T0183438

Lecture Notes of the Institute for Computer Sciences, Social Informatics and Telecommunications Engineering 334

More information about this series at http://www.springer.com/series/8197

Nguyen-Son Vo · Van-Phuc Hoang (Eds.)

Industrial Networks and Intelligent Systems

6th EAI International Conference, INISCOM 2020
Hanoi, Vietnam, August 27–28, 2020
Proceedings

Editors
Nguyen-Son Vo
Faculty of Electrical and Electronics Engineering
Duy Tan University
Da Nang, Vietnam

Van-Phuc Hoang
Le Quy Don Technical University
Hanoi, Vietnam

ISSN 1867-8211 ISSN 1867-822X (electronic)
Lecture Notes of the Institute for Computer Sciences, Social Informatics and Telecommunications Engineering
ISBN 978-3-030-63082-9 ISBN 978-3-030-63083-6 (eBook)
https://doi.org/10.1007/978-3-030-63083-6

This Springer imprint is published by the registered company Springer Nature Switzerland AG
The registered company address is: Gewerbestrasse 11, 6330 Cham, Switzerland

Preface

We are delighted to introduce the proceedings of the 2020 European Alliance for Innovation (EAI) International Conference on Industrial Networks and Intelligent Systems (INISCOM 2020). This conference has brought together researchers, developers, and practitioners from around the world who are leveraging and developing industrial networks and intelligent systems. The theme of INISCOM 2020 was "Computing, Telecommunications Technologies and Applications of 5G-IoT, AI and Cyber-Security to Improve Citizens' Lives."

The technical program of INISCOM 2020 consisted of 25 full papers in oral presentation sessions at the main conference tracks. The conference tracks were: Track 1 – Telecommunications Systems and Networks; Track 2 – Hardware, Software, and Application Designs; Track 3 – Information Processing and Data Analysis; Track 4 – Industrial Networks and Intelligent Systems; and Track 5 – Security and Privacy. Aside from the high-quality technical paper presentations, the technical program also featured one keynote speech. The keynote speaker was Prof. Dong-Seong Kim, from Kumoh National Institute of Technology, South Korea.

Coordination with the steering chairs, Prof. Imrich Chlamtac, Dr. Vien Ngo, and Dr. Ta Chi Hieu, was essential for the success of the conference. We sincerely appreciate their constant support and guidance. It was also a great pleasure to work with such an excellent Organizing Committee team and we thank them for their hard work in organizing and supporting the conference. In particular, the Technical Program Committee (TPC), led by our TPC co-chairs, Dr. Nguyen-Son Vo, Dr. Quoc Tuan Vien, and Prof. Trung Q. Duong, who completed the peer-review process of technical papers and made a high-quality technical program. We are also grateful to conference manager Natasha Onofrei for the support and all the authors who submitted their papers to INISCOM 2020.

We strongly believe that INISCOM provides a good forum for all researcher, developers, and practitioners to discuss all science and technology aspects that are relevant to industrial networks and intelligent systems. We also expect that the future INISCOM will be as successful and stimulating as indicated by the contributions presented in this volume.

October 2020

Nguyen-Son Vo
Van-Phuc Hoang

Organization

Steering Committee

Imrich Chlamtac	University of Trento, Italy
Vien Ngo	Queen's University Belfast, UK
Ta Chi Hieu	Le Quy Don Technical University, Vietnam

Organizing Committee

General Chair

Van-Phuc Hoang	Le Quy Don Technical University, Vietnam

General Co-chairs

Cong-Kha Pham	The University of Electro-Communications, Japan
Xuan-Nam Tran	Le Quy Don Technical University, Vietnam

TPC Chair and Co-chairs

Nguyen-Son Vo	Duy Tan University, Vietnam
Quoc Tuan Vien	Middlesex University, UK
Trung Q. Duong	Queen's University Belfast, UK

Sponsorship and Exhibit Chairs

Luong Duy Manh	Le Quy Don Technical University, Vietnam
Hoa Le-Minh	Northumbria University, UK

Local Chairs

Tran Cong Manh	Le Quy Don Technical University, Vietnam
Van-Trung Nguyen	Le Quy Don Technical University, Vietnam
Do Thanh Quan	Le Quy Don Technical University, Vietnam

Workshops Chairs

Koichiro Ishibashi	The University of Electro-Communications, Japan
Sylvain Guilley	Télécom Paris, France
Xuan-Tu Tran	VNU University of Engineering and Technology, Vietnam

Publicity and Social Media Chairs

Tomohiko Taniguchi	Fujitsu Laboratories, Japan
Nguyen Quoc Dinh	Le Quy Don Technical University, Vietnam
Dao Thi Nga	Le Quy Don Technical University, Vietnam

Publications Chairs

Quang Kien Trinh	Le Quy Don Technical University, Vietnam
Mai Ngoc Anh	Le Quy Don Technical University, Vietnam
Tomoyuki Ohkubo	Advanced Institute of Industrial Technology, Japan

Web Chairs

Trong-Thuc Hoang	The University of Electro-Communications, Japan
Vu Hoang Gia	Le Quy Don Technical University, Vietnam

Posters and PhD Track Chairs

Ulrich Kuhne	Télécom Paris, France
Guanghao Sun	The University of Electro-Communications, Japan
Ta Minh Thanh	Le Quy Don Technical University, Vietnam
Le-Nam Tran	University College Dublin, Ireland

Panels Chairs

Mai-Khanh Nguyen Ngoc	The University of Tokyo, Japan
Le Chung Tran	University of Wollongong, Australia
Berk Canberk	Istanbul Technical University, Turkey

Demos Chairs

Zoran Hadzi-Velkov	Ss. Cyril and Methodius University, Macedonia
Van Sang Doan	Kumoh National Institute of Technology, South Korea
Quang Nguyen The	Le Quy Don Technical University, Vietnam

Tutorials Chairs

Jean-Luc Danger	Télécom Paris, France
Duc Anh Le	Center for Open Data in the Humanities, Tokyo, Japan
Tuan Le	Middlesex University, UK

Technical Program Committee

Truong Khoa Phan	University College London, UK
T. Tuan Nguyen	University of Buckingham, UK
Purav Shah	Middlesex University, UK
Tuan Anh Le	Middlesex University, UK
Cong Trang Mai	Queen's University Belfast, UK
Le Chung Tran	University of Wollongong, Australia
Huy T. Nguyen	Nanyang Technological University, Singapore
G. Suseendran	Vels Institute of Science, Technology & Advanced Studies, India
Falowo Olabisi	University of Cape Town, South Africa
Thang Vu	University of Luxembourg, Luxembourg
Yuanfang Chen	Hangzhou Dianzi University, China

Kien Nguyen	Chiba University, Japan
Nguyen Ngoc Mai Khanh	The University of Tokyo, Japan
Guanghao Sun	The University of Electro-Communications, Japan
Duc Anh Le	Center for Open Data in the Humanities, Tokyo, Japan
Tomoyuki Ohkubo	Advanced Institute of Industrial Technology, Japan
Luong Duy Manh	Le Quy Don Technical University, Vietnam
Kien Trinh	Le Quy Don Technical University, Vietnam
Dao Thi Nga	Le Quy Don Technical University, Vietnam
Huu Hung Nguyen	Le Quy Don Technical University, Vietnam
Xuan Tung Truong	Le Quy Don Technical University, Vietnam
Tang Van Ha	Le Quy Don Technical University, Vietnam
Ta Minh Thanh	Le Quy Don Technical University, Vietnam
Doan Van Sang	Kumoh National Institute of Technology, South Korea
Toan Dao	University of Transport and Communications, Vietnam
Huan Vo	Ho Chi Minh City University of Technology and Education, Vietnam
Van-Ca Phan	Ho Chi Minh City University of Technology and Education, Vietnam
Pham Ngoc Son	Ho Chi Minh City University of Technology and Education, Vietnam
Kien Dang	Ho Chi Minh City University of Transport, Vietnam
Toan Doan	Thu Dau Mot University, Vietnam
Dac-Binh Ha	Duy Tan University, Vietnam
Nguyen Gia Nhu	Duy Tan University, Vietnam

Contents

Telecommunications Systems and Networks

Intelligent Channel Utilization Discovery in Drone to Drone Networks for Smart Cities

Muhammed Raşit Erol and Berk Canberk(✉)

Department of Computer Engineering, Istanbul Technical University, 34469 Ayazaga, Istanbul, Turkey
{erolm15,canberk}@itu.edu.tr

Abstract. Drone networks are playing a significant role in a wide variety of applications such as the delivery of goods, surveillance, search and rescue missions, etc. The development of the drone to drone (D2D) networks can increase the success of these applications. One way of improving D2D network performance is the monitoring of the channel utilization of the link between drones. There are many works about monitoring channel utility; however, either they sense channel physically, which is not reliable and effective due to noise in the channel and miss-sense of signals, or they have protocol-based solutions with high time-complexity. Hence, we propose a less time and power-consuming MAC layer protocol based monitoring model, which works on the IEEE 802.11 RTS/CTS protocol for D2D communication. We work on this protocol because it solves the hidden terminal problem, which can be seen widely in drone communication due to the characteristics of wireless networks and mobility of drones. Our model consists of Searching & Finding and Functional Sub-layers. In the Searching & Finding Sub-layer, we locate the other drones in the air with a specific flying pattern; we also sense and collect frame information on the channel. With a Functional Sub-layer, we calculate channel utilization with Network Allocation Vector (NAV) vector sizes, showing the duration of the drone about how long it must defer from accessing the link. Also, we create a visualization map with Voronoi Diagram. In that diagram, according to drone coordinates, each region is generated after the k-means clustering algorithm, which is one of the simplest and popular unsupervised machine learning algorithms. Hence, each Voronoi section shows channel utility in terms of percentage in a more precise and discretized way. Furthermore, with our model, we decrease the sensing time of the channel by about 25%, and we reduce the power consumption of sensing drone approximately 26%. Also, our model uses about 57% less area during the calculation phase.

Keywords: Monitoring of channel utilization · IEEE 802.11 RTS/CTS · Drone to drone networks · Voronoi diagram · NAV vectors

N.-S. Vo and V.-P. Hoang (Eds.): INISCOM 2020, LNICST 334, pp. 3–18, 2020.
https://doi.org/10.1007/978-3-030-63083-6_1

1 Introduction

Low Altitude Platforms (LAVs), also called drones, are rapidly developing and becoming extremely useful in a variety of areas, from civil applications to military missions due to the structural advantages and moving flexibility on air. Surveillance, search and rescue missions, delivery of goods, construction, and natural disaster monitoring are most standing out applications of drones [1]. The achievement of these applications depends on improvements in network performance. Hence, there are a significant number of challenges in aerial networks to increase network performance [2–4]. In this aspect, to provide reliable, efficient, and stable drone to drone networks, monitoring resources of aerial systems is a crucial mission because that minimizes the cost of maintenance of data flow. Thus, we focused on the topic of resource monitoring, which is channel utilization for the drone to drone (D2D) networks.

D2D network complexity is dramatically expanding in terms of services and topology, which causes challenging network management problems on network resources. Hence, the diagnosing channel utilization as resource monitoring takes crucial place in D2D networks. As mentioned in [5], monitoring characteristics of wireless networks is critical to many management tasks such as fault diagnosis and resource management. Also, in that work, monitoring types are introduced as PHY and MAC behaviors. In this aspect, we focus on the discovery of channel utilization for D2D networks in the field of smart city applications. It is known that smart cities enhance life quality with intelligent things. Therefore, drone collaboration and D2D networks play a vital role in supporting a lot of smart city applications such as D2D communication and network resource management [6]. Thus, in this work, we work on monitoring of channel utilization as resource management of D2D networks in smart cities.

There exist many studies in the recent literature about evaluating channel utilization in many ways for D2D networks. In [7], MIT LL has developed a data collection and visualization framework to monitor and analyze the performance of a high-capacity backbone (HCB) network, which is an example of Mobile AdHoc Networks (MANETs). In that work, the monitoring implemented at various layers of the OSI stack. Furthermore, the channel utilization can be measured with PHY(physical) layer methods. In [8], with the proposed Channel Quality Indicator (CQI) feedback scheme, each cellular-UAV can evaluate link quality by the reference signal. Also, in [9], Negative Acknowledgement (NACK)-related regular feedback system is considered. In this work, if Signal to Interference and Noise Ratio (SINR) is less than the threshold of a special Modulation and Coding Scheme (MCS), the user transmits NACK back to the base station. Moreover, [10] provides novel channel feedback schemes that solve the problem of finding the right feedback mechanism to convey channel information. With this scheme, it is possible to measure channel quality for wireless networks.

None of these works presented on PHY layer are accurate and reliable measurement methods for link quality because PHY layer can be affected by other signals or signal cannot reach the destination due to shadowing effect and mobility of UAVs. Also, it is impossible to obtain any information about chan-

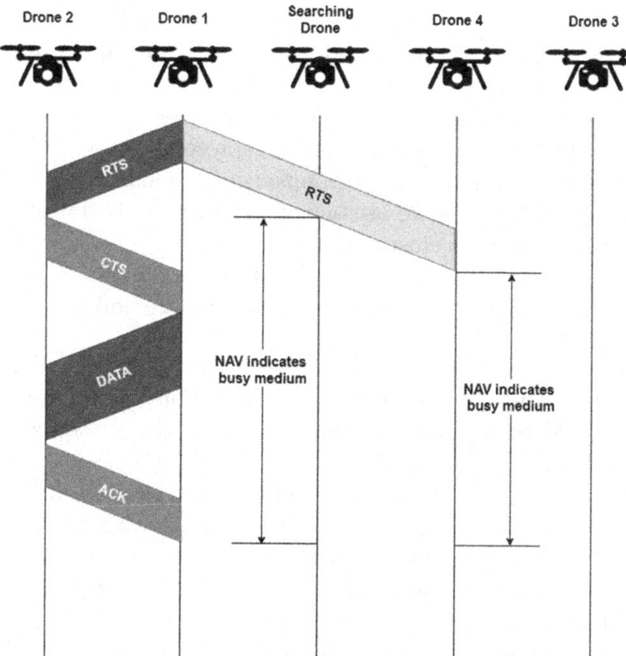

Fig. 1. The RTS/CTS mechanism [11] of scenario Fig. 4

nel quality for the base station, if no NACK is sent. The flow management and logical connection are necessary for more accurate and reliable monitoring of channel utilization; however, these works do not provide these MAC layer properties. Shortly, to be sure about there is a communication in the channel, the MAC layer protocol based approach is needed. Moreover, they have a very complicated implementation of monitoring channel utilization.

Consequently, keeping these studies in mind, we propose a novel monitoring approach of D2D network channel utilization and network traffic type in the field where drones are actively communicating with the IEEE 802.11 protocol. Also, our model works on the MAC layer with flow management and logical connection advantages. Even though the most preferred way of calculating channel utilization is the sensing channel always on the PHY layer, we present a protocol-based method that uses the NAV vector, which is originated from IEEE 802.11 RTS/CTS enabled protocol. In our approach, we calculate channel utilization using the duration field of the frames, which determines the NAV vector size. Furthermore, only one participant can communicate in the channel with IEEE 802.11 RTS/CTS (see Fig. 1); thus, calculating the channel utilization with our method becomes applicable. With our model, we prevent the sensing channel on the PHY layer, which is not an optimal approach due to the power consumption of the searching drone and noise in the channel. The MAC protocol-based system we offered shows there exists absolutely communication in the MAC layer, which is a more exact sensing way rather than blindly sensing the channel. Furthermore, we propose a visualization method using the Voronoi diagram in our work to show channel utilization in the area. Due to a Voronoi map that can be used to find the largest empty circle amid a collection of points, the drone environment where drones are communicating can be represented this method in a more precise way. With this method, drone groups can be visualized more centralized manner within regions because we use one of the unsupervised machine learning algorithms called the k-means clustering method according to drone coordinates. This algorithm clusters drone coordinates and helps to create Voronoi regions. Shortly, the main contributions of this paper include the following:

- We propose a new system model consists of Searching & Finding Sub-layer and Functional Sub-layer Modules, which is responsible for locating drones and creating a Voronoi Diagram. This model works on the MAC layer of IEEE 802.11 RTS/CTS protocol.
- We introduce a practical and more straightforward channel utilization calculation using the properties of the IEEE 802.11 RTS/CTS protocol.
- We present a novel monitoring approach with a k-means clustering algorithm to visualize and analyze network traffic in the Voronoi diagram with a more effective and faster way.
- We can also apply our implementation to future technology WIFI 6, which is the data-driven protocol as we propose. Hence, our model will present a compelling and more uncomplicated novel monitoring method in the future.

The rest of this paper is organized as follows. The network architecture is explained in Sect. 2. In part 3, the system model is indicated. The simulation environment is described in Sect. 4. In Sect. 5, we evaluate the performance of

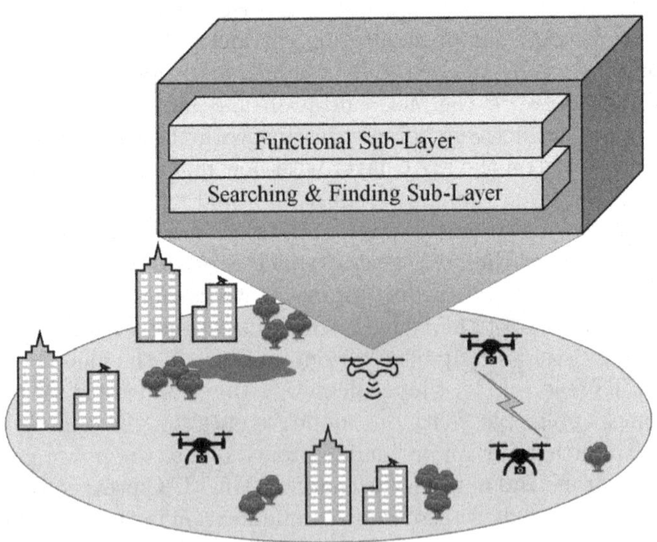

Fig. 2. Network architecture

our proposed model. Finally, we conclude the paper by summarizing the achievements in Sect. 6.

2 Network Architecture

The network topology for our model consists of n drones with one searching drone. These drones are communicating with the IEEE 802.11 wireless networking protocol with RTS/CTS (Request To Send/Clear To Send) mechanism. The RTS/CTS mechanism is created for avoiding the hidden terminal problem in wireless networks and allows only one pair to communicate in the channel. Drones can fly at different heights with the specific moving pattern. However, the searching drone always flies at the pre-determined height. All drones repeat their flying patterns after reaching destination coordinates. Furthermore, we assume that all drones are completed authentication stage for wireless communication. Hence, they communicate directly with each other on the same channel without authentication messages. Moreover, the channel is always busy, and the frame size is randomly generated in the network. We represent the whole network architecture and the component models of the searching drone in Fig. 2.

3 System Model

We divide the proposed system model into two coherent sub-layers titled Searching & Finding Sub-layer and Functional Sub-layer. Searching & Finding Sub-layer is responsible for searching on the area with a specific movement pattern and gathering information from D2D communication. Moreover, we dedicate the Functional Sub-layer to process information belongs to the Searching & Finding Sub-layer. Each of the sub-layers additionally owns some modules. Searching & Finding Sub-layer has two modules entitled Sensing and Data Classification; furthermore, the Functional Sub-layer has two modules entitled Calculation and Visualization. In Fig. 3, we represent the entire system model and the associations between its segments.

3.1 Searching and Finding Sub-layer

Searching & Finding Sub-layer includes Sensing and Data Classification Modules. The Sensing Module determines the movement pattern of the searching drone and executes it. Furthermore, this module performs the classification of data operations and gives the meaning of them.

Sensing Module. This module handles the movement pattern of the searching drone and operations of collecting data from the channel with sensing. This information contains coordinates of communicating drones, source and destination address of the frame, Duration ID in the frame to keep Network Allocation Vector (NAV) timer and frame types such as RTS, CTS, DATA or ACK. After the sensing channel for gathering this information, this module transfers collected data to the next layer called Data Classification.

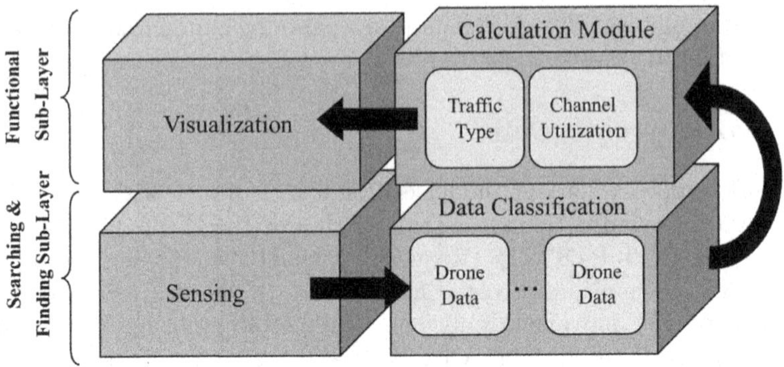

Fig. 3. Layered architecture of proposed system model

Data Classification Module. This module exists at the end of the Searching & Finding Sub-layer. The transferred data from Sensing Module is classified here to match that data with Drone Data sections. The coming information from the below layer is assigned to Drone Data if there exists. In the case of a new drone whose information does not exist in the Drone Data section, is discovered, then the new part is created in the Data Classification Module. All other information about this drone will be assigned this section in the future data gathering. This module's main aim is grouping collected data with corresponding drones to make it easier for calculations in the future. After all these operations, the classified data is transferred to the upper layer named Calculation Module.

3.2 Functional Sub-layer

The functional Sub-layer includes Calculation and Visualization Modules. The Calculation Module calculates channel utilization, and the Visualization Module creates a Voronoi diagram with calculated channel utilization and network traffic type.

Calculation Module. This module exists between Data Classification and Visualization Module. It calculates channel utilization and type of network traffic for each drone. To do that, this module uses the information coming from the Data Classification Module. Furthermore, this module contains submodules called Channel Utilization and Traffic Type Sub-Modules.

Channel Utilization Sub-Module. This module is responsible for computing the channel utilization of the area. To do that, this module uses the properties of the IEEE 802.11 RTS/CTS protocol. Due to this protocol, only one pair can communicate at a certain time. Other drones should wait until the NAV vector reaches zero; after that, if they win back off timer before other drones, then they can transmit their data. The Fig. 4 shows sample scenario. This scenario is an example of our model with less number of drones with the searching drone.

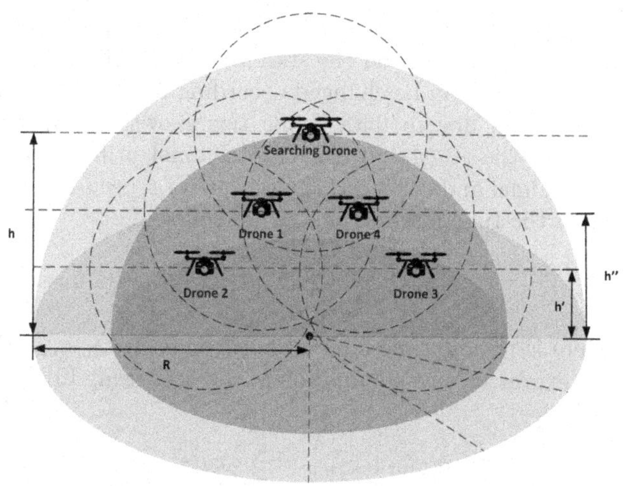

(b) The placement of drones in the area

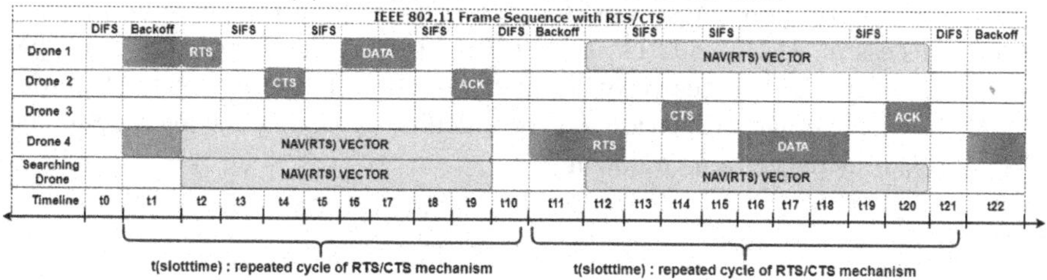

(a) The communication sequence diagram

Fig. 4. The sample scenario of IEEE 802.11 RTS/CTS

The communication sequence diagram of the area can be seen in the below part of the figure. In our model, we concentrate on the repeated cycle of RTS/CTS mechanism denoted as $t_{slottime}$ and natural outcome of IEEE 802.11 RTS/CTS protocol, known as sequentially repeated cycles like in the Fig. 4b. It can be understood that $t_{slottime}$ can be calculated focusing on starting time with the RTS frame after the Backoff timer until the ACK frame is transferred, which showed in Fig. 4b. In this interval, only one drone pair can transmit, and the total transmission time for data always has the same sub-time intervals except for the data frame size. Other time intervals like DIFS duration denoted as t_{DIFS} are constant values determined by the protocol as in work [12]. Hence, $t_{slottime}$ can be denoted as following:

$$t_{slottime} = t_{RTS} + 3 \times t_{SIFS} + t_{CTS} + t_{DATA} + t_{ACK} + t_{DIFS} + t_{Backoff} \quad (1)$$

where $t_{Backoff}$ is a random value between [1, CW], and other time intervals are the part of the IEEE 802.11 RTS/CTS protocol. The CW(contention window) is an integer between $CW_{min} = 32$ and $CW_{max} = 1024$. In our model, we consider this value as average value as $CW = 528$. The other components of $t_{slottime}$ is constant values which depends on the standard of the IEEE 802.11 protocol except t_{DATA}. The t_{DATA} can be changed according to the data frame size.

In this time interval, due to data frame size shows us real traffic in the channel, we focus on the t_{DATA} to calculate channel utilization of the area. Other values are minimal and constant. Hence, rate of t_{DATA} and $t_{slottime}$ gives us channel usage in the time interval $t_{slottime}$ as following:

$$rate = \frac{t_{DATA}}{t_{slottime}} \times 100 \tag{2}$$

In our model, there is no idle time interval between each $t_{slottime}$. Hence, total utilization can be expressed as a summation of the *rate* using Eq. 2 as following:

$$Total_{utilization} = \sum_{n=1}^{N} rate \tag{3}$$

where N is the total number of repeated cycles $t_{slottime}$ in the channel.

The successful IEEE 802.11 RTS/CTS transmission contains multiple $t_{slottime}$ time intervals successively, as in our model. These intervals are independent; hence, average utilization can be calculated, taking the average of the $Total_{utilization}$ using Eq. 3 as following:

$$Average_{utilization} = \frac{Total_{utilization}}{N} \tag{4}$$

Equation 4 provides a powerful and simple calculation of channel utilization only sensing once in the time interval of $t_{slottime}$. The calculation of the channel utilization can be possible when t_{DATA} internal is known. Hence, we present a novel approach for determining t_{DATA} interval using the NAV vector. After sensing once the channel, the NAV vector is created for sensing drone called searching drone in our model and average channel utilization using the Eq. 4 can be calculated with the size of the NAV vector. Thus, we use four types of NAV vector, and the duration of these vectors can be calculated as follows:

$$NAV_{RTS} = t_{RTS} + 3 \times t_{SIFS} + t_{CTS} + t_{DATA} + t_{ACK} \tag{5}$$

$$NAV_{CTS} = t_{CTS} + 2 \times t_{SIFS} + t_{DATA} + t_{ACK} \tag{6}$$

$$NAV_{DATA} = t_{DATA} + t_{SIFS} + t_{ACK} \tag{7}$$

$$NAV_{ACK} = t_{ACK} \tag{8}$$

It can be seen from Eq. 5–8 that all NAV vector types contain t_{DATA} interval except NAV_{ACK}. In our model, the searching drone can only sense the mentioned frames. Except NAV_{ACK} interval, with all sensed NAV vector types, t_{DATA} can be determined because except t_{DATA}, all other time intervals are known by the protocol. Hence, in this module, it is possible to calculate t_{DATA} interval using the proposed algorithm whose flowchart is given in Fig. 5. Using this flowchart, we decide which equation we can use according to the type of frame received and use the proper equation. Each frame type has a different calculation equation given as Eq. 5–8 and using these equations and Eq. 2–4 following equation can be derived:

$$Average_{Utility} = \frac{\sum_{n=1}^{M} (\frac{t_{DATA}}{t_{slottime}} \times 100)}{M} \tag{9}$$

where M is the number of the sensed frame number, and t_{DATA} is calculated using our proposed algorithm, whose flowchart is given in Fig. 5. Also, $t_{slottime}$ is determined according to sensed frame type.

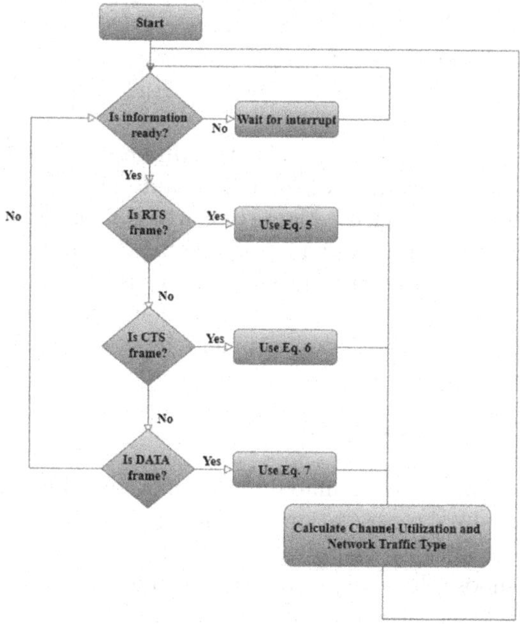

Fig. 5. The flowchart for Calculation Module

Shortly, in this module, according to the frame type, the NAV vector is activated in the searching drone. Using these types of NAV vector, t_{DATA} interval is determined using Eq. 5–7. After that, Eq. 9 gives us individual channel utility for each drone. With Eq. 3–4, known t_{DATA} interval allow us to calculate $Average_{utilization}$. Hence, knowing t_{DATA} interval provides an accurate calculation of the average channel utilization, as mentioned above.

Traffic Type Sub-Module. This module is responsible for determining the traffic type of the region in the Voronoi Diagram. According to the $Average_{utilization}$ of the channel traffic, which is calculated in the Channel Utilization Sub-Module, we determine the traffic type of a region in the Voronoi Diagram in terms of three-level which are low, medium, and high traffic. We define the low level as a percentage of less than 30%, and the medium level as a percentage of between 30% and 60%. Also, we specify the high level as a percentage of more than 60%. After the calculation of these types, this information is transferred to the next module named Visualization Module.

Visualization Module. This module exists after the Calculation Module. After the channel utilization and network traffic type are calculated for each drone, the Voronoi Diagram is created and visualized with each drone's coordinates in this module. This diagram contains regions with average channel utilization and network traffic types. After each iteration, we recreate the map with new data, and we examine the difference in the traffic in this module. The searching drone completes each iteration when it reaches the starting coordinates after visiting all areas.

In this module, as a novel approach, the drone coordinates are grouped, and after that, we construct the Voronoi Diagram. The grouping operation is done with the k-means Clustering Algorithm, which is one of the simplest and popular unsupervised machine learning algorithms. K-means clustering results in a partitioning of the data space into Voronoi cells, which helps us to create Voronoi regions in our model. According to [13], continuous geometric problems can be converted into a discrete graph problem, as in our model.

4 Simulation Environment

The proposed system model is simulated with the pygame library of Python programming language. In the simulation, the drones are placed in the area with Poisson Distribution because it provides centralized distribution with proper lambda value. We choose the lambda value as 10, and the sample size as 1000000. We use the Poisson Distribution for both the x-axis and the y-axis on the map. Though the distribution also provides the same values, we applied the discretization method to these values. However, this method decreases the produced values by Poisson Distribution. Creating values for the x and y-axis with Poisson Distribution is applied multiple times until the coordinates are generated for each drone. Furthermore, drones that are communicating with other already placed drones are also identified with Poisson Distribution. Moreover, we made the placement with some restrictions; the minimum distance with pair drones must be more than ten units.

The model has some assumptions which are mentioned as following. We assume that the searching drone is always flying at the same height; hence, we create the area as a 2D map. Furthermore, we take that all drones are running with a specific movement pattern. The searching operation starts from a

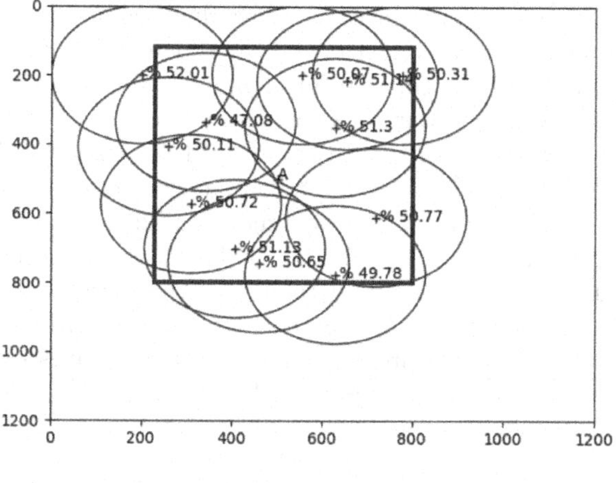

(a) With PHY Solution

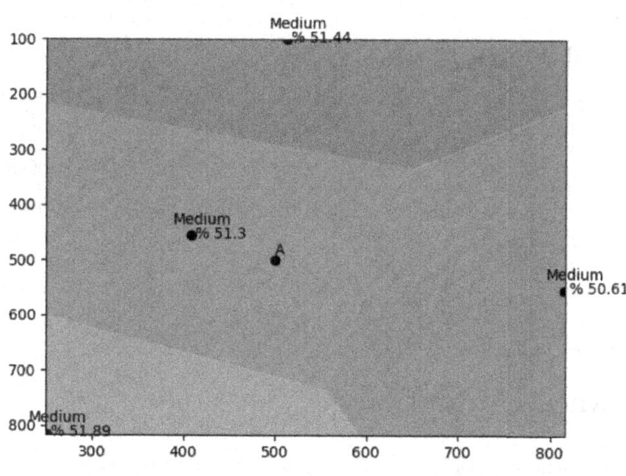

(b) With MAC (Our Model) Solution

Fig. 6. The visualisation model of the area with PHY and MAC

particular location and continues on the map until all regions are visited. These regions have the same size as the radius of all other drones, which decreases miss sensed areas when searching and detecting. When the searching drone visits all parts, the backward movement pattern starts. End of this pattern, the searching drone reaches the starting location, and the total movement pattern begins again.

In the model, each drone communicates only with one drone on the map. The communication between them to be possible, each pair of drones placed within the communication range. The communication range of all drones is the same size unit. The direction of communication is determined randomly. Furthermore, we set the data frame size to random size, and the drones are communicating all the time. Hence, there is no idle time on the channel. All drones have the frame to send when the backoff timer starts decreasing.

There are different monitoring types in terms of the number of drones, such as 8, 12, 16, and 20 in the model. We place drones considering the rate of drone communication range and the map. Hence, a more significant number of drones have a smaller communication range in the model. Also, the searching drone moves with a speed vector that scaled for the map. We discretize the movement of the searching drone and communication time of the other drones in terms of units. Furthermore, the searching drone can move with different sizes of the unit of time; however, communication occurs in each unit of time. Table 1 shows the overall simulation parameters.

Table 1. Simulation Parameters

Map Properties			Poisson Parameters	
Vehicle Number	Radius of drones	Map Size	Lambda	Sample Size
8	250	1000×1000	10	1000000
12	200	1000×1000	10	1000000
16	150	1050×1050	10	1000000
20	100	1000×1000	10	1000000
Drone Receiver Power			0.2 W	

5 Performance Evaluation

We evaluate the performance of our proposed system model based on the visualization of the drone environment. If we create a channel utilization map using only drone coordinates, then there might be miss-sensed drones in the map because, according to IEEE 802.11 RTS/CTS protocol, only one drone can use the channel at a time. Hence, we classify drone locations with the k-means clustering algorithm, and we create Voronoi regions after clustering. Furthermore, we show the comparison of the PHY (Physical Layer) and MAC Layer (our model) solutions with different parameters such that average sensing time, power consumption, and used area during calculation in our simulation results.

Figure 6 shows the results of our novel model as a visualization approach. The figure consists of two parts, which first part shows a map of before our work as PHY, and the second part shows a diagram of using our model as the MAC layer. Each section of the figure has a point A on the same location to offer the same place for evaluation channel utility. Furthermore, we create all maps for the same altitude, in which the searching drone flies denoted as h in the figures. Also, we divide all Voronoi diagrams into four regions.

When we examine Fig. 6, we can see for point A; it is not reliable to find out channel utilization in the first part of the figures. For some samples, there is no specific channel utilization, or there are multiple values that we should consider all of that. With our model, as in part b, we can detect channel utilization more reliably and accurately. Furthermore, we show network traffic types in regions, which cannot be seen in the first part of the figures.

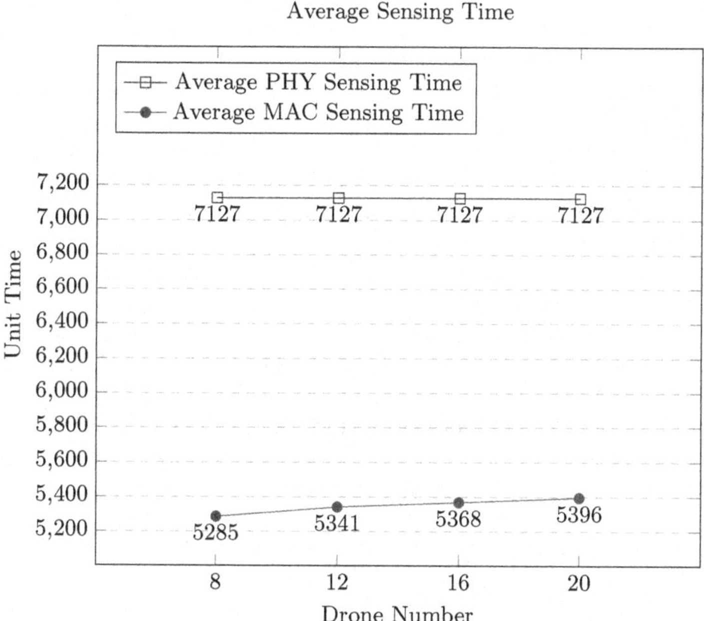

Fig. 7. Average sensing time

In additional visualization evaluation, we experiment average sensing time of the searching drone with MAC solution (our model) and PHY method. Figure 7 shows our model decreases average sensing time about 25%. We also examine power consumption and used area during the calculation of the searching drone as parameters with our simulation. Figure 8 shows the power consumption of the searching drone during the sensing phase, according to PHY and MAC (as our model) layer solutions. It shows that our model decreases power consumption of the searching drone by about 26%. Furthermore, Fig. 9 shows the area used during the calculation phase measured by the searching drone according to PHY and MAC (as our model) layer solutions. It shows that our model decrease used area during the calculation phase about 57%.

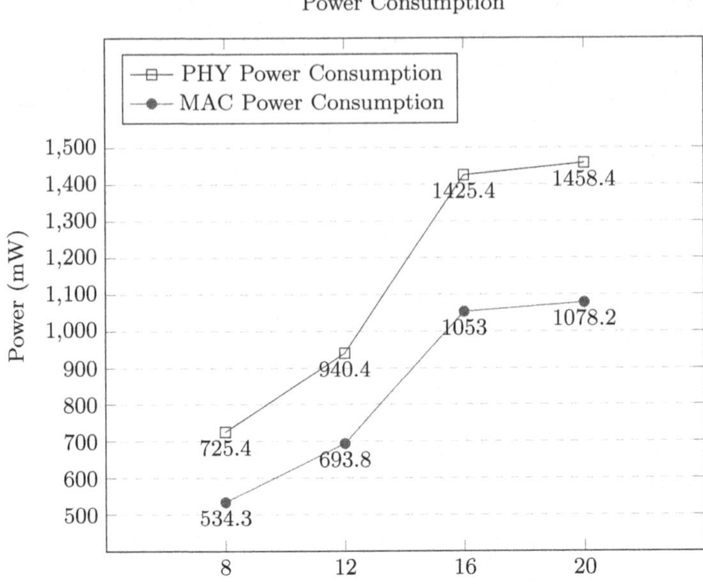

Fig. 8. Power consumption of the searching drone

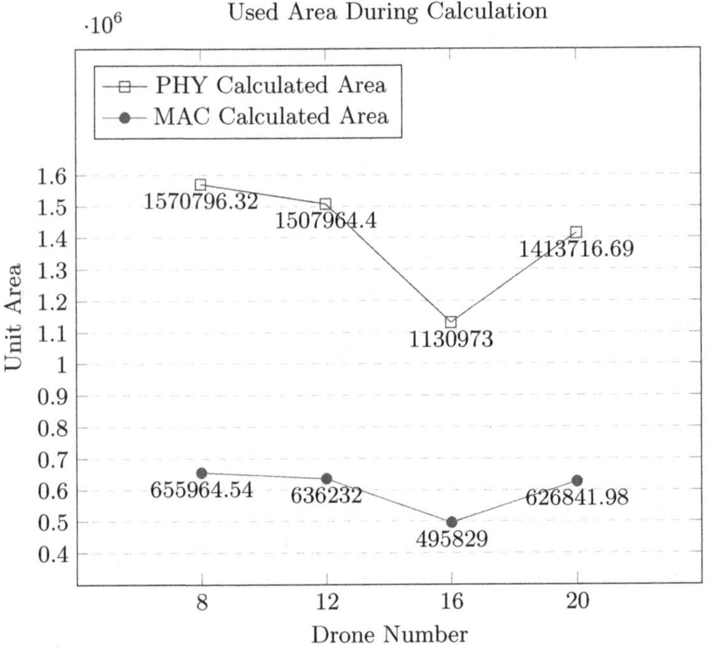

Fig. 9. Used area during calculation

6 Conclusions

In this paper, we propose a novel channel utilization monitoring approach for D2D networks to evaluate the performance of the wireless link. The searching drone flies in the air with a specific flying pattern and senses the channel to locate other drones. Our protocol-based system model process all sensed information and calculate channel utility with our novel method. This method benefits from the IEEE 802.11 RTS/CTS protocol, which allows continuous and discrete communication with one communicating pair at a time in the channel. Furthermore, we provide novel visualization methods with Voronoi Diagram. The power and simplicity of the Voronoi Diagram are applied, and we create the channel utilization map in our work. We divide the map into Voronoi regions with a k-means clustering algorithm for corresponding heights and show channel utility of each area. Furthermore, we decrease the average sensing time of channel 25% with our model. We also reduce the power consumption of the searching drone by about 26%, and we fall used area during the calculation phase, about 57%. In future work, this protocol-based approach for monitoring channel utility can be applied on WIFI 6, which is the data-driven protocol as we propose, and a 3D Voronoi diagram can be created to get a more robust examination of link performance. Artificial Intelligence (AI) based evaluation of channel utility and network traffic type also can be applied to the work.

Acknowledgements. The paper is supported by The Scientific and Technical Research Council of Turkey (TUBITAK) 1001 The Scientific and Technological Research Projects Funding Program with project number 119E434.

References

1. Hayat, S., Yanmaz, E., Muzaffar, R.: Survey on unmanned aerial vehicle networks for civil applications: a communications viewpoint. IEEE Commun. Surv. Tutor. **18**(4), 2624–2661 (2016)
2. Mozaffari, M., Saad, W., Bennis, M., Nam, Y., Debbah, M.: A tutorial on UAVs for wireless networks: applications, challenges, and open problems. IEEE Commun. Surv. Tutor. **21**(3), 2334–2360 (2019)
3. Ariman, M., Seçinti, G., Erel, M., Canberk, B.: Software defined wireless network testbed using raspberry Pi of switches with routing add-on. In: 2015 IEEE Conference on Network Function Virtualization and Software Defined Network (NFV-SDN), pp. 20–21 (2015)
4. Gutierrez-Estevez, D.M., Canberk, B., Akyildiz, I.F.: Spatio-temporal estimation for interference management in femtocell networks. In: 2012 IEEE 23rd International Symposium on Personal, Indoor and Mobile Radio Communications - (PIMRC), pp. 1137–1142 (2012)
5. Nguyen, H., Scalosub, G., Zheng, R.: On quality of monitoring for multichannel wireless infrastructure networks. IEEE Trans. Mob. Comput. **13**(3), 664–677 (2014)
6. Alsamhi, S.H., Ma, O., Ansari, M.S., Almalki, F.A.: Survey on collaborative smart drones and Internet of Things for improving smartness of smart cities. IEEE Access **7**, 128125–128152 (2019)

7. Arbiv, S., et al.: Data collection and analysis framework for mobile ad hoc network research. In: MILCOM 2019 IEEE Military Communications Conference (MILCOM), pp. 1–6 (2019)
8. Park, K., Rhee, J., Shin, M., Hur, W.: A novel CQI feedback channel for cellular UAV system. In: 2019 International Conference on Advanced Technologies for Communications (ATC), pp. 84–88 (2019)
9. Cheng, R., Liu, Y., Cheng, W., Liu, D.: Common feedback channel for multicast and broadcast services. In: 2011 7th International Wireless Communications and Mobile Computing Conference, pp. 1958–1963 (2011)
10. Choi, Y., Bahk, S.: Partial channel feedback schemes maximizing overall efficiency in wireless networks. IEEE Trans. Wirel. Commun. **7**(4), 1306–1314 (2008)
11. Rollins, S.: MAC Layer (2020). https://www.cs.usfca.edu/%7Esrollins/courses/cs686-f08/web/notes/maclayer.html. Accessed 15 April 2020
12. Yanmaz, E., Hayat, S., Scherer, J., Bettstetter, C.: Experimental performance analysis of two-hop aerial 802.11 networks. In: 2014 IEEE Wireless Communications and Networking Conference (WCNC), pp. 3118–3123 (2014)
13. Hammouri, O.M., Matalgah, M.M.: Voronoi path planning technique for recovering communication in UAVs. In: 2008 IEEE/ACS International Conference on Computer Systems and Applications, pp. 403–406 (2008)

Downlink Resource Sharing and Multi-tier Caching Selection Maximized Multicast Video Delivery Capacity in 5G Ultra-Dense Networks

Thanh-Minh Phan[1], Nguyen-Son Vo[2](✉), Minh-Phung Bui[3], Quang-Nhat Tran[3], Hien M. Nguyen[4], and Antonino Masaracchia[5]

[1] Ho Chi Minh City University of Transport, Ho Chi Minh City 700000, Vietnam
ptminh@hcmutrans.edu.vn

[2] Institute of Fundamental and Applied Sciences, Duy Tan University, Ho Chi Minh City 700000, Vietnam
vonguyenson@duytan.edu.vn

[3] Van Lang University, Ho Chi Minh City 700000, Vietnam
{buiminhphung,tranquangnhat}@vanlanguni.edu.vn

[4] Faculty of Electrical-Electronic Engineering, Duy Tan University, Da Nang 550000, Vietnam
nguyenminhhien2501@gmail.com

[5] Queen's University Belfast, Belfast BT7 1NN, UK
a.masaracchiag@qub.ac.uk

Abstract. In this paper, we propose a downlink resource sharing and multi-tier caching selection (DRS-MCS) solution for video streaming applications and services (VASs) in 5G ultra-dense networks (UDNs). The DRS-MCS allows mobile users (MUs) to experience the VASs by multicasting from three-tier caching placements, i.e., macro base station (MBS), femtocell base stations (FBSs), and mobile devices. To do so, the MUs are categorized into three types including 1) sharing users (SUs) that own downlink resources being shared for device-to-device (D2D) communications, 2) caching helpers (CHs) that cache the requested videos for multicasting over D2D communications, and 3) requesting users (RUs) that request the videos. The CHs and the RUs are grouped into different clusters, each cluster has a number of CHs and RUs in close vicinity for D2D multicast communications. We then formulate the DRS-MCS optimization problem. By solving the problem, the DRS-MCS solution can select not only the best pairs of SUs and CHs for D2D multicast communications but also the best caching placements for multicasting in each cluster, so as to maximize the total video capacity delivered to the RUs. Simulation results are shown to demonstrate the benefits of the proposed DRS-MCS solution compared to other conventional multicasting schemes.

Keywords: 5G ultra-dense networks · Caching and clustering · D2D multicast communications · Resource sharing · Video applications and services

N.-S. Vo and V.-P. Hoang (Eds.): INISCOM 2020, LNICST 334, pp. 19–31, 2020.
https://doi.org/10.1007/978-3-030-63083-6_2

1 Introduction

By 2022, it anticipates that together with the rapid increase of mobile users (MUs), the video applications and services (VASs) will bloom and use up about 79% of the mobile data traffic [1]. It is certain in VASs that there are a number of MUs in close vicinity of each other that have the same interest of video contents. In this context, by exploiting the benefits of common interest-sharing nature of dense MUs and broadcast nature of wireless medium, multicasting techniques play an important role in emerging 5G networks since they can provide the system with a high energy- and spectrum-efficiency solution and a high video delivery capacity [2,3].

Most of the multicasting techniques have been studied to apply to device-to-device (D2D) communications [4,5] with the assistance of the MUs that have cached the videos for streaming, namely caching helpers (CHs), and have available downlink resources for sharing, namely sharing users (SUs) [6–12]. The results achieved include tractable model for analysis and optimization design of coverage probability and system capacity [6], reduction in streaming cost at better fairness [7], minimum video delivery delay [8–10], and maximum energy efficiency [11] and sum effective throughput [12]. Other multicasting techniques have further exploited both physical communications features and social attributes of MUs to gain higher system performance [13–20]. However, these studies cannot provide a flexible multicasting strategy to serve the requesting users (RUs) that request the videos by fully utilizing the three-tier caching placements, i.e., macro base station (MBS), femtocell base stations (FBSs), and CHs over D2D communications, in 5G ultra-dense networks (UDNs).

Few of multicasting techniques have been proposed to serve the RUs flexibly by the MBS over conventional cellular transmission and by the CHs over D2D communications with downlink resources shared by the SUs [21,22]. In particular, the authors in [21] have designed a downlink resource sharing and caching helper selection solution to maximize the multicast video delivery in dense D2D 5G networks. The proposed solution has been insightfully studied by considering the social attributes between the CHs and the RUs as well as the constraint on the skewed fairness of RUs, so as to further satisfy the RUs [22]. The existing problem of the works in [21,22] is that they do not exploit the caching placement at the FBSs to fully provide the RUs with three-tier caching selection for the highest system capacity.

In this paper, we utilize the three-tier caching placements at the MBS, the FBSs, and the CHs as well as the downlink resources available at the SUs to propose a downlink resource sharing and multi-tier caching selection (DRS-MCS) solution for VASs in 5G UDNs. To do so, we formulate the DRS-MCS optimization problem for finding the best pairs of the CHs and the SUs in order to multicast the requested videos from the CHs to the RUs over D2D communications that reuse the downlink resources of the SUs. In addition, the DRS-MCS is able to select the best caching placements, i.e., the MBS, the FBSs, or the CHs, to serve the RUs in different clusters at maximum system capacity. The DRS-MCS optimization problem also considers a constraint on the target signal

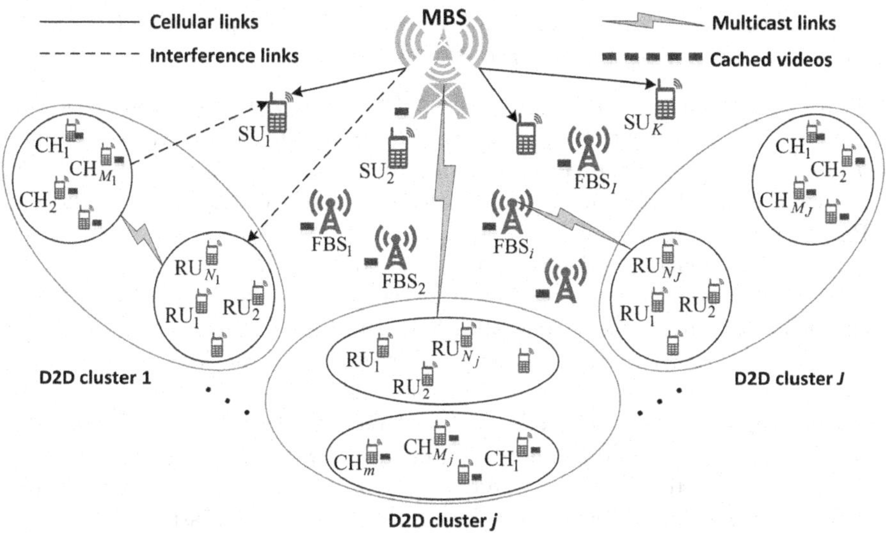

Fig. 1. Multicast Video Streaming in 5G UDNs with DRS-MCS.

to interference plus noise ratio (SINR) of the SUs to guarantee their quality of service (QoS) which is certainly degraded due to the transmissions of the CHs when reusing the downlink resources of the SUs. Simulation results are shown with insightful analysis and discussion to demonstrate the benefits of the proposed DRS-MCS solution compared to other schemes such as conventionally multicasted by the MBS and by both the MBS and the FBSs but without downlink resource sharing.

The rest of this paper is organized as follows. We introduce the system model of multicast video streaming in 5G UDNs with DRS-MCS and describe how it works in Sect. 2. In Sect. 3, we derive the system formulations that enable us to propose the DRS-MCS optimization problem and solution in Sect. 4. The performance evaluation is presented in Sect. 5. Finally, we conclude the paper in Sect. 6.

2 System Model

In this paper, we consider a system of multicast video streaming in 5G UDNs with DRS-MCS as shown in Fig. 1. The system consists of one MBS, I FBSs, K SUs, and J clusters. The cluster j, $j = 1, 2, ..., J$, has M_j CHs that have cached the requested videos and N_j RUs that request the videos. The system provides three ways to flexibly multicast the videos to the RUs that are 1) multicast from the MBS over conventional cellular transmissions, 2) multicast from the FBSs by applying the channel splitting and F-ALOHA schemes that are able to guarantee no interference [23], and 3) multicast over D2D communications by reusing the downlink resources shared by the SUs. At the appropriate time, if there are a number of RUs requesting a particular video, the MBS deploys the DRS-MCS strategy including three steps presented as follows:

- Step 1 - Clustering: To deploy the DRS-MCS strategy, it is necessary to group the CHs and the RUs that are in close vicinity for D2D communications into J clusters. We apply the D2D clustering technique proposed in [24], in order to expand the coverage area of each cluster so that there are M_j CHs and N_j RUs in the cluster j.
- Step 2 - Formulating and solving the DRS-MCS optimization problem: The MBS further collects the system parameters such as the number of SUs and channel information from the MBS, FBSs, and CHs to the RUs and from the MBS and CHs to the SUs. These parameters enable the MBS to formulate the DRS-MCS optimization problem and solve it for the optimal downlink resource sharing index $v_j^{m,k}$. If $v_j^{m,k}$ evaluates to 1 (or 0), the SU k does (or does not) share its downlink resource for D2D multicast communications from the CH m to N_j RUs in the cluster j, $m = 1, 2, ...M_j$. For the purpose of limiting the interference impact caused by the transmissions of CHs on the SUs, i.e., guaranteeing the QoS of the SUs, the DRS-MCS optimization problem considers the constraints such that an SU can share its downlink resource with up to only one CH in the whole system and the target SINR of the SUs is greater than or equal to a given threshold.
- Step 3 - Multicasting: After solving the DRS-MCS optimization problem, the MBS decides which one, i.e., itself, an FBS, or a CH, multicasts the video to all the RUs in the cluster j. In other words, the RUs in the cluster j are served by the MBS, the FBS, or the CH, depending on from which the channel quality is better so that the total multicast video capacity delivered to all RUs in the system is maximized.

3 System Formulations

3.1 Wireless Channel

In this paper, the wireless channel gains are modeled as $G_j^{\mathsf{x},\mathsf{y}} = h_j^{\mathsf{x},\mathsf{y}} g_j^{\mathsf{x},\mathsf{y}}$ [25,26], here $\mathsf{x} \in \{0, i, m\}$, $\mathsf{y} \in \{n, k\}$, $i = 1, 2, ..., I$, $h_j^{\mathsf{x},\mathsf{y}}$ is the exponential power fading coefficient with unit mean, i.e., $\sim$exp(1), and $g_j^{\mathsf{x},\mathsf{y}} = \|d_j^{\mathsf{x},\mathsf{y}}\|^{-\eta}$ is the standard power law path loss function with path loss exponent η, $d_j^{\mathsf{x},\mathsf{y}}$ is the distances from the MBS (x=0), the FBS i (x=i), and the CH m (x=m) to the RU n (y=n) and the SU k (y=k).

3.2 Capacity at RUs

To obtain the video capacity delivered to the RUs, it is required to compute the SINR from the CHs to the RUs and the signal to noise ratio (SNR) from the MBS and the FBSs to the RUs which are respectively presented in the sequel.

In the cluster j, if the CH m is selected to multicast the video to the RU n by reusing the downlink resource shared by the SU k, the SINR from the CH m to the RU n, which is affected by the interference generated from the conventional cellular transmission of the MBS to the SU k, is given by

$$\gamma_j^{m,k,n} = \frac{v_j^{m,k} P_j^m G_j^{m,n}}{N_0 + P_0^k G_j^{0,n}}, \tag{1}$$

where P_j^m and P_0^k are the transmission powers of the CH m in the cluster j and of the MBS (indicated by 0) to the SU k; $G_j^{m,n}$ and $G_j^{0,n}$ are the channel gains from the CH m and the MBS to the RU n in the cluster j; and N_0 is the power of additive white Gaussian noise.

In case there is not any SUs sharing the downlink resources with the CHs, the FBSs are considered multicasting the video to the RUs. The SNR from the FBS i to the RU n in the cluster j is given by

$$\gamma_j^{i,n} = \frac{P_j^{i,n} G_j^{i,n}}{N_0}, \tag{2}$$

where $P_j^{i,n}$ is the transmission power of the FBS i to the RU n and $G_j^{i,n}$ is the channel gain from the FBS i to the RU n in the cluster j.

In addition, we further compute the SNR from the MBS to the RU n in the cluster j which is expressed as

$$\gamma_j^{0,n} = \frac{P_0^j G_j^{0,n}}{N_0}, \tag{3}$$

where P_0^j is the transmission power of the MBS to the RUs in the cluster j.

So far, the capacity at the RU n in the cluster j delivered from the CH m, the FBS i, and the MBS is respectively given as below

$$C_j^{m,k,n} = W \log_2(1 + \gamma_j^{m,k,n}), \tag{4}$$

$$C_j^{i,n} = W \log_2(1 + \gamma_j^{i,n}), \tag{5}$$

and

$$C_j^{0,n} = W \log_2(1 + \gamma_j^{0,n}), \tag{6}$$

where W is the system bandwidth.

Finally, the total system capacity delivered from the MBS, FBSs, and CHs to the RUs in all clusters is expressed as

$$C = \sum_{j=1}^{J} \max\Big\{ \sum_{n=1}^{N_j} C_j^{0,n}, \max\big\{ \sum_{n=1}^{N_j} C_j^{i,n}, i = 1, 2, ..., I\big\}, \sum_{k=1}^{K} \sum_{m=1}^{M_j} \sum_{n=1}^{N_j} C_j^{m,k,n} \Big\}, \tag{7}$$

In Eq. (7), C is so-called the objective function in the DRS-MCS optimization problem.

Algorithm 1. EBSA for DRS-MCS optimization problem

Input: Initial system parameters given in Table 1
Output: $\mathcal{V}^*$, C^*

Generating J search space matrices
$\mathcal{V}_1 = \{V^1_{M_1 \times K}, V^2_{M_1 \times K}, ..., V^{2^{M_1 \times K}}_{M_1 \times K}\}$
$\mathcal{V}_2 = \{V^1_{M_2 \times K}, V^2_{M_2 \times K}, ..., V^{2^{M_2 \times K}}_{M_2 \times K}\}$
...
$\mathcal{V}_J = \{V^1_{M_J \times K}, V^2_{M_J \times K}, ..., V^{2^{M_J \times K}}_{M_J \times K}\}$
$C^* \leftarrow 0$
for each matrix v_1 in $\mathcal{V}_1$, $v_1 = 1, 2, ..., 2^{M_1 \times K}$ **do**
  for each matrix v_2 in $\mathcal{V}_2$, $v_2 = 1, 2, ..., 2^{M_2 \times K}$ **do**
    ...
    for each matrix v_J in $\mathcal{V}_J$, $v_J = 1, 2, ..., 2^{M_J \times K}$ **do**
      if J matrices satisfy (9b), (9c), and (9d) **then**
        Computing C in (7)
        if $C > C^*$ **then**
          $C^* \leftarrow C$
          $\mathcal{V}^* \leftarrow \{V^{v_1}_{M_1 \times K}, V^{v_2}_{M_2 \times K}, ..., V^{v_J}_{M_J \times K}\}$
        end if
      end if
    end for
    ...
  end for
end for

3.3 SINR at SUs

In the DRS-MCS strategy, the SUs have to share the downlink resources with the CHs for D2D multicast communications. This in turn makes the QoS of the SUs degraded due to the interference from the transmissions of the CHs when reusing the downlink resources. To limit the interference impact on the SUs for a high QoS guarantee, it is necessary to compute the SINR at the SU k which is given by

$$\gamma^k = \frac{P_0^k G_0^k}{N_0 + v_j^{m,k} P_j^m G_j^{m,k}}, \tag{8}$$

where G_0^k and $G_j^{m,k}$ are the channel gains from the MBS and the CH m in the cluster j to the SU k.

4 DRS-MCS Optimization Problem and Solution

The DRS-MCS optimization problem aims at maximizing the objective function C (7). We further take into account the constraints on $v_j^{m,k}$ so that an SU can share its downlink resource with up to only one CH in the whole system (9b) and there is up to only one SU sharing its downlink resource with one CH in a cluster (9c). In addition, the target SINR γ_0 is considered to guarantee the QoS of the SUs (9d). The DRS-MCS optimization problem is formulated as below

$$\max_{v_j^{m,k}} C \tag{9a}$$

$$\text{s.t. } \sum_{k=1}^{K} \sum_{m=1}^{M_j} v_j^{m,k} \leq 1, j = 1, 2, ..., J, \tag{9b}$$

$$\sum_{j=1}^{J} \sum_{m=1}^{M_j} v_j^{m,k} \leq 1, k = 1, 2, ..., K, \tag{9c}$$

$$v_j^{m,k} P_j^m G_j^{m,k} \leq \frac{P_0^k G_0^k}{\gamma_0} - N_0, k = 1, 2, ..., K, \tag{9d}$$

$$j = 1, 2, ..., J, m = 1, 2, ..., M_j.$$

where the constraint (9d) is derived from Eq. (8) by letting $\gamma^k \geq \gamma_0$.

The DRS-MCS optimization problem is solved by using exhaustive binary searching algorithm (EBSA) [27] presented in **Algorithm** 1. To solve the DRS-MCS optimization problem, we separate $v_j^{m,k}$ into J variables associated with J clusters, the variable j, i.e., $V_j = V_{M_j \times K}$, is an $M_j \times K$ matrix. So, finding $v_j^{m,k} = 1$ (or 0) is equivalent to finding the element at the row m and the column k of the matrix V_j evaluates to 1 (or 0). The variable V_j has its own search space of $\mathcal{V}_j = \{V_{M_j \times K}^1, V_{M_j \times K}^2, ..., V_{M_j \times K}^{v_j}, ..., V_{M_j \times K}^{2^{M_j \times K}}\}$, $v_j = 1, 2, ..., 2^{M_j \times K}$.

In **Algorithm** 1, line 1 generates J search spaces of J variables. The search space j has $2^{M_j \times K}$ matrices. Then, the output maximum value C^* is initially set at 0 in line 2. In lines 3–6, each permutation of J matrices created by selecting a matrix in each search space is considered checking if it satisfies the constraints (9b), (9c), and (9d) or not (line 7). If satisfied, the objective function C is computed (line 8) to obtain a higher value C^* and find the corresponding result $\mathcal{V}^*$ (lines 9–11). The EBSA terminates when it completes the computation of all permutations for finding the maximum value C^* and the optimal result $\mathcal{V}^*$. It is noted that the EBSA introduces a high memory and time complexity of $\mathcal{O}(2^{K \times \sum_{j=1}^{J} M_j})$. However, we apply the EBSA to solving the DRS-MCS optimization problem thanks to its simple implementation for the exact optimal results. Finding other proper algorithms that achieve exact or approximated optimal results at lower memory and time complexity is beyond the scope of the paper.

5 Performance Evaluation

In this paper, the system parameters used to deploy the DRS-MCS strategy for multicast video streaming in 5G UDNs are listed in Table 1. The distances between the MBS and the SUs/RUs, the FBSs and the RUs, the CHs and the SUs, and the CHs and the RUs are randomly uniform distributed in the ranges of [100, 1,000] m, [50, 200] m, [50, 100] m, and [1, 50] m, respectively. To evaluate the performance of the proposed DRS-MCS solution, we compare it to other schemes including average capacity (AVE), minimum capacity (MIN), without

Table 1. Parameters Setting

Symbols	Specifications
I	5 FBSs
J	5 Clusters
K	5 SUs
$\{M_j\}$	{2, 4, 6, 8, 10} CHs
$\{N_j\}$	{5, 10, 15, 20, 25} RUs
W	5 MHz
P_0^j, P_0^k	Fixed to 5W
$P_j^{i,n}$	Fixed to 0.1W
P_j^m	Randomly uniform distributed in the range of [0.001, 0.01]W
N_0	10^{-13}W
η	4 (path loss exponent)
γ_0	5 dB

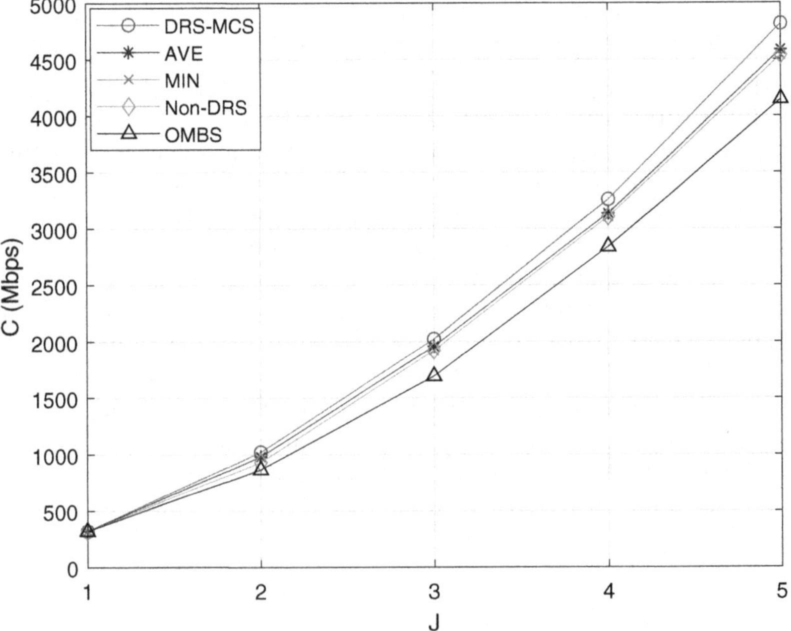

Fig. 2. System performance versus the number of clusters J.

downlink resource sharing (Non-DRS), and only MBS (OMBS). In AVE and MIN, we compute the average capacity and the minimum capacity of all the permutations in J search spaces that satisfy the constraints (9b), (9c), and (9d),

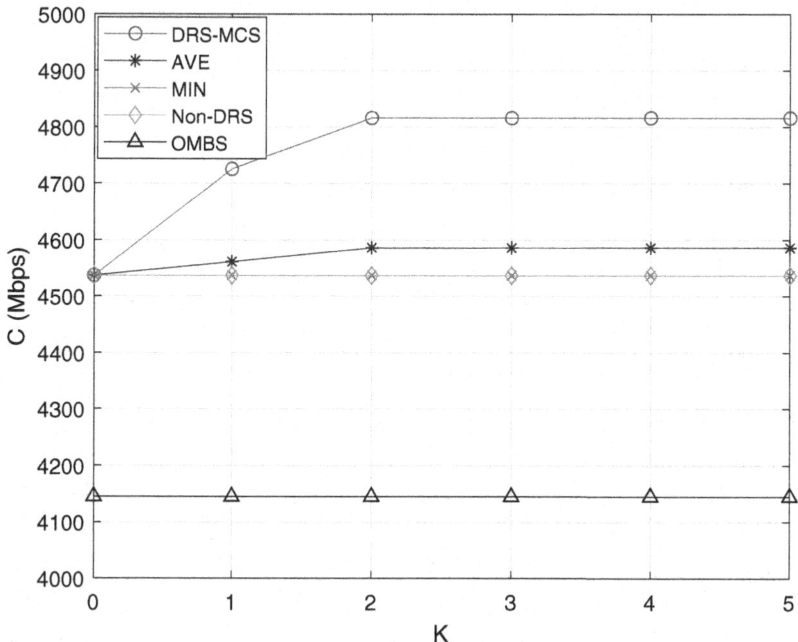

Fig. 3. System performance versus the number of SUs K.

instead of computing the maximum capacity as given in **Algorithm** 1. In Non-DRS, we do not consider sharing the downlink resources of the SUs. And in OMBS, the RUs are traditionally served by the MBS.

We evaluate the system performance of DRS-MCS, AVE, MIN, Non-DRS, and OMBS versus the number of clusters (J) as shown in Fig. 2. If $J = 1$, the performance of all schemes are the same since the best caching placement selected to serve the RUs is the MBS. Increasing J yields the higher number of RUs served by MBS, FBSs, and D2D multicast communications, and thus providing the RUs with higher system capacity. In comparison, the DRS-MCS outperforms the others thanks to more caching placement selection opportunities. The AVE, MIN and Non-DRS gain higher performance than the OMBS since they can further exploit the FBSs to serve the RUs. It is noted that the MIN and Non-DRS provide the RUs with the same performance since the minimum capacity of DRS-MCS is equivalent to the context of Non-DRS. The OMBS serves the RUs the worst system capacity due to no FBSs nor DRS assisted.

The system performance of DRS-MCS, AVE, MIN, Non-DRS, and OMBS versus the number of SUs (K) is illustrated in Fig. 3. We can observe that if there is no SU ($K = 0$) to share the downlink resources, the performance of DRS-MCS, AVE, MIN, Non-DRS are the same, but it is higher than that of the OMBS thanks to the assistance of FBSs. Increasing K provides more downlink resource sharing opportunities to increase the performance of DRS-MCS

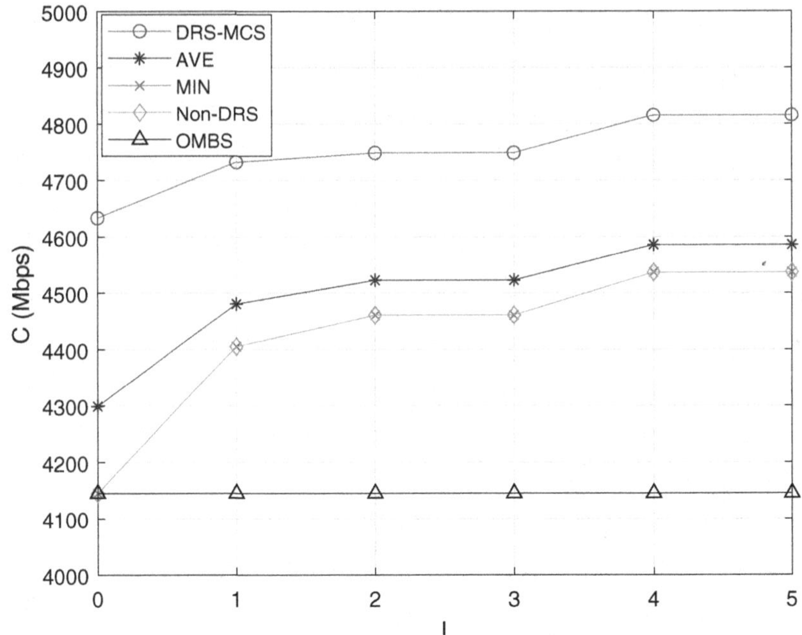

Fig. 4. System performance versus the number of FBSs I.

and AVE. Meanwhile, the performance of MIN and Non-DRS is the same and keeps unchanged with respect to K. Interestingly, the system performance gets saturated when K is high enough, i.e., $K = 3$. This finding helps the system designers consider selecting a proper number of SUs for high system capacity at reasonable computation cost of EBSA.

Figure 4 plots the system performance versus the number of FBSs (I). The results show that if there is no FBS ($I = 0$), the MIN and Non-DRS obviously become the OMBS due to no FBSs nor DRS assisted. The performance of DRS-MCS, AVE, MIN, and Non-DRS increases in accordance with the increase of I, but getting saturated if I is high enough ($I = 4$) or the new FBSs added are not better in terms of providing higher system capacity than the existing ones. In comparison, the DRS-MCS always outperforms the other AVE, MIN, Non-DRS, and OMBS schemes. In addition, similar to selecting the number of SUs, implementing DRS-MCS strategy must carefully consider selecting a proper number of FBSs to gain high system capacity at reasonable cost of computational resource and system architecture modification.

In Fig. 5, we further investigate the effect of the target SINR (γ_0) of SUs on the performance of DRS-MCS, AVE, MIN, Non-DRS, and OMBS. It is certain that increasing γ_0 to guarantee the QoS of SUs reduces the downlink resource sharing opportunities. As a result, the performance of DRS-MCS and AVE is reduced to that of MIN and Non-DRS when $\gamma_0 = 25\,\text{dB}$. It is important to

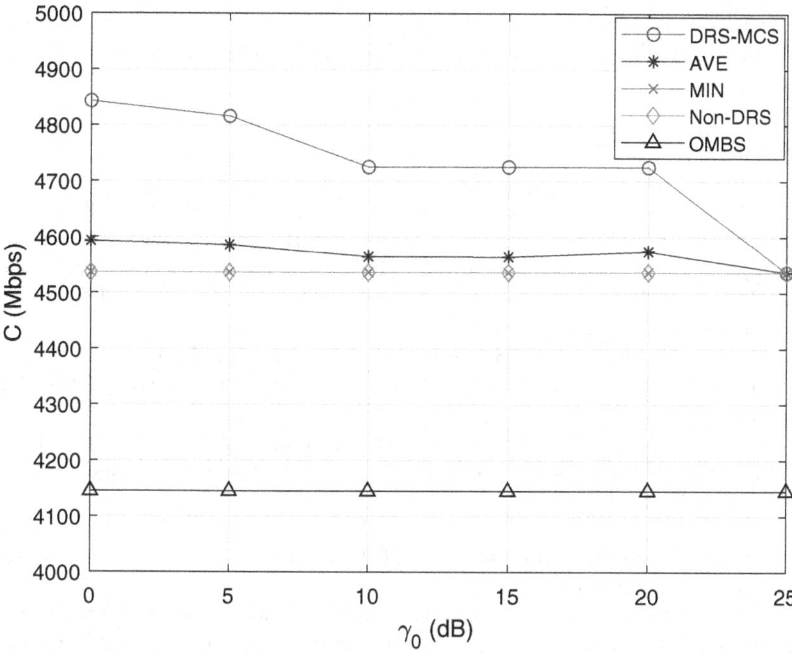

Fig. 5. System performance versus the number of FBSs γ_0.

observe that the proper value of γ_0 should be carefully selected so that the system capacity is high enough while guaranteeing the SUs a high QoS, $\gamma_0 = 5\,\text{dB}$ for example. Depending on the QoS demands of SUs, different values of γ_0 can be selected to make the DRS-MCS gain different system performances that are always higher than the system performances of other schemes.

6 Conclusion

In this paper, we have proposed the DRS-MCS solution for video streaming applications and services in 5G UDNs. The proposed DRS-MCS not only allows the RUs to receive the videos flexibly multicasted from the three-tier caching placements, i.e., MBS, FBSs, and CHs, but also enables to pair the SUs that have available downlink resources with the CHs that cache the videos, for D2D multicast communications by reusing the shared downlink resources. The objective of the DRS-MCS solution is to serve the RUs the highest system capacity while guaranteeing the QoS of the SUs by limiting the interference transmitted by the CHs that reuse the downlink resources of the SUs. Simulation results are insightfully analyzed to demonstrate the benefits of the proposed DRS-MCS solution compared to the other schemes. In addition, useful suggestions for system design and modification are provided to achieve the most effective DRS-MCS solution.

Acknowledgement. This research is funded by Vietnam National Foundation for Science and Technology Development (NAFOSTED) under grant number 102.04-2018.308.

References

1. Cisco, "Cisco Visual Networking Index: Global Mobile Data Traffic Forecast Update," in 2017–2022 White Paper, February 2019. https://www.cisco.com
2. Lecompte, D., Gabin, F.: Evolved multimedia broadcast/multicast service (eMBMS) in LTE-advanced: overview and Rel-11 enhancements. IEEE Commun. Mag. **50**(11), 68–74 (2012)
3. Araniti, G., Condoluci, M., Scopelliti, P., Molinaro, A., Iera, A.: Multicasting over emerging 5G networks: challenges and perspectives. IEEE Netw. **31**(2), 80–89 (2017)
4. Thanh, T.L., Hoang, T.M.: Cooperative spectrum-sharing with two-way AF relaying in the presence of direct communications. EAI Endorsed Trans. Ind. Netw. Intell. Syst. **5**(14), 1–9 (2018)
5. Nguyen, T.H., Nguyen, D.Q., Nguyen, V.D.: Quality of service provisioning for D2D users in heterogeneous networks. EAI Endorsed Trans. Ind. Netw. Intell. Syst. **6**(21), 1–7 (2019)
6. Lin, X., Ratasuk, R., Ghosh, A., Andrews, J.G.: Modeling, analysis, and optimization of multicast device-to-device transmissions. IEEE Trans. Wirel. Commun. **13**(8), 4346–4359 (2014)
7. Zulhasnine, M., Huang, C., Srinivasan, A.: Exploiting cluster multicast for P2P streaming application in cellular system. In: Proceedings of IEEE Wireless Communications and Networking Conference, Shanghai, China, April 2013, pp. 1–6 (2013)
8. Peng, B., Peng, T., Liu, Z., Yang, Y., Hu, C.: Cluster-based multicast transmission for device-to-device (D2D) communication. In: Proceedings of IEEE 78th Vehicular Technology Conference, Las Vegas, NV, September 2013, pp. 1–5 (2013)
9. Militano, L., Condoluci, M., Araniti, G., Molinaro, A., Iera, A., Muntean, G.-M.: Single frequency-based device-to-device-enhanced video delivery for evolved multimedia broadcast and multicast services. IEEE Trans. Broadcast. **61**(2), 263–278 (2015)
10. Zhu, Y., Qin, X., Zhang, P.: An efficient multicast clustering scheme for D2D assisted offloading in cellular networks. In: Proceedings of IEEE/CIC International Conference on Communications in China, Beijing, China, August 2018, pp. 1–5 (2018)
11. Hmila, M., Fernandez-Veiga, M., Rodriguez-Perez, M.: Distributed resource allocation approach for device-to-device multicast communications. In: Proceedings of International Conference on Wireless and Mobile Computing, Networking and Communications, Limassol, Cyprus, October 2018, pp. 1–8 (2018)
12. Kim, J.-H., Joung, J., Lee, J.W.: Resource allocation for multiple device-to-device cluster multicast communications underlay cellular networks. IEEE Commun. Lett. **22**(2), 412–415 (2018)
13. Cao, Y., Jiang, T., Chen, X., Zhang, J.: Social-aware video multicast based on device-to-device communications. IEEE Trans. Mob. Comput. **15**(6), 1528–1539 (2016)
14. Zhao, P., Feng, L., Yu, P., Li, W., Qiu, X.: A social-aware resource allocation for 5G device-to-device multicast communication. IEEE Access **5**, 15717–15730 (2017)

15. Moghaddam, S.S., Ghasemi, M.: Efficient clustering for multicast device-to-device communications. In: Proceedings of International Conference on Computer and Communication Engineering, Kuala Lumpur, Malaysia, September 2018, pp. 1–6 (2018)
16. Yang, L., Wu, D., Cai, Y.: A distributed social-aware clustering approach in D2D multicast communications. In: Proceedings of International Wireless Communications and Mobile Computing Conference, Limassol, Cyprus, pp. 1–6 (2018)
17. Wu, Y., Wu, D., Yang, L., Shi, X., Ao, L., Fu, Q.: Matching-coalition based cluster formation for D2D multicast content sharing. IEEE Access **7**, 73913–73928 (2019)
18. Yang, L., Wu, D., Xu, S., Zhang, G., Cai, Y.: Social-energy-aware user clustering for content sharing based on D2D multicast communications. IEEE Access **6**, 36092–36104 (2018)
19. Wu, Y., Wu, D., Yang, L., Xu, S.: Incentive-based cluster formation for D2D multicast content sharing. In: Proceedings of Asia-Pacific Conference on Communications, Ningbo, China, November 2018, pp. 1–6 (2018)
20. Feng, L., Zhao, P., Zhou, F., Yin, M., Yu, P., Li, W., Qiu, X.: Resource allocation for 5G D2D multicast content sharing in social-aware cellular networks. IEEE Commun. Mag. **56**(3), 112–118 (2018)
21. Phan, T.-M., Vo, N.-S., Bui, M.-P., Dang, X.-K., Ha, D.-B.: Downlink resource sharing and caching helper selection control maximized multicast video delivery capacity in dense D2D 5G networks. J. Sci. Technol. **18**(4.2), 12–20 (2020)
22. Vo, N.-S., Phan, T.-M., Bui, M.-P., Dang, X.-K., Viet, N.T., Yin, C.: Social-aware spectrum sharing and caching helper selection strategy optimized multicast video streaming in dense D2D 5G networks. IEEE Syst. J. 1–12 (2020)
23. Cheung, W., Quek, T., Kountouris, M.: Throughput optimization, spectrum allocation, and access control in two-tier femtocell networks. IEEE J. Sel. Areas Commun. **30**(3), 561–574 (2012)
24. Gyawali, S., Xu, S., Ye, F., Hu, R.Q., Qian, Y.: A D2D based clustering scheme for public safety communications. In: Proceedings of IEEE 87th Vehicular Technology Conference, Porto, Portugal, pp. 1–5 (2018)
25. Vo, N.-S., Duong, T.Q., Tuan, H.D., Kortun, A.: Optimal video streaming in dense 5G networks with D2D communications. IEEE Access **6**, 209–223 (2017)
26. Bhardwaj, A., Agnihotri, S.: Energy- and spectral-efficiency trade-off for D2D-multicasts in underlay cellular networks. IEEE Wirel. Commun. Lett. **7**(4), 546–549 (2018)
27. Vo, N.-S., Duong, T.Q., Guizani, M., Kortun, A.: 5G optimized caching and downlink resource sharing for smart cities. IEEE Access **6**, 31457–31468 (2018)

Performance Analysis of Relay Selection on Cooperative Uplink NOMA Network with Wireless Power Transfer

Van-Long Nguyen[1(✉)], Van-Truong Truong[2], Dac-Binh Ha[2,3], Tan-Loc Vo[4], and Yoonill Lee[5]

[1] Graduate School, Duy Tan University, Danang 550000, Vietnam
vanlong.itqn@gmail.com

[2] Faculty of Electrical-Electronic Engineering, Duy Tan University, Danang 550000, Vietnam
truongvantruong@dtu.edu.vn, hadacbinh@duytan.edu.vn

[3] Institute of Research and Development, Duy Tan University, Danang 550000, Vietnam

[4] Pham Van Dong University, Quang Ngai, Vietnam
vtlocqng@gmail.com

[5] Department of Engineering Technology, Purdue University Northwest, Hammond, IN, USA
lee2273@pnw.edu

Abstract. Wireless power transmission in the next-generation wireless networks is the subject that attracts a lot of attention from academia and industry. In this work, we study and analyze the performance of relay selection on uplink non-orthogonal multiple access (NOMA) networks with wireless power transmission. Specifically, the considered system consists of one base station, multiple power-constrained relays and a pair of NOMA users. The best relay (with highest energy harvested from the base station) is chosen to cooperate with two users which use NOMA scheme to send messages to the base station. To analyze the performance, based on the statistical characteristics of signal-to-noise ratio (SNR) and signal-to-interference-plus-noise ratio (SINR), using the Gaussian-Chebyshev quadrature method, the closed-form expressions of outage probability and throughput for two users are derived. In order to understand more details about the behavior of this considered system, the numerical results on outage probability and throughput of a given system are provided following the system key parameters, such as the transmit power, the number of relays, time switching ratio and energy conversion efficiency. In the end, the theoretical result is also verified by using the Monte-Carlo simulation. The simulation results demonstrate that the performance of the system is improved by increasing the number of relays.

Keywords: Non-orthogonal multiple access · Wireless power transfer · Cooperative network · RF energy harvesting · Outage probability · Throughput

N.-S. Vo and V.-P. Hoang (Eds.): INISCOM 2020, LNICST 334, pp. 32–44, 2020.
https://doi.org/10.1007/978-3-030-63083-6_3

1 Introduction

In recent years, we have witnessed an explosion in the number of wireless devices: smartphones, wireless sensors, wireless devices, etc. These devices are increasingly integrated with many functions, leading to the urgent need for the power supply. One of the methods that attracted the attention of both academics and the industry is radio frequency (RF) energy harvesting (EH), which enables the converting of the received RF signals into electricity energy [1–5]. We also realize that the growingtrend of wireless networks is to serve a greater number of users while still have faster data transmission speed and higher reliability. In recent years, there has been many researches on 5G networks to reach that goal, in which, NOMA has emerged as the strongest candidate with the ability to serve multiple users using the same amount of time and frequency resources [6–8]. Most of articles currently focus on NOMA downlink networks [9,10], while uplink networks [11–14] are equally important in many systems, such as wireless sensors networks. The paper [11] proposed uplink NOMA network for two users, in which, users with better channel condition communicated directly with Base Station (BS), while the remaining user with bad channel condition was supported by half-duplex decode-and-forward (DF) relay. The uplink NOMA model in which the cell-center user directly connected to the BS and DF relay, while the cell-edge user communicated with the BS with the help of DF relay, is proposed in [12]. The outage probability and average sum rate were analyzed to study the performance of this considered system. The [13] studied NOMA uplink system with massive connectivity requirements, including Internet of Things (IoT) nodes, mobile devices, or unmanned aerial vehicles (UAVs). In particular, users can select buffer-aided (BA) relay in the cluster of relays, which is called flex-NOMA, to minimize packet delay. Outage probability of system is presented for comparing with OMA model. Another study of two-user uplink NOMA is presented in [14], in which, user near the BS acts as a transducer which can be switched in half-duplex or full-duplex to aid communication between remote users with BS. Two scenarios are given by the author to evaluate the system: (1) a direct channel between the BS and remote users and (2) no direct channel between BS and remote users.

As such, the combination between NOMA and relay techniques and RF EH will be considered for the new generation network.

Different from the works above, our work focus on the performance analysis of relay selection on cooperative uplink NOMA network with wireless power transfer. Our article contributes to the ideas below:

1. Proposing the relay selection on cooperative uplink NOMA network with wireless power transfer scheme.
2. Deriving exact closed-form expressions of outage probability and throughput for each user.
3. Evaluating the performance using tools of outage probability and throughput expressions.
4. The behavior of the considered system is assessed concerning different key parameters, such as transmit power, number of the relay, time switching ratio and energy conversion efficiency.

The rest of this paper is organized as follows. The system model is presented in Sect. 2. The performance of the considered system is analyzed in Sect. 3. The numerical results are shown in Sect. 4. Finally, Sect. 5 draws the conclusion of our paper.

2 Network and Channel Models

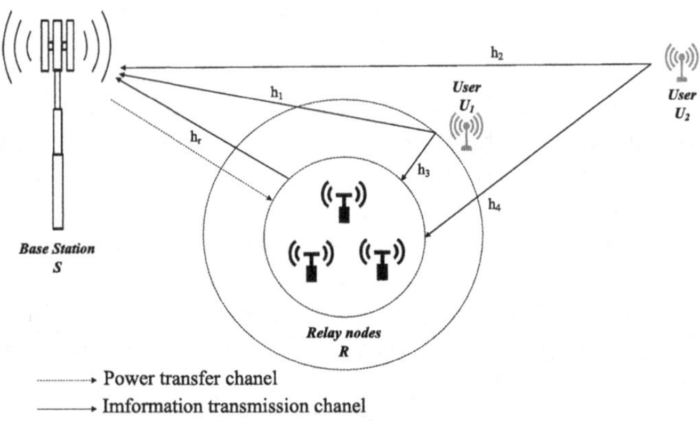

Fig. 1. System model for uplink NOMA AF relaying network

Figure 1 is a model of an uplink cooperative NOMA system in which the signals are transmitted from the two NOMA users (near user U_1 and far user U_2) to the base station (S) with the cooperation of energy-constrained amplify-and-forward relay nodes (R). Assuming that all users and relay nodes are single antenna devices and operate in half-duplex mode. The relay nodes are located closer to the base station than the users. With two users, the one that has shorter distance to the base station is named as U_1, the farther one is U_2.

At the beginning of each transmission, the channel information from S to the relays is collected by the base station using pilot signals. Thereby, S identifies the relay with the best ability to receive energy to specify it the relay node from the user to the base station. The protocol for this system is proposed as follows:

1. In the first phase (Power transfer phase): Based on Rayleigh fading channel coefficient from link S to relay nodes R (h_r), S chose the relay node (R^*), which had the best Rayleigh fading channel coefficient (h_{r^*}) to transmit RF energy with power P_0 in the time αT where $\alpha (0 < \alpha < 1)$ denotes the time switching ratio and T is the block time.
2. In the second phase (Information transmission phase): Users U_1 and U_2 simultaneously broadcast their signals (s_1, s_2) to S and R on the same frequency blocks, following the NOMA scheme within the duration of $(1 - \alpha)T/2$.

3. In the third phase (Relay phase): R^* amplifies the received signal and forwards to S in the remaining time of $(1-\alpha)T/2$ with the energy harvested from S by using the best Rayleigh fading channel coefficient (h_{r^*}).

Finally, S uses the selection combining (SC) scheme and successive interference cancellation (SIC) to combines two received signals, i.e., the direct signal from users and the relaying signal from R^* [15].

2.1 Power Transfer Phase

In this phase, S transmits energy to the relay nodes with a power of P_0, the energy gained at the relay during αT can be described as follow:

$$E_{R^*} = \frac{\eta P_0 |h_{r^*}|^2 \alpha T}{d_{r^*}^{\theta}}, \tag{1}$$

where η indicates the energy conversion efficiency ($0 < \eta < 1$), $|h_r^*|^2$ denotes the Rayleigh fading channel coefficient of $S - R^*$, d_{r^*} is the Euclidean distance of link $S - R^*$ and θ is denoted as the path-loss exponent.

2.2 Information Transmission Phase

In the second phase, both users U_1 and U_2 simultaneously broadcast their own information to base station and relays in the duration of $(1-\alpha)T/2$. It is assumed S applying SIC to decode s_1 and s_2 from the direct signal. The instantaneous SINR for detecting the signal from U_1 is:

$$\gamma_d^{s_1} = \frac{P_1 |h_1|^2 / d_1^{\theta}}{P_2 |h_2|^2 / d_2^{\theta} + \sigma^2} = \frac{\gamma_1 |h_1|^2}{\gamma_2 |h_2|^2 + 1} \tag{2}$$

The SNR for detecting the signal from U_2 is:

$$\gamma_d^{s_2} = \gamma_2 |h_2|^2 \tag{3}$$

where $\gamma_1 = \frac{P_1}{\sigma^2 d_1^{\theta}}, \gamma_2 = \frac{P_2}{\sigma^2 d_2^{\theta}}$ are denoted as SNR at U_1 and U_2, P_i ($i = \{1, 2\}$) denote the transmit powers of users U_1 and U_2, respectively. h_i denotes Rayleigh fading channel coefficients between users and S. Note that in this paper, the received signal at S and R^* are y_1 and y_2, respectively, similar to the formulas (2) and (5) in the paper [15]

2.3 Relay Phase

In this phase, R^* amplifies and forwards the received signal to S by using the harvested energy E_{R^*} in the duration of $(1-\alpha)T/2$. We ignore the processing power required by the transmit/receive the circuitry of R. The transmit power of R^* is given by

$$P_{R^*} = \frac{2\eta\alpha P_0 |h_r^*|^2}{(1-\alpha) d_{r^*}^{\theta}} = \frac{a P_0 |h_{r^*}|^2}{d_{r^*}^{\theta}}, \tag{4}$$

where $a = \frac{2\eta\alpha}{1-\alpha}$

Using the AF scheme, the transmit signal at R^* with transmit power P_{R^*} is given by

$$y_{R^*} = Gy_2, \tag{5}$$

where G is the relaying gain in the AF scheme applied at R. G is expressed as

$$G \triangleq \sqrt{\frac{P_{R^*}}{P_1|h_3|^2/d_3^\theta + P_2|h_4|^2/d_4^\theta + \sigma^2}}. \tag{6}$$

The received signal at S in this phase is written as

$$y_3 = \frac{Gh_{r^*}}{\sqrt{d_{r^*}^\theta}}\left(\sqrt{\frac{P_1}{d_3^\theta}}h_3 s_1 + \sqrt{\frac{P_2}{d_4^\theta}}h_4 s_2 + n_2\right) + n_3, \tag{7}$$

where $n_3 \sim \mathcal{CN}(0, \sigma^2)$

Assuming that S only uses the relaying signal to detect s_1 and s_2 by applying SIC, the instantaneous SINR to detect s_1 at S in this phase is expressed as

$$\begin{aligned}\gamma_r^{s_1} &= \frac{P_1 G^2|h_3|^2|h_{r^*}|^2/d_3^\theta d_{r^*}\theta}{G^2(P_2|h_4|^2/d_4^\theta + \sigma^2)|h_{r^*}|^2/d_{r^*}^\theta + \sigma^2}\\ &= \frac{a\gamma_3\gamma_{r^*}|h_3|^2|h_{r^*}|^4}{\gamma_3|h_3|^2 + (\gamma_4|h_4|^2+1)(a\gamma_{r^*}|h_{r^*}|^4+1)}.\end{aligned} \tag{8}$$

The instantaneous SNR to detect s_2 at S in this phase is obtained by

$$\gamma_r^{s_2} = \frac{a\gamma_4\gamma_{r^*}|h_3|^2|h_{r^*}|^4}{\gamma_3|h_3|^2 + \gamma_4|h_4|^2 + a\gamma_{r^*}|h_{r^*}|^4 + 1} \tag{9}$$

where $\gamma_{r^*} = \frac{P_0}{\sigma^2 d_{r^*}^\theta}, \gamma_3 = \frac{P_1}{\sigma^2 d_3^\theta}, \gamma_4 = \frac{P_2}{\sigma^2 d_4^\theta}$ are denoted as transmit SNR at S, U_1 and U_2, respectively.

By applying the SC scheme at S, the instantaneous SINR and SNR to detect s_1 and s_2 at S in this phase are respectively given by

$$\gamma_{s_1} = max\left\{\gamma_r^{s_1}, \gamma_d^{s_1}\right\}, \tag{10}$$

$$\gamma_{s_2} = max\left\{\gamma_r^{s_2}, \gamma_d^{s_2}\right\}. \tag{11}$$

Notice that $|h_i|^2$ ($i \in \{1,2,3,4\}$) are assumed as the i.i.d. Rayleigh channel gains following exponential distributions with parameters λ_i Thus, the probability density function (PDF) and the cumulative distribution function (CDF) of these random variables $|h_i|^2$ are written as

$$f_{|h_i|^2}(x) = \frac{1}{\lambda_i}e^{-\frac{x}{\lambda_i}}, \tag{12}$$

$$F_{|h_i|^2}(x) = 1 - e^{-\frac{x}{\lambda_i}}. \tag{13}$$

R^* is the best capable of receiving power, selected from N relays. Therefore, PDF and CDF of the random variables $|h_{r^*}|^2$ are written as

$$f_{|h_{r^*}|^2}(x) = \frac{N}{\lambda_{r^*}} e^{-\frac{x}{\lambda_{r^*}}} \left(1 - e^{-\frac{x}{\lambda_{r^*}}}\right)^{(N-1)}, \tag{14}$$

$$F_{|h_{r^*}|^2}(x) = \left(1 - e^{-\frac{x}{\lambda_{r^*}}}\right)^N = \sum_{k=0}^{N} C_k^N (-1)^{N-k} e^{-\frac{(N-k)x}{\lambda_{r^*}}}. \tag{15}$$

For further calculation, first we derive the expressions of CDF and PDF of $X = \frac{\gamma_j |h_j|^2}{\gamma_k |h_k|^2 + 1}$ as follows

$$F_X(x) = \Pr\left(\frac{\gamma_j |h_j|^2}{\gamma_k |h_k|^2 + 1} < x\right) = 1 - \frac{\lambda_j \gamma_j e^{-\frac{x}{\lambda_j \gamma_j}}}{\lambda_j \gamma_j + \lambda_k \gamma_k x}. \tag{16}$$

$$f_X(x) = \frac{\lambda_j \gamma_j \lambda_k \gamma_k e^{-\frac{x}{\lambda_j \gamma_j}}}{(\lambda_j \gamma_j + \lambda_k \gamma_k x)^2} + \frac{e^{-\frac{x}{\lambda_j \gamma_j}}}{\lambda_j \gamma_j + \lambda_k \gamma_k x}. \tag{17}$$

where $j \in \{1, 3\}, k \in \{2, 4\}$

3 Performance Analysis

In this section, we analyze the performance of this considered system in terms of outage probability (P_{out}) and throughput (τ). The outage probability P_{out} is defined as the probability that the instantaneous capacity,

$$C = \frac{1-\alpha}{2} log_2(1 + SINR). \tag{18}$$

falls below a predetermined rate threshold. We assume that $R_{th} > 0 (bps/Hz)$ is the minimum required data rate for both users. Therefore, P_{out} is expressed as follows

$$\begin{aligned} P_{out} &= Pr\left(\frac{1-\alpha}{2} log_2(1 + SINR) < R_{th}\right) \\ &= Pr\left(SINR < 2^{\frac{2R_{th}}{1-\alpha}} - 1 \triangleq \gamma_{th}\right). \end{aligned} \tag{19}$$

For our considered system, the outage probability $P_{out}^{s_1}$ for detecting s_1 and $P_{out}^{s_2}$ for detecting s_2 are expressed as

$$P_{out}^{s_1} = Pr(\gamma_{s_1} < \gamma_{th}) = Pr(\gamma_d^{s_1} < \gamma_{th}) \cdot Pr(\gamma_r^{s_1} < \gamma_{th}), \tag{20}$$

$$P_{out}^{s_2} = Pr(\gamma_{s_2} < \gamma_{th}) = Pr(\gamma_d^{s_2} < \gamma_{th}) \cdot Pr(\gamma_r^{s_2} < \gamma_{th}), \tag{21}$$

respectively.

Proposition 1. *Under Rayleigh fading, the outage probability* $P_{out}^{s_1}$ *for detecting* s_1 *for detecting is given by (22) and (23), respectively.*

$$Pr(\gamma_d^{s1} < \gamma_{th}) = 1 - \frac{\gamma_1\lambda_1}{\gamma_{th}\gamma_2\lambda_2 + \gamma_1\lambda_1} e^{-\frac{\gamma_t}{\gamma_1\lambda_1}} \tag{22}$$

$$\begin{aligned} Pr(\gamma_r^{s1} < \gamma_{th}) = 1 - \frac{(-1)^N \pi\mu e^{-\frac{\gamma_{th}}{\lambda_3\gamma_3}}}{M} \sum_{k=1}^{N} C_k^N (-1)^{(N-k)} \sum_{i=1}^{M} u_i^{\sqrt[k]{\left(\frac{\mu}{ln^2 u_i}+\gamma_{th}+1\right)}-1} \\ \times \frac{\lambda_3\gamma_3\lambda_4\gamma_4 + \lambda_3\gamma_3 + \lambda_4\gamma_4(\frac{\mu}{ln^2 u_i} + \gamma_{th})}{\left[\lambda_3\gamma_3 + \lambda_4\gamma_4(\frac{\mu}{ln^2 u_i} + \gamma_{th})\right]^2 ln^3 u_i} e^{\frac{-\mu}{\lambda_3\gamma_3 ln^2 u_i}} \sqrt{1-\beta_i^2} \end{aligned} \tag{23}$$

where $u_i = \frac{\beta_i+1}{2}$ *with* $\beta_i = \cos\left(\frac{2i-1}{2M}\pi\right)$, *and* M *is the complexity-vs-accuracy trade-off coefficient.*

Proof. See Appendix B.

Proposition 2. *Under Rayleigh fading, the outage probability* $P_{out}^{s_2}$ *for detecting* s_2 *is given by (24), respectively.*

$$\begin{aligned} P_{out}^{s2} = \left(1 - e^{-\frac{\gamma_{th}}{\lambda_2\gamma_2}}\right) \Bigg[1 + \frac{\pi^2 \mu e^{-\frac{\gamma_{th}}{\lambda_4\gamma_4}}}{2MH\lambda_4\gamma_4} \sum_{k=0}^{N} C_k^N (-1)^{N-k} \sum_{i=1}^{M}\sum_{j=1}^{H} e^{-\frac{\mu}{\lambda_4\gamma_4 ln^2 u_i}} \\ \times \frac{u_i^{\sqrt[N-k]{\left(\frac{\mu}{ln^2 u_i} - \lambda_3\gamma_3 ln v_j + \gamma_{th}+1\right)}}}{u_i ln^3 u_i} \sqrt{(1-\phi_i^2)(1-\varphi_j^2)} \Bigg] \end{aligned} \tag{24}$$

where $u_i = \frac{\phi_i+1}{2}, \phi_i = cos(\frac{2i-1}{2M}\pi), v_j = \frac{\varphi_j+1}{2}, \varphi_j = cos(\frac{2j-1}{2H}\pi), M$ *and* H *are the complexity-vs-accuracy trade-off coefficients.*

Proof. See Appendix A.

Proposition 3. *The throughput expressions for* U_1 *and* U_2 *are respectively given by*

$$\tau_1 = \frac{R_{th}(1-\alpha)(1-P_{out}^{s_1})}{2}, \tag{25}$$

$$\tau_2 = \frac{R_{th}(1-\alpha)(1-P_{out}^{s_2})}{2}. \tag{26}$$

4 Nummerical Results and Disscussion

In this section, the performance of the system is analyzed based on the results of the outage probability and throughput of the system by using Monte-Carlo simulation.

No	Description	Notation	Value	Unit
1	The distances from S to U_1	d_1	3	
2	The distances from S to U_2	d_2	6	
3	The distances from R^* to U_1	d_3	2	
4	The distances from R^* to U_2	d_4	5	
5	The distances from S to R^*	d_{r^*}	1	
6	The rate threshold	R_{th}	1	bps/Hz
7	The variance of additive noises	σ^2	1	
8	The path loss exponent	θ	2	
9	The complexity-vs-accuracy trade-off coefficients	M, H	100	

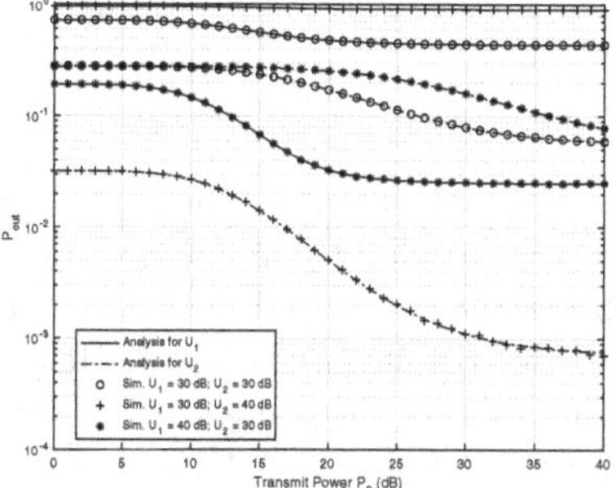

Fig. 2. Outage probability of the U_1 and U_2 versus P_0 with different values of (P_1, P_2) where $\alpha = 0.4, \eta = 0.8$

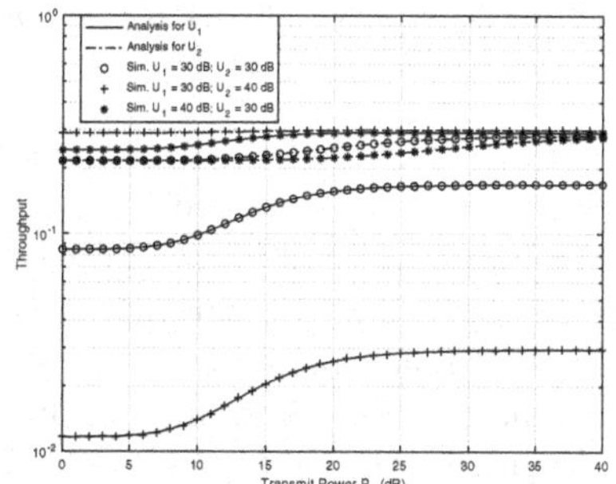

Fig. 3. Throughput of the U_1 and U_2 versus P_0 with different values of (P_1, P_2) where $\alpha = 0.4, \eta = 0.8$

The predefined simulation parameters are set in the table below:

Figure 2 and Fig. 3 show the system performance results, which are, particularly, the outage probability and the throughput at users U_1 and U_2, in comparison with transmit power at the base station (P_0) with different values of the transmit powers at users U_1 (P_1) and U_2 (P_2). Specifically, from Fig. 2, we can see that the outage probability at U_1 and U_2 decreases when increasing P_0. At the same time, when P_2 increases, the user's performance reduces as a result because of the increase in interference. This is explained in formulas (2) and (8). Similar conclusions are also obtained in case of considering the out probability for user U_2 in Fig. 2. Specifically, the system performance at user U_2 improves if we increase P_2 and decrease P_1, as shown in formulas (3) and (9). In Fig. 3, the throughputs of both users U_1 and U_2 improve when increasing P_0. In this result, we also see that, when increasing P_1 and decreasing P_2, the throughput of U_1 increases, in contrast, the throughput of U_2 increases when increasing P_2 and decreasing P_1.

The number of relay nodes is also an important parameter of the system. Figure 4 and Fig. 5 show the effect of the number of relays on the performance

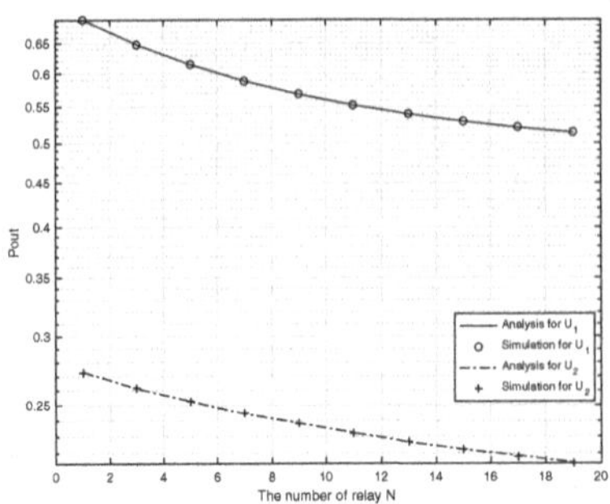

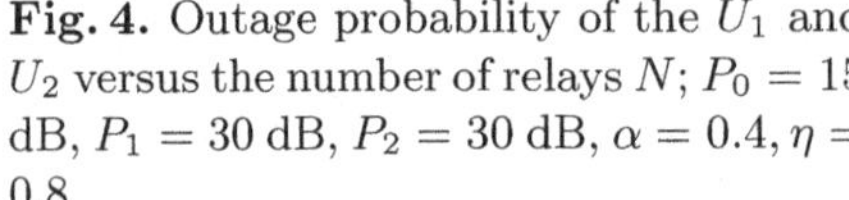

Fig. 4. Outage probability of the U_1 and U_2 versus the number of relays N; $P_0 = 15$ dB, $P_1 = 30$ dB, $P_2 = 30$ dB, $\alpha = 0.4, \eta = 0.8$

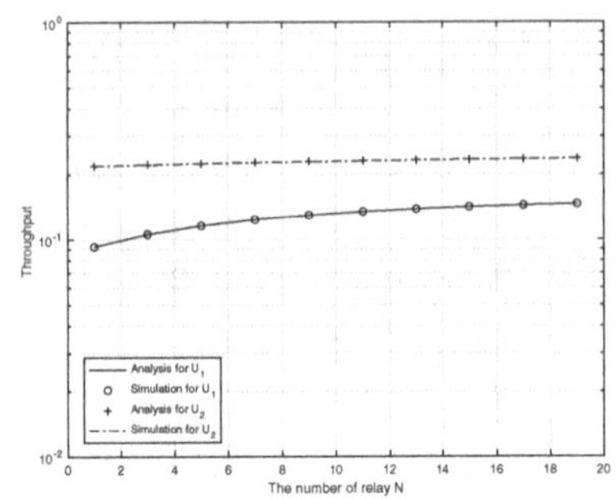

Fig. 5. Throughput of the U_1 and U_2 versus the number of relays N; $P_0 = 15$ dB, $P_1 = 30$ dB, $P_2 = 30$ dB, $\alpha = 0.4, \eta = 0.8$

of the system in terms of outage probability and throughput. In Fig. 4, the outage probability of both users decreases when increasing the number of relays. On the contrary, the throughput of both users increases in Fig. 5 when the number of relays increases. Through the above two results, we conclude that the performance of the system is improved when increasing the number of relays.

Figure 6 is the simulation result of the outage probability of both users U_1 and U_2 following the energy conversion efficiency. We see that the outage probability decreases with the increasing of the energy conversion efficiency. It means that the performance of the system in terms of the outage probability is improved by increasing this parameter. Figure 7 shows that the throughputs of both users U_1 and U_2 increases with an increase in energy conversion efficiency. The reason for the two results in Fig. 6 and Fig. 7 is that as the energy conversion efficiency increases, the energy for amplifying and forwarding at R^* increases, thus the signal transfer efficiency at the uplink of the system is increased.

The impact of time switching ratio on system performance in terms of outage probability is illustrated in Fig. 8. This result shows that, with the simulation parameters described on the table and the values: $P_0 = 15$ dB, $P_1 = 30$ dB, $P_2 = 30$ dB, $\eta = 0.8$, when increasing α $(\alpha > 0)$ until 0.1, the outage probability of both users U_1 and U_2 decreases. Then, with the value of α $(\alpha > 0.1)$ progressing to 1, the outage probability of both users U_1 and U_2 increases. Thereby, we find that there are a value of the time switching ratio at which the average outage probability is minimum. The algorithm for finding values of the time switching ratio to minimize the outage probability of this system will continue to be studied in the future.

The System performance simulation results in terms of throughput versus α were performed in Fig. 9, which performed a throughput simulation of both users U_1 and U_2. It can be observed that the throughput is reduced when we increase the value of the tim switching ratio.

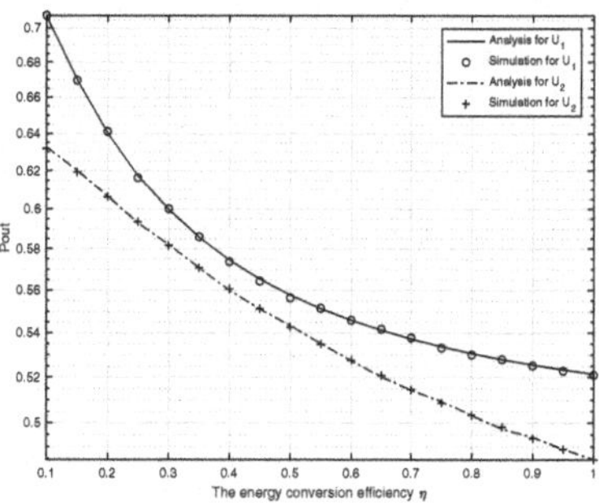

Fig. 6. Outage probability of the U_1 and U_2 versus the energy conversion efficiency; $P_0 = 15$ dB, $P_1 = 30$ dB, $P_2 = 30$ dB, $\alpha = 0.4$

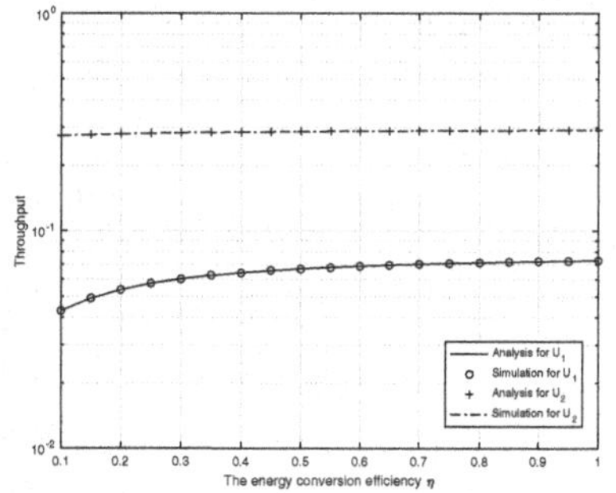

Fig. 7. Throughput of the U_1 and U_2 versus the energy conversion efficiency; $P_0 = 15$ dB, $P_1 = 30$ dB, $P_2 = 30$ dB, $\alpha = 0.4$

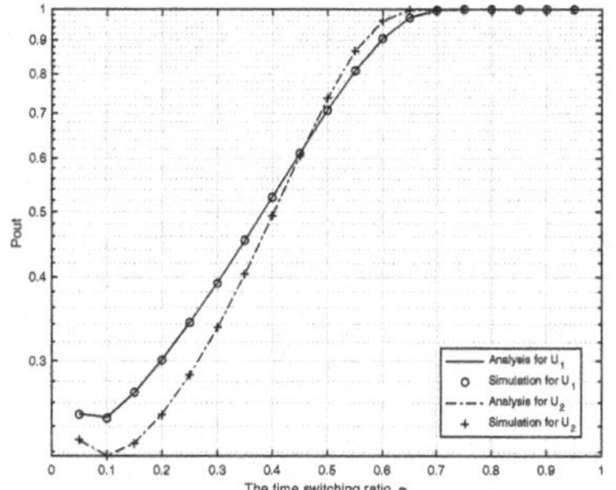

Fig. 8. Outage probability of the U_1 and U_2 versus the time switching ratio; where $P_0 = 15$ dB, $P_1 = 30$ dB, $P_2 = 30$ dB, $\eta = 0.8$

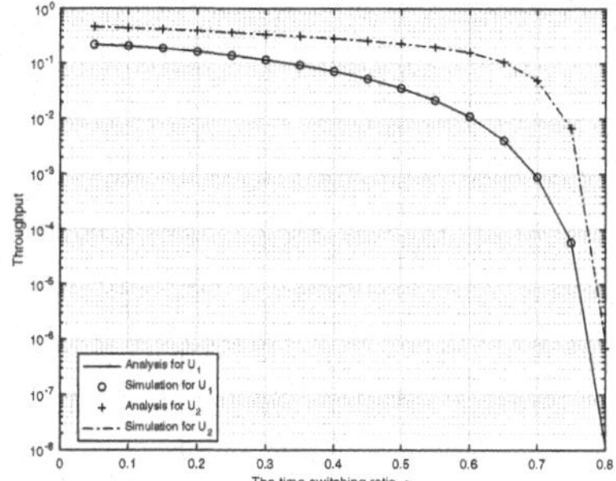

Fig. 9. Throughtput of the U_1 and U_2 versus the time switching ratio; where $P_0 = 15$ dB, $P_1 = 30$ dB, $P_2 = 30$ dB, $\eta = 0.8$

5 Conclusion

In this work, we studied an uplink cooperative NOMA network with wireless power transfer, in which, the best relay (i.e., the relay with the highest power received from the base station) helps the users forward information to the base station. The closed-form expressions of the outage probability and throughput for two users are given to describe the performance of the system. The simulation results showed that the performance of the U_1 can be improved when increasing P_0, P_1, N, and η. On the contrary, the performance of the U_1 decreases when increasing P_2. Similarly, the performance of the U_2 improves when increasing P_0, P_2, N, η and decreasing P_1. Especially, iteratively using numerical methods, we discovered that the system exists a value of time switching ratio that makes the average outage probability to be minimized. Therefore, we need to carefully consider the time switching ratio for the best system performance.

Appendix A

$$
\begin{aligned}
P_{out}^{s2} &= \Pr\left(\gamma_2|h_2|^2 < \gamma_{th}\right) . \Pr\left(\frac{a\gamma_r\gamma_4|h_4|^2|h_{r*}|^4}{a\gamma_{r*}|h_{r*}|^4 + \gamma_3|h_3|^2 + \gamma_4|h_4|^2 + 1} < \gamma_{th}\right) \\
&= F_{|h_2|^2}\left(\frac{\gamma_{th}}{\gamma_2}\right)\left[F_{|h_4|^2}\left(\frac{\gamma_{th}}{\gamma_4}\right) + \int_0^\infty \int_{\frac{\gamma_{th}}{\gamma_4}}^\infty F_{|h_{r*}|^2}\right. \\
&\quad \left.\left(\sqrt{\frac{\gamma_{th}(\gamma_3 x + \gamma_4 y + 1)}{a\gamma_{r*}(\gamma_4 y - \gamma_{th})}}\right) f_{|h_4|^2}(y) dy f_{|h_3|^2}(x) dx\right] \\
&\overset{(a)}{=} \left(1 - e^{-\frac{\gamma_{th}}{\lambda_2\gamma_2}}\right)\left[1 + \frac{2\mu e^{-\frac{\gamma_{th}}{\lambda_4\gamma_4}}}{\lambda_3\lambda_4\gamma_4}\sum_{k=0}^{N} C_k^N (-1)^{N-k}\right. \\
&\quad \left.\int_0^\infty \int_0^1 \frac{z^{\sqrt[N-k]{\left(\gamma_3 x + \frac{\mu}{ln^2 z} + \gamma_{th} + 1\right)}} . e^{-\frac{\mu}{\lambda_4\gamma_4 ln^2 z} - \frac{x}{\lambda_3}}}{z ln^3 z} dz dx\right] \\
&\overset{(b)}{=} \left(1 - e^{-\frac{\gamma_{th}}{\lambda_2\gamma_2}}\right)\left[1 + \frac{2\mu e^{-\frac{\gamma_{th}}{\lambda_4\gamma_4}}}{\lambda_3\lambda_4\gamma_4}\sum_{k=0}^{N} C_k^N (-1)^{N-k}\right. \\
&\quad \left.\times \int_0^\infty \frac{\pi}{2M}\sum_{i=1}^{M} \frac{u_i^{\sqrt[N-k]{\left(\gamma_3 x + \frac{\mu}{ln^2 u_i} + \gamma_{th} + 1\right)}} . e^{-\frac{\mu}{\lambda_4\gamma_4 ln^2 u_i} - \frac{x}{\lambda_3}}\sqrt{1 - \phi_i^2}}{u_i ln^3 u_i} dx\right] \\
&= \left(1 - e^{-\frac{\gamma_{th}}{\lambda_2\gamma_2}}\right)\left[1 + \frac{\pi\mu e^{-\frac{\gamma_{th}}{\lambda_4\gamma_4}}}{M\lambda_3\lambda_4\gamma_4}\sum_{k=0}^{N} C_k^N (-1)^{N-k}\sum_{i=1}^{M}\frac{e^{-\frac{\mu}{\lambda_4\gamma_4 ln^2 u_i}}\sqrt{1 - \phi_i^2}}{u_i ln^3 u_i}\right. \\
&\quad \left.\times \int_0^\infty u_i^{\sqrt[N-k]{\left(\gamma_3 x + \frac{\mu}{ln^2 u_i} + \gamma_{th} + 1\right)}} . e^{-\frac{x}{\lambda_3}} dx\right] \\
&\overset{(c)}{=} \left(1 - e^{-\frac{\gamma_{th}}{\lambda_2\gamma_2}}\right)\left[1 + \frac{\pi^2\mu e^{-\frac{\gamma_{th}}{\lambda_4\gamma_4}}}{2MH\lambda_4\gamma_4}\sum_{k=0}^{N} C_k^N (-1)^{N-k}\right. \\
&\quad \left.\times \sum_{i=1}^{M}\sum_{j=1}^{H} \frac{e^{-\frac{\mu}{\lambda_4\gamma_4 ln^2 u_i}} u_i^{\sqrt[N-k]{\left(\frac{\mu}{ln^2 u_i} - \lambda_3\gamma_3 ln v_j + \gamma_{th} + 1\right)}}}{u_i ln^3 u_i}\sqrt{(1 - \phi_i^2)(1 - \varphi_j^2)}\right]
\end{aligned}
$$

where step (a) is obtained by letting $z = e^{-\sqrt{\frac{\mu}{(\gamma_4 y - \gamma_{th})}}}$ in which $\mu = \frac{\gamma_{th}}{a\lambda_{r*}\gamma_{r*}}$, step (b) and (c) are obtained by applying the Gaussian-Chebyshev quadrature method in which $u_i = \frac{\phi_i + 1}{2}, \phi_i = cos(\frac{2i-1}{2M}\pi), v_j = \frac{\varphi_j + 1}{2}, \varphi_j = cos(\frac{2j-1}{2H}\pi), M$ and H are the complexity-vs-accuracy trade-off coefficients.

Appendix B

$$
\begin{aligned}
Pr(\gamma_d^{s1} < \gamma_{th}) &= Pr\left(\frac{\gamma_1|h_1|^2}{\gamma_2|h_2|^2+1} < \gamma_{th}\right) = Pr\left(|h_1|^2 < \frac{\gamma_{th}(\gamma_2|h_2|^2+1)}{\gamma_1}\right) \\
&= \int_0^{\infty} F_{|h_1|^2}\left(\frac{\gamma_{th}(\gamma_2 x+1)}{\gamma_1}\right) f_{|h_2|^2}(x)dx \\
&= 1 - \frac{1}{\lambda_2} e^{-\frac{\gamma_{th}}{\lambda_1\gamma_1}} \int_0^{\infty} e^{-x\frac{\gamma_{th}\gamma_2\lambda_2+\gamma_1\lambda_1}{\gamma_1\lambda_1\lambda_2}} dx = 1 - \frac{\gamma_1\lambda_1}{\gamma_{th}\gamma_2\lambda_2+\gamma_1\lambda_1} e^{-\frac{\gamma_t}{\gamma_1\lambda_1}}
\end{aligned}
$$

$$
\begin{aligned}
Pr(\gamma_r^{s1} < \gamma_{th}) &= \Pr\left(\frac{a\gamma_3\gamma_r|h_3|^2|h_{r*}|^4}{(\gamma_4|h_4|^2+1)a\gamma_{r*}|h_{r*}|^4+\gamma_3|h_3|^2+\gamma_4|h_4|^2+1} < \gamma_{th}\right) \\
&= \Pr\left(a\gamma_r|h_{r*}|^4\left[\gamma_3|h_3|^2 - \gamma_{th}\left(\gamma_4|h_4|^2+1\right)\right] < \gamma_{th}\left[\gamma_3|h_3|^2+\gamma_4|h_4|^2+1\right]\right) \\
&= \Pr\left(\gamma_3|h_3|^2 - \gamma_{th}(\gamma_4|h_4|^2+1) < 0\right) \\
&+ \Pr\left(|h_{r*}|^4 < \frac{\gamma_{th}(\gamma_3|h_3|^2+\gamma_4|h_4|^2+1)}{a\gamma_{r*}\left[\gamma_3|h_3|^2 - \gamma_{th}(\gamma_4|h_4|^2+1)\right]}, \gamma_3|h_3|^2 - \gamma_{th}(\gamma_4|h_4|^2+1) > 0\right) \\
&= \Pr\left(\frac{\gamma_3|h_3|^2}{\gamma_4|h_4|^2+1} < \gamma_{th}\right) + \Pr\left(|h_{r*}|^4 < \frac{\gamma_{th}\left(\frac{\gamma_3|h_3|^2}{\gamma_4|h_4|^2+1}\right)}{a\gamma_{r*}\left(\frac{\gamma_3|h_3|^2}{\gamma_4|h_4|^2+1} - \gamma_{th}\right)}, \right. \\
&\left. \frac{\gamma_3|h_3|^2}{\gamma_4|h_4|^2+1} > \gamma_{th}\right) \\
&\overset{(a)}{=} 1 + (-1)^N \sum_{k=1}^{N} C_k^N (-1)^{(N-k)} \int_0^{\infty} e^{-\frac{k}{\lambda_{r*}}\sqrt{\frac{\gamma_{th}(y+\gamma_{th}+1)}{a\gamma_{th}y}}} \\
&\times \left[\frac{\lambda_3\gamma_3\lambda_4\gamma_4 e^{-\frac{y+\gamma_{th}}{\lambda_3\gamma_3}}}{(\lambda_3\gamma_3+\lambda_4\gamma_4(y+\gamma_{th}))^2} + \frac{e^{-\frac{y+\gamma_{th}}{\lambda_3\gamma_3}}}{(\lambda_3\gamma_3+\lambda_4\gamma_4(y+\gamma_{th}))}\right] dy \\
&\overset{(b)}{=} 1 + 2(-1)^N \mu e^{-\frac{\gamma_{th}}{\lambda_3\gamma_3}} \sum_{k=1}^{N} C_k^N (-1)^{(N-k)} \int_0^1 z^{\sqrt[k]{\left(\frac{\mu}{ln^2 z}+\gamma_{th}+1\right)}-1} \\
&\times \frac{\lambda_3\gamma_3\lambda_4\gamma_4 + \lambda_3\gamma_3\lambda_4\gamma_4\left(\frac{\mu}{ln^2 z}+\gamma_{th}\right)}{\left[\lambda_3\gamma_3+\lambda_4\gamma_4\left(\frac{\mu}{ln^2 z}+\gamma_{th}\right)\right]^2 ln^3 z} e^{-\frac{\mu}{\lambda_3\gamma_3 ln^2 z}} dz \\
&\overset{(c)}{=} 1 - \frac{(-1)^N \pi\mu e^{-\frac{\gamma_{th}}{\lambda_3\gamma_3}}}{M} \sum_{k=1}^{N} C_k^N (-1)^{(N-k)} \sum_{i=1}^{M} u_i^{\sqrt[k]{\left(\frac{\mu}{ln^2 u_i}+\gamma_{th}+1\right)}-1} \\
&\times \frac{\lambda_3\gamma_3\lambda_4\gamma_4 + \lambda_3\gamma_3 + \lambda_4\gamma_4(\frac{\mu}{ln^2 u_i}+\gamma_{th})}{\left[\lambda_3\gamma_3+\lambda_4\gamma_4(\frac{\mu}{ln^2 u_i}+\gamma_{th})\right]^2 ln^3 u_i} e^{\frac{-\mu}{\lambda_3\gamma_3 ln^2 u_i}} \sqrt{1-\beta_i^2}
\end{aligned}
$$

where step (a) is obtained by using Eqs. (13) and (17), step (b) is obtained by letting $z = e^{-\frac{\mu}{\sqrt{y}}}$ in which $\mu = \frac{\gamma_{th}}{a\lambda_{r*}^2\gamma_{r*}}$, step (c) is obtained by applying the Gaussian-Chebyshev quadrature method in which $u_i = \frac{\beta_i+1}{2}$ with $\beta_i = \cos\left(\frac{2i-1}{2M}\pi\right)$, and M is the complexity-vs-accuracy trade-off coefficient.

References

1. Zhang, J.W., Bai, X., Han, W.Y., Zhao, B.H., Xu, L.J., Wei, J.J.: The design of radio frequency energy harvesting and radio frequency-based wireless power transfer system for battery-less self-sustaining applications. Int. J. RF Microwave Comput. Aided Eng. **29**(1), e21658 (2019)

2. Lu, X., Wang, P., Niyato, D., Kim, D.I., Han, Z.: Wireless networks with RF energy harvesting: a contemporary survey. IEEE Commun. Surv. Tutorial **17**(2), 757–789 (2014)
3. Ha, D.B., Agrawal, J.P.: Performance analysis for NOMA relaying system in next-generation networks with RF energy harvesting. In: Wireless Energy Transfer Technology. IntechOpen (2019)
4. Huang, K., Lau, V.K.: Enabling wireless power transfer in cellular networks: architecture, modeling and deployment. IEEE Trans. Wireless Commun. **13**(2), 902–912 (2019)
5. Divakaran, S.K., Krishna, D.D.: RF energy harvesting systems: an overview and design issues. Int. J. RF Microwave Comput. Aided Eng. **29**(1), e21633 (2019)
6. Tabassum, H., Ali, M.S., Hossain, E., Hossain, M., Kim, D.I.: Non-orthogonal multiple access (NOMA) in cellular uplink and downlink: challenges and enabling techniques. arXiv preprint arXiv:1608.05783 (2016)
7. Zaidi, S.K., Hasan, S.F., Gui, X.: Evaluating the ergodic rate in SWIPT-aided hybrid NOMA. IEEE Commun. Lett. **22**(9), 1870–1873 (2018)
8. Liu, Y., Ding, Z., Elkashlan, M., Poor, H.V.: Cooperative non-orthogonal multiple access with simultaneous wireless information and power transfer. IEEE J. Sel. Areas Commun. **34**(4), 938–953 (2016)
9. Zhu, J., Wang, J., Huang, Y., Navaie, K., Ding, Z., Yang, L.: On optimal beamforming design for downlink MISO NOMA systems. IEEE Trans. Veh. Technol. **69**, 3008–3020 (2020)
10. Wang, L., Xu, D.: Resource allocation in downlink SWIPT-based cooperative NOMA systems. KSII Trans. Internet Inf. Syst. **14**(1), 20–39 (2020)
11. Kim, J.O., Uddin, M.B., Shin, S.Y.: Outage analysis of NOMA exploited coordinated direct and relay assisted uplink transmission. In: Proceedings of the 7th International Conference on Information and Communication Technology (ICoICT 2019), pp. 1–6. IEEE (2019)
12. Xu, Y., Wang, G., Li, B., Jia, S.: Performance of D2D aided uplink coordinated direct and relay transmission using NOMA. IEEE Access **7**, 151090–151102 (2019)
13. Nomikos, N., Michailidis, E.T., Trakadas, P., Vouyioukas, D., Zahariadis, T., Krikidis, I.: Flex-NOMA: exploiting buffer-aided relay selection for massive connectivity in the 5G uplink. IEEE Access **7**, 88743–88755 (2019)
14. Yue, X., Liu, Y., Kang, S., Nallanathan, A., Ding, Z.: Exploiting full/half-duplex user relaying in NOMA systems. IEEE Trans. Commun. **66**(2), 560–575 (2017)
15. HA, B.D., Tran, D.D., Vo, S.N.: Cooperative transmission in uplink NOMA networks with wireless power transfer. J. Sci. Technol. Issue Inf. Commun. Technol. **17**(122), 20–27 (2019)

Convolutional Neural Network-Based DOA Estimation Using Non-uniform Linear Array for Multipath Channels

Van-Sang Doan[1], Thien Huynh-The[1], Van-Phuc Hoang[2], and Dong-Seong Kim[1](✉)

[1] Kumoh National Institute of Technology, Gumi, South Korea
{vansang.doan,thienht,dskim}@kumoh.ac.kr
[2] Le Quy Don Technical University, Hanoi, Vietnam
phuchv@lqdtu.edu.vn

Abstract. In this paper, a novel convolutional neural network (CNN) was designed for DOA estimation, which could deploy in radio-electronics systems for improving the accuracy and operation efficiency. The proposed model was evaluated with different hyper-parameter configurations for optimization, and then a suitable model was compared with other existing models to demonstrate its preeminence. Regarding dataset generation, our work considered the influence of both Gaussian noise and multipath channels to DOA estimation accuracy. According to the analysis, in frame of this study, the model with 5 conv-blocks, 48 filters, and a filter size of 1×7 achieved the best performance in terms of accuracy (75.27% at +5 dB SNR) and prediction time (10.1 ms) that notably outperformed two other state-of-the-art CNN model-based DOA estimation techniques.

Keywords: Convolution neural network · DOA estimation · Multipath channels · Antenna array

1 Introduction

Many years, direction of arrival (DOA) estimation has been an active research topic for application in various areas, including radar, sonar, and communication. In military radar and sonar systems, DOA estimation facilitates determining the location of targets to help for surveillance, tracking and control in the ground, air, and water areas of nations. In communications, the quality of wireless connections is greatly improved if DOA estimation is employed to target the user. In many contexts, the DOA estimation plays a role as a spatial filter. In addition, the arrival angle information of a signal source in the electronic warfare

Supported by ICTCRC, Kumoh National Institute of Technology, South Korea.

N.-S. Vo and V.-P. Hoang (Eds.): INISCOM 2020, LNICST 334, pp. 45–56, 2020.
https://doi.org/10.1007/978-3-030-63083-6_4

operation is very useful to explore further data about the enemy situation that enable us to build a plan or decide a proper activity.

In almost applications, an antenna array with multiple elements is often used for steering a beam pattern toward the defined direction. As well-known methods, multiple signal classification (MUSIC) [1] and estimation of signal parameters via rotation invariance (ESPRIT) exploit the Eigen-decomposition of the covariance matrix of received signals to determine the arrival angles [2]. In decades, researchers attempt to solve the DOA estimation for coherent signals. Therefore, plenty of array pre-processing techniques are proposed, such as forward spatial smoothing (FSS) [3], forward/backward spatial smoothing (FBSS) [4], joint spatial-temporal method [5], and Toeplitz approximation [6]. They all were deeply analyzed and evaluated.

Recently, the machine learning technique is introduced as an effective approach to solving the DOA estimation problem, for example, support vector machine (SVM) is used to estimate DOA of multiple plane waves in [7–9]. As part of machine learning, deep learning is a well-known universal approximation theorem that can learn features deeply by designing neural networks with multiple hidden layers [10]. Accordingly, a deep neural network (DNN) proposed in [11] was designed for multitask autoencoder and a sequence of parallel multiple-layer classifiers. As a result, the DNN achieved a DOA estimation performance with higher accuracy than SVM and MUSIC techniques. Despite enhancing DOA estimation accuracy, the multilayer perception consumes a high computational complexity, expensive architecture, and time delay. Besides, convolutional neural network (CNN) allow learning feature automatically without expert knowledge in a specific domain. In 2017, *Adavanne et al.* demon-strated a stacked convolutional recurrent neural network, namely DOAnet, for esti-mating DOA in both azimuth and elevation angles [12]. In that work, the input data of DOAnet was signal frames in time domain. The study evaluated the DOA estimation possibility of DOAnet on anechoic, matched and unmatched reverberation dataset, and indicated that the approach performed better than MUSIC in most scenarios. In 2019, *Chakrabarty et al.* proposed a CNN-based method for DOA estima-tion of multiple speakers, here, the short-time Fourier transform (STFT) of received signals was used as a pre-processing to generate the input feature map, which is di-rectly fed into the CNN [13]. In another approach, the input feature map was columns of a covariance matrix, which played a role as a spatial spectrum [14]. However, the method did not consider multipath propagation conditions.

Despite achieving the remarkable performance of DOA estimation based on deep learning techniques, the above-mentioned researches have concerned the assessment of uniform linear array (ULA) but not non-uniform linear array (NLA); therefore, it still remains room for exploration. In addition, no signal dataset with the multipath effect is synthesized in those publications for learning CNN models, and this, therefore, motivates us to generate a new dataset for performance evaluation, in which diverse factors affecting the incoming signals are considered exhaustively. Accordingly, this paper presented a CNN model, namely DOA-ConvNet, which was designed according to a combination of dense

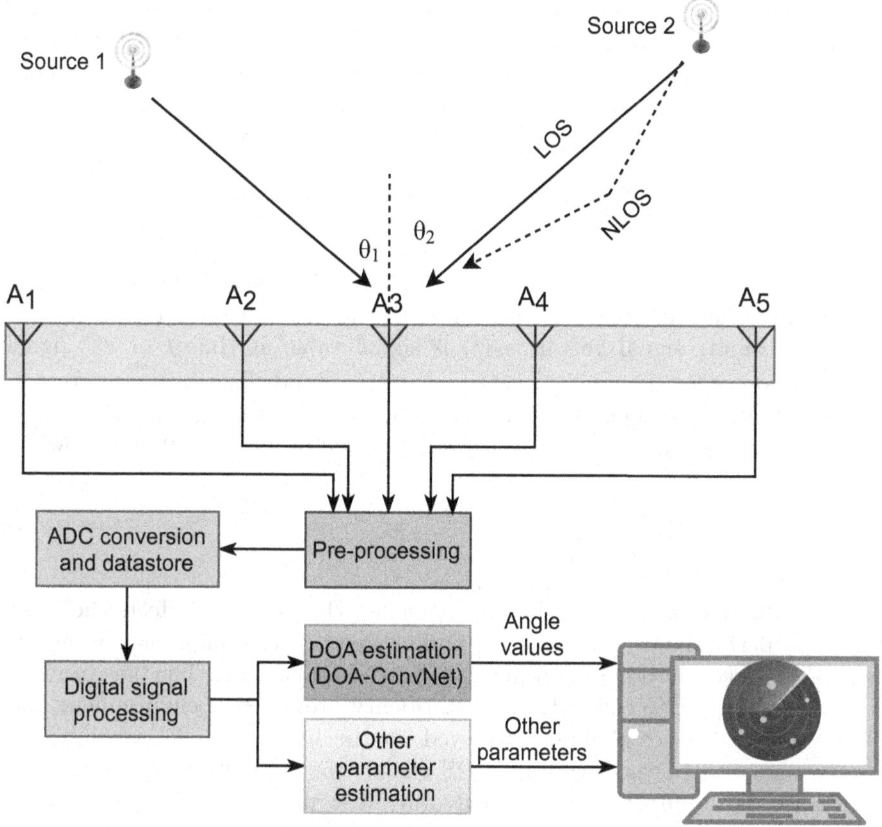

Fig. 1. Direction finding system with 5-element NLA for DOA estimation using DOA-ConvNet model.

and residual network architecture. The proposed model was evaluated to achieve the optimal performance in terms of accuracy and time-consuming.

2 Signal Model and Dataset Generation

2.1 Signal Model of Antenna Array

As shown in Fig. 1, a signal model of a linear antenna array is illustrated. In principle, an incoming signal is captured by the antenna array. Then, the received signals of elements are pre-processed at pre-processing block and converted to digital form at ADC conversion and datastore block. In the digital signal processing block, the signals are re-constructed and prepared for DOA-ConvNet. The output of DOA-ConvNet is a DOA value of the received signal. Along with other signal parameters, the estimated DOA is indicated in a display for surveillance and tracking.

Accordingly, assume that a signal travels from one transmitter to a non-uniform linear array (NLA) with M elements in K paths of the multipath propagation. The light-of-sight (LOS) path is in the directions of θ_1, and other non-light-of-sight (NLOS) ones are in the direction of θ_2, θ_3, ..., θ_K. Then, the output signals of NLA expressed as a matrix $\mathbf{x}$ in time domain are written as:

$$\mathbf{x} = \mathbf{A} \cdot \mathbf{s} + \mathbf{n} \tag{1}$$

where $\mathbf{x} = [x_1(t), x_2(t), ..., x_M(t)]^T$ is output signal vector of the array, $x_m(t)$ is signal value at time t of mth antenna element; $\mathbf{s} = [s_1(t), s_2(t), ..., s_K(t)]^T$ presents incoming signal vector, $s_k(t)$ is signal value at time t of kth signal from direction of θ_k; $\mathbf{n} = [n_1(t), n_2(t), ..., n_M(t)]^T$ stands for white noise vector, whose elements are independent; $\mathbf{A} = [\mathbf{a}_1(\theta_1), \mathbf{a}_2(\theta_2), ..., \mathbf{a}_K(\theta_K)]$ is defined as a steering matrix of the array, $\mathbf{a}_k(\theta_k)$ is a steering vector for angle θ_k as following:

$$\mathbf{a}_k(\theta_k) = \left[1, e^{-j\frac{2\pi d_1 \sin\theta_k}{\lambda_k}}, ..., e^{-j\frac{2\pi d_{M-1} \sin\theta_k}{\lambda_k}}\right]^T \tag{2}$$

The multipath propagation is a phenomenon that strongly affects the DOA estimation performance of a direction finding system. The angle bias occurs by the non-useful correlation of non-line-of-sight (NLOS) signals. The bias problem sometimes is solved by calibration for stationary propagation environment, but not for mobile devices that are deployed on the battlefield. As a result, the antenna array output will contain LOS signal, their NLOS components, and white noise. By acquisition of the array output signals with multiple samples, the direction-finding system can estimate the arrival angle of strong incoming signals. After converting output signals to digital, the matrix $\mathbf{x}$ has a size of $M \times N$, with M is number of antenna elements, and N is number of signal samples. In the case of I/Q (In-phase/Quadratic-phase) receiver, the matrix $\mathbf{x}$ is assigned with the size of $M \times N \times 2$. The matrix $\mathbf{x}$ now becomes input data for the proposed DOA-ConvNet.

2.2 Dataset Generation

In this study, an NLA with configuration of $\mathbf{d} = \{0, d_1, d_2, d_3, d_4\} = \{0, 3, 5, 7, 10\}\lambda/2$ is considered for evaluating the performance of DOA-ConvNet for DOA estimation task. The NLA configuration has a symmetric property that provides a high accuracy of DOA estimation of multiple coherent signals, as demonstrated in [18]. Regard to the data structure, the signal frame for each acquisition is of size 256 samples. By using the array signal model as presented in the previous subsection, a dataset with consideration of multiple noise levels and randomized multipath propagation is generated for training and validating the proposed network. Accordingly, the dataset is produced according to parameters summarized in Table 1. Herein, the input feature map is defined by the size of matrix $\mathbf{x}$, concretely is of size $5 \times 256 \times 2$. All signal feature maps in the

Table 1. Summary of signal dataset generation parameters.

Parameter	Value
Array	NLA, $d = [0, 3, 5, 7, 10]\lambda/2$
Carrier Frequency (LOS)	35 MHz
Sampling frequency	350 MHz
Signal strength	1 V
Number of NLOS signals	6
Multipath delay (NLOS)	$\mathbf{U}\{[1, 1000]\}$ ns
Maximum Doppler shift	10 Hz
Multipath attenuation (NLOS)	$\mathbf{U}\{[-100, -10]\}$ dB
SNR	−20 dB to +25 dB, step 5 dB
Angle of arrival	−89° to +89°, step 1°
Number of angle classes	179 class names
Data size of one signal window	$5 \times 256 \times 2$

dataset are labeled corresponding to angles from −89° to +89° with the step of 1°, so there is a total of 179 label names that correspond to 179 output classes of the proposed network. In detail, 10,000 signal feature maps are generated for each angle with various SNRs from −15 dB to +15 dB with the step of 5 dB. Each SNR has 179,000 signal feature maps. Overall, the dataset has 1,790,000 observation maps, in which 80% is used for training and 20% is for validation.

3 CNN-Based DOA Estimation Model

An ELINT receiver always operates with two channels I and Q, which provide full amplitude and phase information of received signals. Therefore, the input data size should be of $M \times N \times 2$, where M is number of antenna elements, N is number of samples per signal window, and number 2 presents two signal channels, I and Q. According to the problem in this work, the input data size is assigned by $5 \times 256 \times 2$. Inspired by the ResNet architecture [15], our deep neural network model (namely DOA-ConvNet) is designed with a primary flow (main path) and skip-connections (residual paths), as shown in Fig. 2a and listed in Table 2. Accordingly, the primary flow is constructed by a series of conv-blocks connected consecutively from input to output layers. Each conv-block, as shown in Fig. 2b, consists of two branches, whose outputs are synthesized by an addition layer. Each branch contains two consecutive one-dimensional (1D) convolution layers, one max-pool layer, and one ELU activation function layer. In 1D-convolution layer, the convolution operation is executed between an input data and a kernel so that its formula can express as follows:

$$\mathbf{y}(i) = \sum_{m} \mathbf{c}(m) \cdot \mathbf{x}(i - m) \tag{3}$$

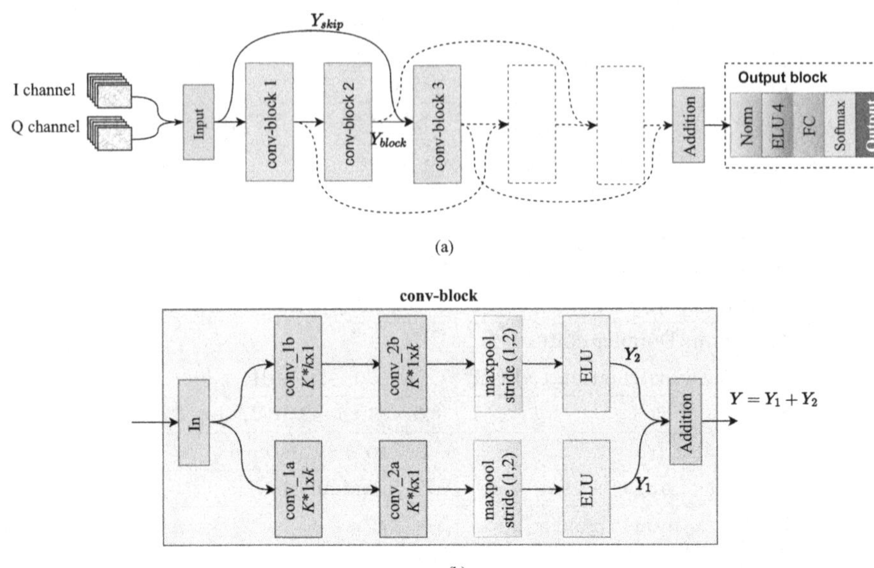

Fig. 2. DOA-ConvNet model structure: (a) *overall*; (b) *conv-block*.

where $\mathbf{y}$ is an output matrix, $\mathbf{x}$ is an input matrix, $\mathbf{c}$ is a kernel matrix, and m, i, are element indices.

It can be observed that two 1D-convolution layers with filter sizes of $1 \times k$ and $k \times 1$ are employed instead of one convolution layer with a filter size of $k \times k$. This way helps to reduce the number of learnable parameters that produce the model with a light-weight structure [16,17]. In the second branch, two convolution layers with two mentioned kernel sizes are exchanged their positions. The second convolution layer is followed by a maxpool layer whose hyper-parameters are designed by a pool-size of 1×3 and stride of $(1, 2)$. The role of maxpool layer is to perform down-sampling to reduce the size of feature maps and extract the robust features before going to the activation layer. The activation function plays a crucial role in CNNs as a brain neuron, which activates the useful features and impairs the redundant ones. In our model, we use an exponential linear unit (ELU) function, a smooth activation function, which provides better signal propagation through the network than ReLU one. The ELU function is expressed as follows:

$$y = \begin{cases} x & \text{if } x \geq 0 \\ \alpha(e^x - 1) & \text{if } x < 0 \end{cases} \tag{4}$$

where y denotes the output, x stands for input value and α is a scalar number between 0.1 and 0.3. In this work, it is chosen by 0.2.

4 Experiment Result and Discussion

Our model is intended with changeable hyper-parameters, including number of conv-blocks, number of filters in a convolution layer, and size of filters. Different configurations of the model will be evaluated, and then a suitable one will be chosen for competing with other existing models in terms of accuracy and time-consuming. For each configuration, the model must be trained with the synthesized dataset, which is described in Sect. 2 before the experiment and assessment.

Table 2. Summary of signal dataset generation parameters.

Block	Layer	Output	Description
Input	Input	$5 \times 256 \times 2$	256 Sample/frame
			5 antenna element
			2 signal channels I/Q
p^{th} conv-blocks	*Branch 1:*		
($p = 1, 2, \ldots, P$), P is	- Conv	$5 \times 256/2^{p-1} \times K$	K filters, $1 \times k$, stride $(1, 1)$
number of conv-blocks	- Conv	$5 \times 256/2^{p-1} \times K$	K filters, $k \times 1$, stride $(1, 1)$
	- Maxpool	$5 \times 256/2^{p} \times K$	Poolsize 1×3, stride $(1, 2)$
	- Activation	$5 \times 256/2^{p} \times K$	ELU: exponential linear unit
	Branch 2:		
	- Conv	$5 \times 256/2^{p-1} \times K$	K filters, $k \times 1$, stride $(1, 1)$
	- Conv	$5 \times 256/2^{p-1} \times K$	K filters, $1 \times k$, stride $(1, 1)$
	- Maxpool	$5 \times 256/2^{p} \times K$	Poolsize 1×3, stride $(1, 2)$
	- Activation	$5 \times 256/2^{p} \times K$	ELU: exponential linear unit
	- *Addition*	$5 \times 256/2^{p} \times K$	Branch 1 + branch 2 + skip-connection (except $n = 1$)
p^{th} Skipconnection	- Conv	$5 \times 256/2^{p+1} \times K$	K filters, 1×1, stride $(1, 1)$
($p = 1, 2, \ldots, P - 1$)	- Maxpool	$5 \times 256/2^{p-1} \times K$	Poolsize 1×6, stride $(1, 4)$
Output block	Norm	$5 \times 256/2^{P} \times K$	Batch normalization
	Activation	$5 \times 256/2^{P} \times K$	ELU
	FC	$1 \times 1 \times 179$	179 classes
	Softmax	$1 \times 1 \times 179$	Softmax function
	Classification		Prediction execution

4.1 DOA-ConvNet with Different Filter Size

As mentioned in the previous section, the pair of consecutive convolution layers in each conv-block branch of DOA-ConvNet are assigned with K filters, and filter sizes of $1 \times k$ and $k \times 1$, respectively. For evaluating the impact of filter size on the performance, the network is configured with following hyper-parameters: 5 conv-blocks, 64 filters per convolution layer, and various filter sizes with $k = [3, 5, 7, 9]$.

The experimental result presented in Fig. 3 shows the comparison of angle classification accuracy depending on various SNR levels and different convolution filter sizes. It shows that in spite of giving the best accuracy at low SNR levels (from -20 dB to 0 dB), the model with the filter size of 1×9 has the worst

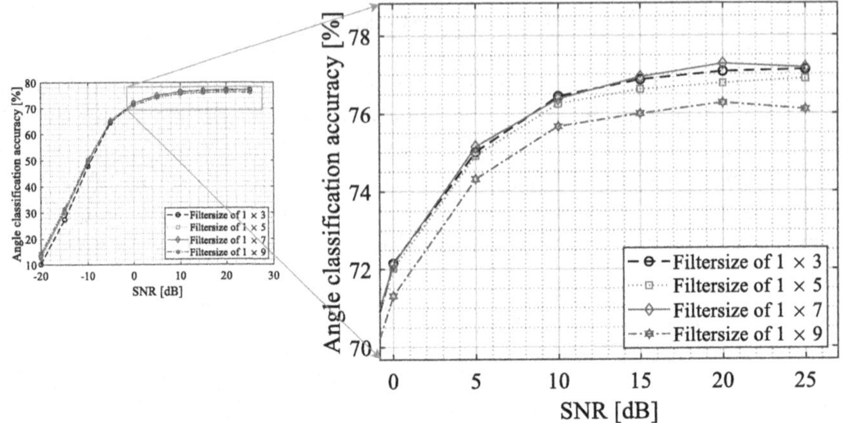

Fig. 3. Angle classification performance of DOA-ConvNet with different filter sizes depending on various SNRs.

performance at high SNRs (> 0 dB). Whereas, the model with the filter size of 1 × 3 seem lowest correct angle classification rate for SNR < 0 dB but it outperforms other ones at higher SNRs excepting 1 × 7. By observation, the model with the filter size of 1 × 7 obtains very good performance at all SNR values, especially it achieves the best accuracy with SNR > 0 dB.

4.2 DOA-ConvNet with Different Number of Filters

In Subsect. 4.1, the filter size with $k = 7$ is selected as a suitable configuration of the convolution dimension; thus, the second experiment is performed with different numbers of filters in a convolution layer while 5 conv-blocks are still used in this evaluation.

As a result, Fig. 4 shows the different performances of the model due to various filter numbers. Logically, more filters result in more neurons in the model; thus, more representative information should be extracted for better prediction performance. However, we observe that the model with a greater number of filters does not gain higher angle classification accuracy that the number of filters should be satisfied with dataset properties, that only be discovered by experiments. In particular, in this framework the model with 48 filters estimates DOA with the highest accuracy, whereas the models with 64 and 96 filters obtain lower performance.

4.3 DOA-ConvNet with Different Conv-Blocks

"*How many convolution layers are optimal for a neural network model-based DOA estimation*" is always a difficult question for answering. Indeed, how big is the network, it depends on the dataset properties and the number of output classes. Therefore, a classifier must be fitted with a particular dataset. For the

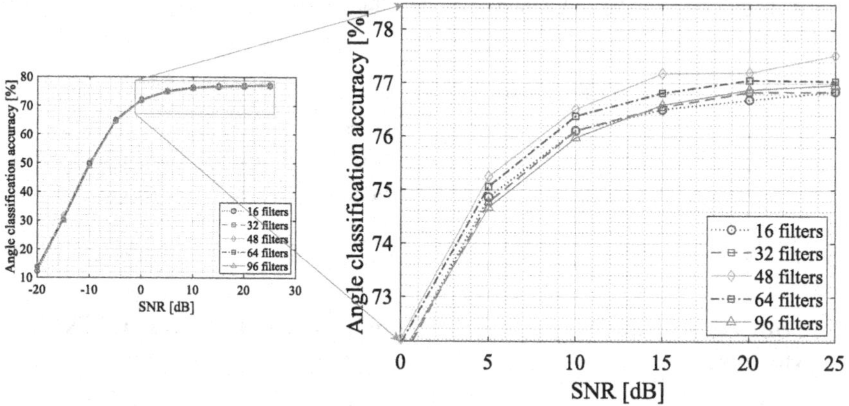

Fig. 4. Angle classification performance of DOA-ConvNet with different numbers of filters depending on various SNRs.

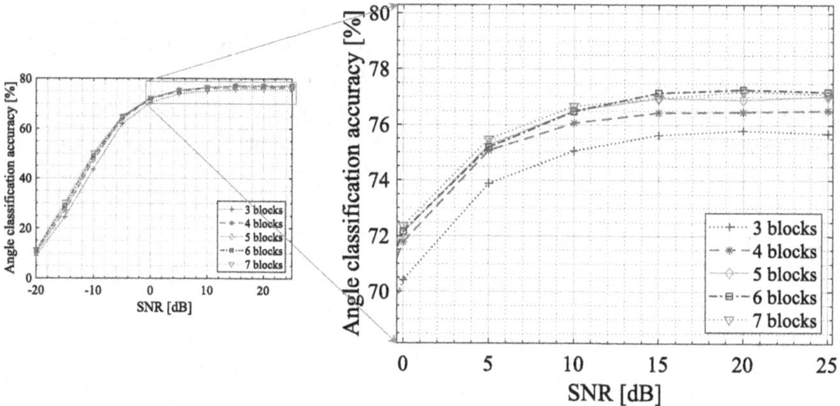

Fig. 5. Angle classification performance of DOA-ConvNet with different numbers of conv-blocks depending on various SNRs.

DOA estimation problem, DOA-ConvNet al.so has to be optimized with a suitable number of convolution layers, which must obtain good accuracy and fast execution. Herein, the number of convolution layers is represented by the number of conv-blocks.

By experiment, the accuracy performance depending on SNRs and different numbers of conv-blocks is plotted in Fig. 5, which shows that DOA-ConvNet with 5, 6, and 7 conv-blocks achieves the almost same classification efficiency and significantly better than that with 3 and 4 conv-blocks. Moreover, the execution time performance also is taken into account for assessment. This result is listed in Table 3, where the numerical results enable us to suggest that the model with 5 conv-blocks is chosen to satisfy the accuracy, structural, and time cost requirements.

Table 3. Performance of DOA-CovNet with different numbers of conv-blocks at SNR = +5 dB.

Number of Conv-blocks	Learnable parameters	Classification accuracy	Time-consuming
3	1.9M	73.88%	7.3 ms
4	1.1M	75.07%	8.8 ms
5	0.7M	75.27%	10.1 ms
6	0.5M	75.19%	11.8 ms
7	0.4M	75.48%	13.6 ms

4.4 Comparison of DOA-ConvNet with Other Existing CNN Models

As a result of the aforementioned analysis, DOA-ConvNet with 5 conv-blocks,48 filters, and a filter size of $k = 7$ is our designated CNN model to compete with two other existing ones (namely DOA-CNN-1 and DOA-CNN-2) for DOA estimation task. DOA-CNN-1 consists of 4 consecutive pairs of Conv+ReLU layers, following by three fully connected layers with output sizes of 512 for two first ones and 179 classes for the last one, and finalized by softmax layer and output decision layer [13]. DOA-CNN-2 is constructed by four consecutive pairs of Conv+ReLU layers also, but its input data is a spatial spectrum, which is pre-processed by beamforming technique [14]. Both mentioned CNN models were

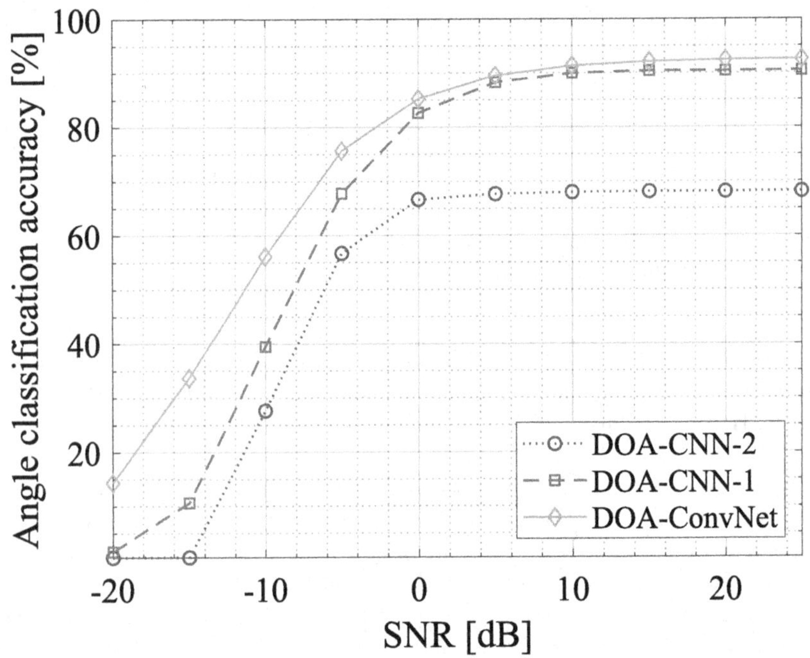

Fig. 6. Comparison of DOA-ConvNet with DOA-CNN-1 and DOA-CNN-2.

demonstrated remarkable outcomes in their own case studies. In this comparison, we applied those two CNN models for our signal dataset design, based on 5-element NLA along with Gaussian noise and multipath channels, to compare with DOA-ConvNet, while other setting up parameters remain as original. The comparison result is plotted in Fig. 6.

Regarding the structural volume, DOA-CNN-1, and DOA-CNN-2 have respectively 8,662,451 and 960,488 learnable parameters, whereas DOA-ConvNet has the number of parameters of 646,619 only. However, in the accuracy comparison, DOA-ConvNet gains the highest performance at all SNR values, the second place for DOA-CNN-2, and the worst for DOA-CNN-1. As we mentioned above, DOA-ConvNet is designed for re-usage of former feature maps, which can improve the robust features, therefore, it achieves higher accuracy.

5 Conclusion

This paper has demonstrated a novel CNN model-based DOA estimation method (namely DOA-ConvNet). The proposed model was evaluated based on different structural hyper-parameters by experiments to gain an optimal one with the good trade-off in terms of accuracy, structural, and time cost. In addition, a signal dataset was synthesized with consideration of both Gaussian noise and multipath channels. A non-uniform linear array with the configuration of $\mathbf{d} = [0, 3, 5, 7, 10]\lambda/2$, along with the optimal DOA-ConvNet, is proposed to outperform other existing models for the same DOA estimation task. In the future work, we intend to develop the network for different array geometries such as circular, rectangle planar or spherical arrays, which can estimate signal DOA in both azimuth and elevation planes.

Acknowledgment. This work was supported under the framework of international cooperation program managed by National Research Foundation of Korea (NRF-2019K2A9A1A09081533) and Priority Research Centers Program through the National Research Foundation of Korea funded by the Ministry of Education, Science and Technology (NRF-2018R1A6A1A03024003).

References

1. Schmidt, R.: Multiple emitter location and signal parameter estimation. IEEE Trans. Antennas Propag. **34**, 276–280 (1986)
2. Roy, R., Kailath, T.: ESPRIT-estimation of signal parameters via rotational invariance techniques. IEEE Trans. Acoust. Speech Signal Process. **37**, 984–995 (1989)
3. Reddy, K.M., Reddy, V.U.: Analysis of spatial smoothing with uniform circular arrays. IEEE Trans. Signal Process. **47**, 1726–1730 (1999)
4. Pillai, S.U., Kwon, B.H.: Forward/backward spatial smoothing techniques for coherent signal identification. IEEE Trans. Acoust. Speech Signal Process. **37**, 8–15 (1989)
5. Lin, J.-D., Fang, W.-H., Wu, M.-L.: Joint spatial-temporal channel parameter estimation using tree-structured MUSIC. In: IEEE 55th Vehicular Technology Conference. VTC Spring 2002 (Cat. No.02CH37367) (2002)

6. Wang, Q., Chen, H., Zhao, G., Chen, B., Wang, P.: An improved direction finding algorithm based on Toeplitz approximation. Sensors **13**, 746–757 (2013)
7. Pastorino, M., Randazzo, A.: A smart antenna system for direction of arrival estimation based on a support vector regression. IEEE Trans. Antennas Propag. **53**, 2161–2168 (2005)
8. Gonnouni, A.E., Martinez-Ramon, M., Rojo-Alvarez, J.L., Camps-Valls, G., Figueiras-Vidal, A.R., Christodoulou, C.G.: A support vector machine MUSIC algorithm. IEEE Trans. Antennas Propag. **60**(10), 4901–4910 (2012)
9. Sun, F.-Y., Tian, Y.-B., Hu, G.-B., Shen, Q.-Y.: DOA estimation based on support vector machine ensemble. Int. J. Numer. Model. Electron. Netw. Dev. Fields, e2614 (2019)
10. Huang, H., Yang, J., Huang, H., Song, Y., Gui, G.: Deep learning for super-resolution channel estimation and DOA estimation based massive MIMO system. IEEE Trans. Veh. Technol. **67**(9), 8549–8560 (2018)
11. Liu, Z.-M., Zhang, C., Yu, P.S.: Direction-of-arrival estimation based on deep neural networks with robustness to array imperfections. IEEE Trans. Antennas Propag. **66**(12), 7315–7327 (2018)
12. Adavanne, S., Politis, A., Virtanen, T.: Direction of arrival estimation for multiple sound sources using convolutional recurrent neural network, CoRR, vol. abs/1710.10059 (2017)
13. Chakrabarty, S., Habets, E.A.P.: Multi-speaker DOA estimation using deep convolutional networks trained with noise signals. IEEE J. Sel. Top. Signal Process. **13**(1), 8–21 (2019)
14. Wu, L., Liu, Z.-M., Huang, Z.-T.: Deep convolution network for direction of arrival estimation with sparse prior. IEEE Signal Process. Lett. **26**(11), 1688–1692 (2019)
15. He, K., Zhang, X., Ren, S., Sun, J.: Deep residual learning for image recognition. In: 2016 IEEE Conference on Computer Vision and Pattern Recognition (CVPR), pp. 770–778 (2016)
16. Huynh-The, T., Hua, C.-H., Pham, Q.-V., Kim, D.-S.: MCNet: an efficient CNN architecture for robust automatic modulation classification. IEEE Commun. Lett. **24**(4), 811–815 (2020)
17. Huynh-The, T., Hua, C., Kim, J., Kim, S., Kim, D.: Exploiting a low-cost CNN with skip connection for robust automatic modulation classification. In: 2020 IEEE Wireless Communications and Networking Conference (WCNC), pp. 1–6, Seoul, Korea (South) (2020)
18. Doan, V.-S., Kim, D.-S.: DOA estimation of multiple non-coherent and coherent signals using element transposition of covariance matrix. ICT Express **6**(2), 67–75 (2020)

An UAV and Distributed STBC for Wireless Relay Networks in Search and Rescue Operations

Cong-Hoang Diem[1(✉)] and Takeo Fujii[2]

[1] Department of Computer Networks, Faculty of Information Technology, Hanoi University of Mining and Geology (HUMG), Hanoi, Vietnam
diemconghoang@humg.edu.vn

[2] Advanced Wireless and Communication Research Center (AWCC), The University of Electro-Communications (UEC), Tokyo, Japan

Abstract. This paper proposes a transmission method using unmanned aerial vehicle (UAV) with distributed Space-Time Block Code (STBC) for multi-hop wireless relay networks in search and rescue operations. First, an UAV is considered to add to the hop with the minimum output signal-noise-ratio (SNR) and operates as a relay node to maintain the links between adjacency nodes in network, expand the transmission coverage area and improve the transmission performance. In addition, in order to overcome the difficulty in assigning the STBC patterns to the distributed relays and also alleviate the complexity of system design and implementation, the original STBC pattern is modified while keeping the same cooperative diversity gain. Finally, an algorithm is proposed to find out the optimal location of the added UAV in the hop, where the UAV has the best contribution to the data transmission performance between the transmitter and the receiver. It can be seen from the simulation results, the optimal location of the added UAV depends on not only the environment of real scenarios but also the distributed cooperative diversity gain. We can confirm that the proposed method achieves the significant performance improvement while keeping the simple operation of system for UAV communications in search and rescue operations.

Keywords: STBC · UAV communications · MANET · Wireless relay networks · Optimal location

1 Introduction

The use of unmanned aerial vehicles (UAVs) is rapidly growing in the past few decades due to their broad range of application domains that include telecommunications, delivery of medical supplies, and search and rescue (SAR) operations

---

Partly supported by Hanoi University of Mining and Geology (HUMG), Vietnam and Advanced Wireless and Communication Research Center (AWCC), The University of Electro-Communications (UEC), Tokyo, Japan.

N.-S. Vo and V.-P. Hoang (Eds.): INISCOM 2020, LNICST 334, pp. 57–73, 2020.
https://doi.org/10.1007/978-3-030-63083-6_5

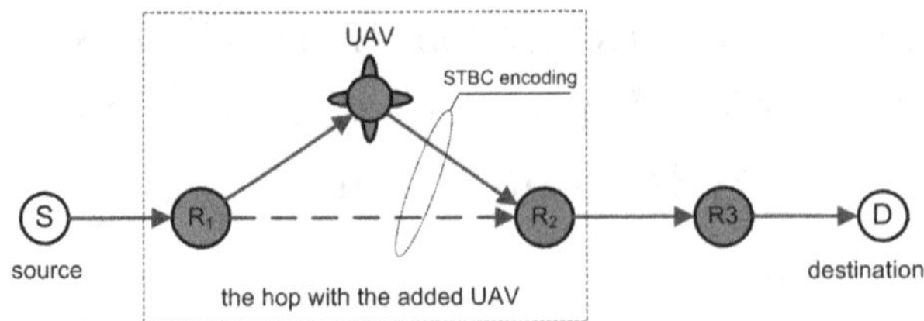

Fig. 1. Multi-hop wireless relay network with the added UAV.

[1]. Wireless networks with UAVs, commonly known as UAV communications, is also a subject that has attracted researchers' attention in recent years. This is because UAV communications can provide reliable and cost-effective wireless connectivity for nodes without infrastructure coverage [2]. Multi-hop Wireless networks with UAVs are seen as a sub form of the well-known concept of Multi-hop Mobile ad-hoc network (MANET). Therefore, wireless networks with UAVs also share common features with MANET. Thanks to the distinctive characteristics such as independent, dynamic and self-adaptive natures, MANET is generally used in emergency communications. MANET is generally applicable to ensure connectivity in disaster relief situations that are usually hampered by the absence of a network or communication infrastructures [3]. The integration of communication systems with wireless medium and the growth of Internet technology has made MANET capable to operate and function during rapid emergency deployment without relying upon infrastructure communication systems. These characteristics make MANET suitable for the efficient communication during natural disaster and search and rescue operations [4,5].

Besides sharing the common features with MANET above, there are several unique characteristics that make UAV communications more suitable with search and rescue operations, namely, mobility, topology changes, and radio propagation. UAVs are expected to be an important component of UAV communications for achieving high-speed wireless communications [6]. UAVs can adjust their altitude to avoid obstacles and enhance the likelihood of establishing line-of-sight (LoS) communication links to ground users. UAVs can also operate as wireless relays to provide wireless connectivity between two distant users or user groups without reliable direct links for improving transmission performance and coverage of ground wireless devices [2]. In addition, using UAVs as relay nodes for search and rescue operations in wireless communications can bring the many benefits as surveying the environment and collecting evidence about position of a mission person [7]. However, UAV communications are also faced with several new challenges such as highly dynamic network topologies, sparse and intermittent network connectivity, intelligent energy usage and replenishment, effective interference management, and so on. Therefore, new communication protocols, basic networking architecture, main channel characteristics, and performance enhancing techniques for UAV communications should be investigated [8].

In UAV communications, orthogonal frequency division multiplexing (OFDM) scheme has proposed as a good candidate to obtain reliable and high performance wireless communication links [9,10]. The guard interval of OFDM has tolerance to not only the influence of the delay spread but also the transmitting timing offset among distributed nodes. In wireless networks, a cooperative diversity is known as a technique allowing single-antenna nodes to reap some of the benefits of Multiple-Input Multiple-Output (MIMO) systems [11,12]. Space-Time Block Code (STBC) encoding scheme is also often used to obtain the higher cooperative diversity gain without channel state information (CSI) which is referred as channel properties of a communication link at the transmitting node [13]. In addition, the STBC encoding scheme can achieve both full diversity order and full data transmission rate with simple decoding algorithm at the receiver. The diversity can also be obtained by using distributed relays in cooperative systems, where each pattern of STBC encoding is transmitted by the different relays [14]. Therefore, in this paper, the distributed cooperative diversity using STBC scheme is considered to use in UAV communications.

As mentioned above, once network infrastructure is destroyed by natural disaster, the connections between end users and communication system are disrupted. With the dynamic and self-configuring nature, UAV communications are a suitable solution for emergency communications in search and rescue operations. However, due to the communication distance or obstacles such as trees, hills and mountains and so on, the several links between adjacency nodes in multi-hop wireless networks can be disconnected. In multi-hop wireless networks, the transmission channel can be modeled as an equivalent single hop with the minimum output signal-noise-ratio (SNR). Therefore, the end-to-end performance of multi-hop communication systems can be derived based on the performance of the equivalent single hop [15]. As a result, when the performance of the single hop with the minimum output SNR is improved, the end-to-end performance of the system can also be improved.

Therefore, in this paper, an UAV is first considered to add to the hop with the minimum output SNR and operates as a relay to maintain the links between adjacency nodes in network, expand the transmission coverage area and improve the transmission performance as shown in Fig. 1. In addition, the distributed STBC cooperative systems have difficulty in assigning the STBC patterns to the distributed relays. In order to overcome this problem and also alleviate the complexity of system design and implementation, the original STBC pattern is modified. For the modified STBC pattern, the transmission signal is changed and encoded only at the UAV while keeping the same cooperative diversity gain in comparison with the original STBC pattern. The transmission signal from relay nodes is same to that from source node. Finally, an algorithm is proposed to find out the optimal location of the added UAV at the hop with the minimum output SNR. The proposed algorithm is based on the Received-Signal-Strength Indicator (RSSI) of the beacon packets to find out the optimal location of the UAV, where the UAV has the best contribution to the data transmission performance between the transmitter and the receiver. From the simulation results, we can confirm

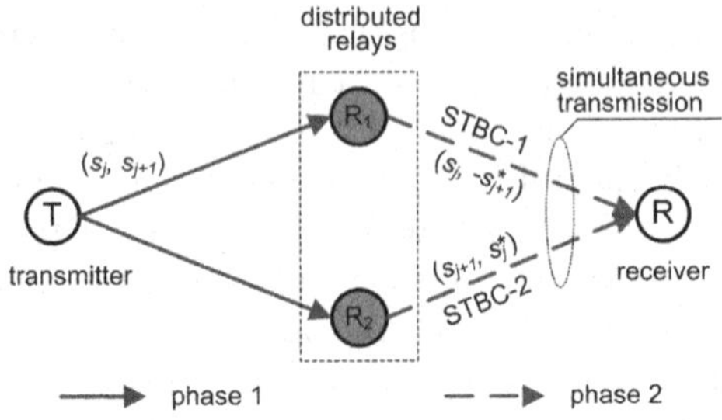

Fig. 2. Distributed STBC cooperative diversity.

that the proposed method can achieve the significant performance improvement while keeping the simple operation of system for UAV communications in search and rescue operations.

The remainder of the paper is organized as follows. A system model and the proposed method are described in Sect. 2 and Sect. 3, respectively. Next, the performance is evaluated through simulation and results are presented and analyzed in Sect. 4. Finally, Sect. 5 concludes this paper.

2 Distributed STBC Cooperative Diversity System

The system under consideration consists of one transmitter, two distributed relays and one receiver, which are all single-antenna nodes, operating over slow, flat, fading channels as shown in Fig. 2. It is assumed that perfect channel state information (CSI) is available at receivers but not at transmitters. A time-division channel allocation is used for medium access, inter-relay interference therefore is not considered in the signal model. It is also assumed that the decode-and-forward (DF) protocol and the distributed STBC model with two transmit nodes (Tx) and one receive node (Rx) is considered to use in this paper. At the relays, the received signal from the transmitter is first decoded, and then re-encoded and forwarded to the receiver [16]. By using the different STBC patterns at the relays, the cooperative diversity gain of the maximum ratio combing (MRC) can be obtained in the receiver.

We assume that the input signal for the j-th symbol is s_j, and the next symbol is s_{j+1}, where $j = 2a$ and a is the even number. The s_j and s_{j+1} symbols

Table 1. The original STBC patterns.

	Symbol j (t)	Symbol $j+1$ ($t+T$)
Branch 1	s_j	$-s^*_{j+1}$
Branch 2	s_{j+1}	s^*_j

are encoded and simultaneously transmitted from the transmitter and the UAV. The transmitted STBC patterns are shown in Table 1, where * is the complex conjugate operation. At the receiver, the signals are received and combined with the different path losses and fading fluctuations. In order to simplify the explanation, we focus on an OFDM sub-carrier signal for explanation. At the first transmission or phase 1, the received signal of the s_j symbol at the i-th relay is given by,

$$y_{ji} = \sqrt{P_t}\beta_{ti}h_{ti}s_j + n_{ji}, \tag{1}$$

where P_t is transmission power. h_{ti} and β_{ti} are fading channel coefficient and path loss gain between the transmitter and the i-th relay, respectively. The path loss gain is calculated with α the path loss exponential, a constant whose measured value range from 1.6 to 6 [17]. n_{ji} is noise component of the s_j symbol. At the next transmission or phase 2, both relays simultaneously transmit the data packet with the different STBC patterns as shown in Table 1. The received signals of the s_j and s_{j+1} symbols at the i-th receiver are expressed as follows,

$$\begin{aligned} y_{ji} &= H_{r_1 i}s_j + H_{r_2 i}s_{j+1} + n_{ji}, \\ y_{(j+1)i} &= -H_{r_1 i}s^*_{j+1} + H_{r_2 i}s^*_j + n_{(j+1)i}, \end{aligned} \tag{2}$$

where $H_{ri} = \sqrt{P_t}\beta_{ri}h_{ri}$ is channel response between the r-th relay in the current hop and the i-th relay in the next hop. n_{ji} and $n_{(j+1)i}$ are noise components. The received signals can be represented in term of vector $\mathbf{y}$ as follows,

$$\mathbf{y} = \begin{bmatrix} y_{ji} \\ y_{(j+1)i} \end{bmatrix} = \begin{bmatrix} s_j & s_{j+1} \\ -s^*_{j+1} & s^*_j \end{bmatrix} \begin{bmatrix} H_{r_1 i} \\ H_{r_2 i} \end{bmatrix} + \begin{bmatrix} n_{ji} \\ n_{(j+1)i} \end{bmatrix}. \tag{3}$$

The Eq. (3) can also be written in term of matrix as follows,

$$\mathbf{y} = \mathbf{sH} + \mathbf{n}, \tag{4}$$

where $\mathbf{y}$, $\mathbf{H}$, and $\mathbf{n}$ are 2×1 matrices, $\mathbf{s}$ is a 2×2 matrix. Without loss of generality, after some elementary manipulations and conjugating the second row of (3), the received signals can be equivalently expressed as follows,

$$\begin{bmatrix} y_{ji} \\ y^*_{(j+1)i} \end{bmatrix} = \begin{bmatrix} H_{r_1 i} & H_{r_2 i} \\ H^*_{r_2 i} & -H^*_{r_1 i} \end{bmatrix} \begin{bmatrix} s_j \\ s_{j+1} \end{bmatrix} + \begin{bmatrix} n_{ji} \\ n^*_{(j+1)i} \end{bmatrix} \tag{5}$$

It can also be represented in term of matrix form,

$$\tilde{\mathbf{y}} = \mathbf{Hs} + \tilde{\mathbf{n}}. \tag{6}$$

where $\tilde{\mathbf{y}}$, $\mathbf{s}$, and $\tilde{\mathbf{n}}$ are 2×1 matrices, $\mathbf{H}$ is a 2×2 matrix. At the receiver, it is assumed that the channel response is known exactly, the combined signals can be obtained by multiplying both sides of Eq. (6) by $\mathbf{H}^H$ as follows,

$$\tilde{\mathbf{s}} = \mathbf{H}^H\tilde{\mathbf{y}}. \tag{7}$$

Substituting (6) into (7), these combined signals can be written as follows,

$$\begin{aligned}\tilde{s}_j &= H^*_{r_1 i} y_{ji} + H_{r_2 i} y^*_{(j+1)i} = (|H_{r_1 i}|^2 + |H_{r_2 i}|^2) s_j + w_j, \\ \tilde{s}_{j+1} &= H^*_{r_2 i} y_{ji} - H_{r_1 i} y^*_{(j+1)i} = (|H_{r_1 i}|^2 + |H_{r_2 i}|^2) s_{j+1} + w_{j+1},\end{aligned} \tag{8}$$

where $w_j = H^*_{r_1 i} n_{ji} + H_{r_2 i} n^*_{(j+1)i}$ and $w_{j+1} = H^*_{r_2 i} n_{ji} - H_{r_1 i} n^*_{(j+1)i}$ are the noise components. These combined signals are then sent to the maximum likelihood detector to choose which symbol was actually transmitted from the transmitter by applying least squares (LS) detection,

$$\hat{s}_k = arg \min_{s_k \in S_M} |\tilde{s}_k - \alpha s_k|^2, \tag{9}$$

where $k \in \{j, j+1\}$, $\alpha = |H_{r_{11} i}|^2 + |H_{r_2 i}|^2$, S_M is the set of M transmitted symbols which is known at both the transmitter and receiver, and $|.|$ denotes a magnitude operator. As a result, the diversity gain of the maximum ratio combing (MRC) can be obtained if the channel response is known exactly at the receiver. By using the cooperative diversity with distributed STBC encoding, the network diversity gain can be obtained in this system. Then the decoded signals $\tilde{s}_j$ and $\tilde{s}_{j+1}$ can be detected.

3 Proposed Method

In this method, it is assumed that the location of relays and wireless transmission environment are known by the added UAV. The Received-Signal-Strength Indicator (RSSI) between nodes can be determined by the beacon packets in network. In the multi-hop wireless networks, the transmission channel can be modeled as an equivalent single hop with the minimum output signal-noise-ratio (SNR). Therefore, the end-to-end performance of multi-hop communication systems can be derived based on the performance of the equivalent single hop [15]. As a result, when the performance of the single hop with the minimum output SNR is improved, the end-to-end performance of the system can also be improved. In this method, it is also assumed that the hop with the minimum output SNR of the multi-hop wireless network is determined based on the information of RSSI between relays at each hop in network. The RSSIs from the added UAV to the transmitter and the receiver at the hop are also known as shown in Fig. 3.

Table 2. The modified STBC patterns.

	Symbol j (t)	Symbol $j+1$ ($t+T$)
Transmitter	s_j	s_{j+1}
UAV	$-s^*_{j+1}$	s^*_j

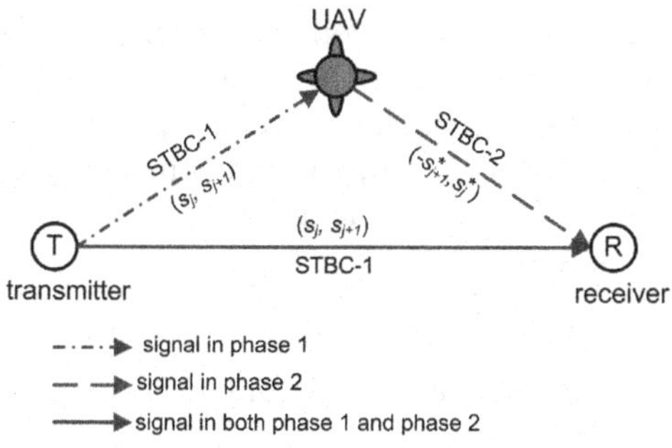

Fig. 3. STBC cooperative diversity.

An UAV is first jointed at the hop and operates as a distributed relay in order to improve the transmission performance of the hop, which in turn leads to improving the end-to-end performance of system. Next, in order to overcome the difficulty in assigning the STBC patterns to the distributed relays and also alleviate the complexity of system design and implementation, the original STBC pattern is modified as shown in Table 2.

It can be seen from the table that the transmission signal is changed only at the added UAV and the same at the transmitter. An algorithm is then also proposed to find out the optimal location of the added UAV at the hop with the minimum output SNR of the multi-hop wireless relay networks. This algorithm is based on the RSSI of the beacon packets, where the UAV has the best contribution to the data transmission performance of Bit Error Rate (BER) between the transmitter and the receiver.

In this method, only the hop with the minimum output SNR in the multi-hop wireless network is considered. At the hop, the distance from transmitter to the receiver is larger than transmission range and there are no any existing relays between the transmitter and receiver as shown in Fig. 4. However, the signal from the transmitter can still be sensed or received weakly at the receiver. This is because the signal is much attenuated by the obstacles such as tree, building, hill or mountain and the distance between the transmitter and the receiver. As a result, this hop is identified as broken and has the minimum output SNR in multi-hop wireless networks. Therefore, an UAV is added in order to maintain the connection and improve the transmission performance of the hop. The signal description for this case as follow. Similarly to Eq. (1), at the first time slot or phase 1, the received signal of the s_j symbol at the UAV and the receiver are given by Eqs. (10) and (11), respectively,

$$y_{ju} = h_{tu}s_j + n_{ju}, \tag{10}$$

$$y_{jr} = h_{tr}s_j + n_{jr}, \tag{11}$$

where h_{tu} and h_{tr} are channel response between the transmitter and UAV, and between the transmitter and the receiver, respectively. At the UAV, the received signal is first decoded, and then re-encoded according to the modified STBC pattern in Table 2 and forwarded to the receiver in phase 2. The signal is also transmitted from the transmitter to the receiver in phase 2. At the receiver, the received signals of the s_j and s_{j+1} symbols from the transmitter and the UAV in phase 2 are combined and expressed as follows,

$$\begin{aligned} y_{jr} &= h_{tr}s_j - h_{ur}s_{j+1}^* + n_{jr}, \\ y_{(j+1)r} &= h_{tr}s_{j+1} + h_{ur}s_j^* + n_{(j+1)r}. \end{aligned} \tag{12}$$

The received signals can be represented in term of vector $\mathbf{y}$ as follows,

$$\mathbf{y} = \begin{bmatrix} y_{jr} \\ y_{(j+1)r} \end{bmatrix} = \begin{bmatrix} s_j & -s_{j+1}^* \\ s_{j+1} & s_j^* \end{bmatrix} \begin{bmatrix} h_{tr} \\ h_{ur} \end{bmatrix} + \begin{bmatrix} n_{jr} \\ n_{(j+1)r} \end{bmatrix}. \tag{13}$$

The Eq. (13) can also be written in term of matrix as the Eq. (3). The orthogonality of the modified STBC encoding pattern can be verified by,

$$\mathbf{s}_1^H \mathbf{s}_2 = \begin{bmatrix} s_j^* \; s_{j+1}^* \end{bmatrix} \begin{bmatrix} -s_{j+1}^* \\ s_j^* \end{bmatrix} = s_j^* s_{j+1}^* - s_j^* s_{j+1}^* = 0, \tag{14}$$

where $(.)^H$ denotes Hermitian conjugate, $\mathbf{s}_1$ and $\mathbf{s}_2$ are the first and second column vector of matrix $\mathbf{s}$, respectively. As a result, in comparison with the original STBC encoding patterns, the cooperative diversity gain of the modified STBC encoding patterns can be obtained is the same. For the modified STBC patterns, the transmitter or source node transmits the s_j and s_{j+1} original symbols, which is similar to the symbols transmitted in multi-hop Single-Input and Single-Output (SISO) transmissions. The symbols are encoded by the only UAV in order to alleviate complexity of system. At the receiver, if the channel response is known exactly, the decoded signals of the s_j and s_{j+1} symbols are derived as follows,

$$\begin{aligned} \tilde{s}_j &= h_{tr}^* y_j + h_{ur} y_{j+1}^* = (|h_{tr}|^2 + |h_{ur}|^2) s_j, \\ \tilde{s}_{j+1} &= -h_{ur} y_j^* + (h_{tr}^*) y_{j+1} = (|h_{tr}|^2 + |h_{ur}|^2) s_{j+1}. \end{aligned} \tag{15}$$

In these equations, the noise components are omitted to keep the presentation simple.

In the case that, the received signal from the transmitter in phase 1 is stored at the receiver. In phase 2, the signal retransmitted from the transmitter is combined with the signal from the UAV and the signal stored in phase 1, the received signal at the receiver can be expressed as follows,

$$\begin{aligned} y_{jr} &= \sum_{m=1,2} h_{tr}^{(m)} s_j - h_{ur} s_{j+1}^* + \sum_{m=1,2} n_{jr}^{(m)}, \\ y_{(j+1)r} &= \sum_{m=1,2} h_{tr}^{(m)} s_{j+1} + h_{ur} s_j^* + \sum_{m=1,2} n_{(j+1)r}^{(m)}, \end{aligned} \tag{16}$$

where $m = 1, 2$ means that the signal from transmitter in phase 1 and phase 2, respectively. Similarly to Eq. (15), the decoded signals of the s_j and s_{j+1} symbols are derived as follows,

$$\tilde{s}_j = (\left| \sum_{m=1,2} h_{tr}^{(m)} \right|^2 + |h_{ur}|^2) s_j, \tag{17}$$

$$\tilde{s}_{j+1} = (\left| \sum_{m=1,2} h_{tr}^{(m)} \right|^2 + |h_{ur}|^2) s_{j+1}. \tag{18}$$

The SNR of the received signal at the receiver (γ) for both s_j and s_{j+1} symbols from the transmitter can be derived by using Eqs. (17) and (18),

$$\gamma_{tr} = \frac{E_r}{N_0} = \frac{E_t(\left| \sum_{m=1,2} h_{tr}^{(m)} \right|^2)}{N_0}, \tag{19}$$

where E_t and E_r is the average signal energy at the transmitter and the receiver, respectively. N_0 is the noise power. The SNR values for the channels from the transmitter to the UAV (γ_{tu}) and from the UAV to the receiver (γ_{ur}) without using STBC encoding can be expressed as follows,

$$\gamma_{tu} = \frac{\gamma_{tr}}{(x^2 + l^2)^{\alpha/2}}, \tag{20}$$

$$\gamma_{ur} = \frac{\gamma_{tr}}{((1-x)^2 + l^2)^{\alpha/2}}, \tag{21}$$

where α is the path loss exponential. When the distance between the transmitter and the receiver is denoted d_{tr} and normalized to 1, $0 \leq x \leq 1$ indicates the relative location of the UAV with respect to the receiver [18]. $x = 0$ means that the UAV is located at the transmitter according to horizontal axis. The height of the UAV is denoted by l. The value of l is determined based on real scenarios so that l minimizes and ensures the links from the UAV to the transmitter and the receiver are Light of Sight (LoS). As shown in Fig. 4, the distance from the transmitter to the UAV and from the UAV to the receiver are denoted by $d_{tu} = (x^2 + l^2)^{1/2}$ and $d_{ur} = ((1-x)^2 + l^2)^{1/2}$, respectively. The SNR of the received signal at the receiver when UAV transmitting the symbols with using STBC can be derived as follows,

$$\gamma_{tur}^{(STBC)} = G\gamma_{tr}, \tag{22}$$

where G is the cooperative diversity gain of using STBC encoding. The maximum diversity gain (G) can be expressed as the number of independent channels in the multiple antennas systems as follows, $G = N_t \times N_r$, where N_t and N_r are the number of transmit and receive antennas, respectively [19]. In order to determine the optimal location of the UAV, it is assumed that the transmit power of all nodes and the location of the transmitter and the receiver are fixed. We also assume that the channels from the transmitter to the UAV and from the UAV to the receiver are symmetric fading channels. Since the RSSI decreases when the distance between the transmitter and the receiver increases, which leads to the increasing of packet error rate (PER), the optimal location of the UAV must be on the vertical plane that passes through the transmitter and the receiver as

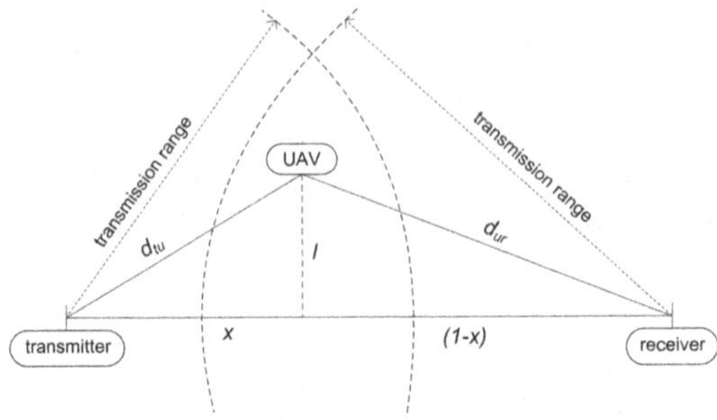

Fig. 4. The hop with the added UAV in network.

shown in Fig. 4. However, in real scenarios, there can be obstacles between the transmitter and the receiver. The packet transmission error probability of the flat fading channel from the transmitter and the receiver can be expressed by,

$$P_{e_{tr}} = \int P_{R_r}(E)P_0(E)dE \tag{23}$$

where E is the amplitude of received signal at the receiver, P_{R_r} is the probability density function (pdf) of received signal envelope of fading distribution. P_0 is PER of a receiver with input signal power of $\frac{E^2}{2}$. The packet transmission error probability with the UAV operating as an intermediary node and using STBC encoding can be expressed by,

$$P_{e_{tur}} = P_{e_{tr}}(1 - P_{s_{tu}} P_{s_{tur}}^{(STBC)}) \tag{24}$$

$$P_{s_{tu}} = 1 - \int P_{R_u}(E)P_0(E)dE \tag{25}$$

$$P_{R_u}(E) = \frac{E}{{\delta_u}^2} exp(-\frac{E^2}{2{\delta_u}^2}) \tag{26}$$

where $P_{s_{tu}}$ is the packet transmission success probability from the transmitter to the UAV in phase 1. $P_{s_{tur}}^{(STBC)}$ is the packet transmission success probability by combining the received signals from the transmitter and the UAV to the receiver in phase 2 and the stored signal from the transmitter in phase 1. $P_{e_{tur}}$ is the packet transmission error probability for both phase 1 and 2. Since the location of the transmitter, the receiver and the transmit power are fixed, the value of $P_{e_{tr}}$ does not change in this case.

$$\min_{x,l}\{P_{e_{tur}}\} = P_{e_{tr}}(1 - \max_{x,l}\{P_{s_{tu}} P_{s_{tur}}^{(STBC)}) \tag{27}$$

Next, an optimization algorithm is proposed in order to find the optimal location of the added UAV which maximizes the final data transmission performance at the receiver. From the Eq. (24), the packet transmission error probability $P_{e_{tur}}$ reaches the minimum value when $\{P_{s_{tu}} P_{s_{tur}}^{(STBC)}\}$ is maximum. Both $P_{s_{tu}}$ and

$P_{s_{tur}}^{(STBC)}$ depend on the UAV location. More specifically, when the distance from the UAV to the transmitter is small, the distance from the UAV to the receiver is large and vice versa. As a result, $P_{s_{tu}}$ is large but $P_{s_{tur}}^{(STBC)}$ is small, and vice versa. In other words, the value of $P_{e_{tur}}$ depends on the height of UAV l and the location of UAV x. With the assumptions as the above mentioned and a fixed transmission power level for nodes, $P_{e_{tur}}$ can be expressed by the empirical function as follow,

$$P_{e_{tur}} = f(\gamma_{\{x,l\}}). \tag{28}$$

The value of $\gamma_{\{x,l\}}$ is determined by the Eq. (29),

$$\gamma_{\{x,l\}} = \max_{x,l}\{\gamma_{tu}, \gamma_{tur}^{(STBC)}\}, \tag{29}$$

where γ_{tu} and $\gamma_{tur}^{(STBC)}$ are calculated according to Eqs. (19) and (22), respectively. Due to function f of the Eq. (28) is a decreasing function, $P_{e_{tur}}$ value reaches the minimum when $\gamma_{\{x,l\}}$ is maximum.

The algorithm is proposed in order to find out the optimal location of the UAV so that $\gamma_{\{x,l\}}$ is maximum. The proposed algorithm is described in detail in Algorithm 1. In this algorithm, the first condition is that the added UAV is within the transmission range of both transmitter and receiver. In other words, the received SNR values γ_{tu} and γ_{ur} are greater than the received SNR threshold γ_{th}. The height of UAV l is then determined so that the links from both transmitter and receiver to the UAV are Light of Sight (LoS) or least affected by the obstacles and nearest to the straight line connected directly between the transmitter and the receiver. Next, the search boundaries and the number of steps are set. The

Algorithm 1. For the optimal location of UAV

Input:the locations of the transmitter and receiver
Output: $\{x, l\}$
Set search boundaries: $\{x_{start}, x_{end}\}$, $\{l_{start}, l_{end}\}$
Set the number of steps: N_x, N_l
Calculate the step size of x: $(x_{start} - x_{end})/N_x$
Calculate the step size of l: $(l_{start} - l_{end})/N_l$
Calculate $\gamma_{\{x,l\}} = \min_{x,l}\{\gamma_{tu}, \gamma_{tur}^{(STBC)}\}$
for $j = l_{start} : N_l :$ to l_{end} **do**
    for $i = x_{start} : N_x :$ to x_{end} **do**
        Calculate $\gamma_{\{i,j\}} = \max_{i,j}\{\gamma_{tu}, \gamma_{tur}^{(STBC)}\}$
        if $\gamma_{\{i,j\}}$ greater than $\gamma_{\{x,l\}}$ **then**
            $\gamma_{\{x,l\}} = \gamma_{\{i,j\}}$
            $x = i$; $l = j$
        else
            $\gamma_{\{x,l\}} = \gamma_{\{x,l\}}$
        end if
    end for
end for

Table 3. Simulation conditions

Modulation method	OFDM QPSK
Number of sub-carriers	52
FFT size	64
Length of guard interval	16 samples
Number of pilot symbols	4 symbols
Number of data symbols	16 symbols
Reference distance	100 [m]
Rice channel model	$K_R = 10$
Noise level at the receiver	−85 [dBm]
Channel estimation	Perfect
Antenna gain of each nodes	0 [dBi]
Direction of antenna	Omnidirectional
Transmit power	15 [dBm]
Number of generated packets	10000 packets

values x and l are divided into small steps within the search boundaries. Finally, $\gamma_{\{x,l\}}$ is calculated based on the beacon packets and compared in order to find the maximum value.

4 Performance Evaluation

4.1 Simulation Conditions

In order to evaluate the performance of the proposed method, three nodes are configured as shown in Fig. 4. In this scenario, the distance from the transmitter to the receiver is 300 [m]. As mentioned above, this distance is normalized to 1, x indicates the relative location of the UAV in comparison with the transmitter and the receiver, $0 \leq x \leq 1$. $x = 0$ and 1 mean that the UAV is located at the transmitter and the receiver according to horizontal axis, respectively. The transmission range is set to 250 [m]. Therefore, the horizontal search boundary is set from $x_{start} = 0.15$ to $x_{end} = 0.85$. The vertical search boundary is set from 1.5 [m] to 100 [m], corresponding to $l_{start} = 0$ and $l_{end} = 1$ for vertical axis. The signal is transmitted at the frequency band of 2.4 GHz [20]. The free space path loss exponent is set to 2 in this simulation [21]. The other simulation conditions are listed in Table 3. In this simulation, it is also assumed that the UAV is static and the delay of all paths from nodes is received within the guard interval of OFDM, therefore, the inter symbol interference (ISI) is not considered. In this method, the modified STBC patterns shown in Table 2 are assigned for the UAV in each sub-carrier.

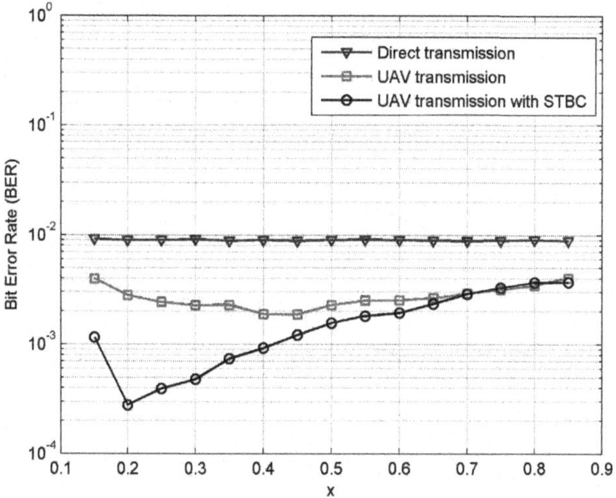

Fig. 5. The transmission performance with the different location of UAV (x) on horizontal axis.

4.2 Simulation Results

First, the bit error rate (BER) of the proposed method with the different locations of the UAV (x) on horizontal axis is shown in Fig. 5. In this simulation, the horizontal location of the UAV is varied from $x = 0.15$ to $x = 0.85$. The vertical location of the UAV is fixed $l = 0.5$. Transmit power is set to be 15 [dBm]. From the figure, direct transmission means that the signal is received directly from the transmitter. Since the receiver is out of the transmission range of the transmitter, direct transmission does not achieve a good performance, only approximately 10^{-2}. This is because that the radio signal is attenuated by radiation in space and affected by fading. When an UAV operates as a relay and STBC encoding is not used, which is named UAV transmission without STBC, the performance is improved significantly in comparison with that of direct transmission. It can be seen that the best performance can be obtained when the location of the UAV is in the approximately middle, $x = 0.45$. This is because that the signal from the transmitter is received and decoded by the UAV then forwarded to the receiver. The signal transmitted in phase 1 from the transmitter to the UAV and phase 2 from the UAV to the receiver is the same. As mentioned in Sect. 3, the channels from the transmitter to the UAV and from the UAV to the receiver are symmetric fading channels. In this simulation, the signal in phase 2 is also retransmitted to the receiver to get better performance. Therefore, when the UAV is in the approximately middle, the performance of system can reach to the best value. In order to get cooperative diversity gain, STBC encoding is used at the UAV. Due to the STBC cooperative diversity gain, UAV transmission with STBC encoding has much better performance than UAV transmission without STBC encoding. From the figure, the best performance can be obtained in this

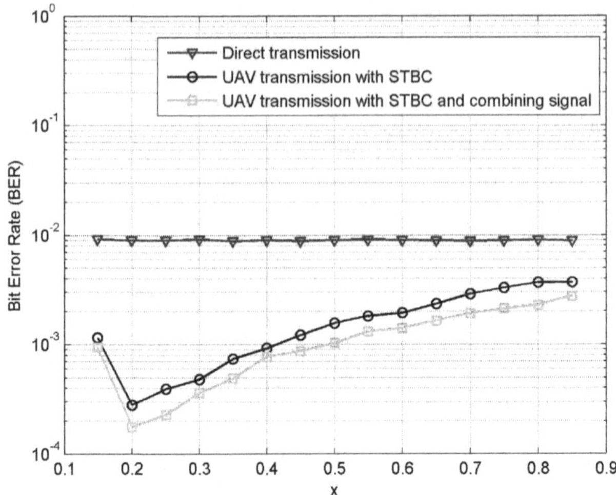

Fig. 6. The transmission performance with combining signal.

case when the location of UAV is about 60 [m] from the transmitter or $x = 0.2$. This is because that the signal is encoded according to Table 2 and transmitted from both the UAV and the transmitter to the receiver in phase 2. Therefore, STBC cooperative diversity gain can be obtained, which in turn leads to the significant improvement of performance in phase 2. As a result, the optimal location of the UAV on horizontal axis will move to the transmitter depending on the cooperative diversity gain. Next, in order to improve the performance of system, the signal transmitted from the transmitter to the receiver in phase 1 is stored at the receiver and then combined with the signal transmitted from both the transmitter and the UAV in phase 2. From the figure, the performance of the UAV transmission with STBC and combining signal is better performance than that of UAV transmission with STBC. However, the optimal location of the UAV on horizontal axis is the same. This is because the combining signal leads to higher SNR at the receiver in phase 2 but the STBC cooperative diversity gain does not increase. Therefore, the optimal location of UAV does not change in this case as shown in Fig. 6.

Next, the effect of the different location of the UAV (l) according to vertical axis on the performance of system is shown in Fig. 7. In this simulation, the vertical location of the UAV is varied from $l = 0.1$ to $l = 0.9$. The horizontal location of the UAV is fixed $x = 0.2$. Transmit power is also set to be 15 [dBm]. It can be seen from the figure that the performance decreases slightly when l value increases. The reason is that when l value increases, the distance from the transmitter to the UAV and also from the UAV to the receiver is greater. However, the different distance is quite small. Therefore, the SNR at the receiver only decreases slightly. This leads to a slight decrease in the performance of

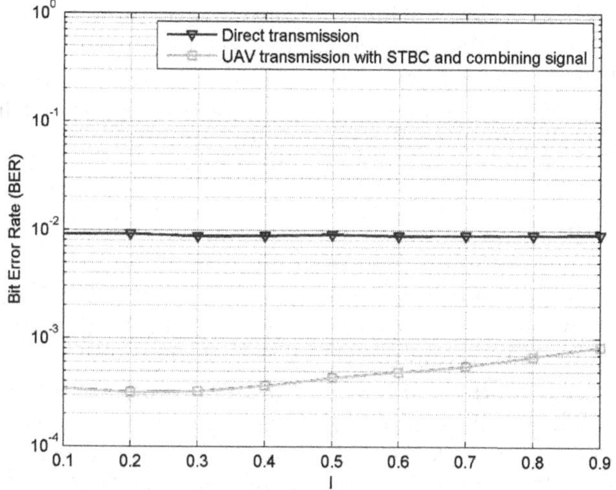

Fig. 7. The transmission performance with the different location of UAV (l) on vertical axis.

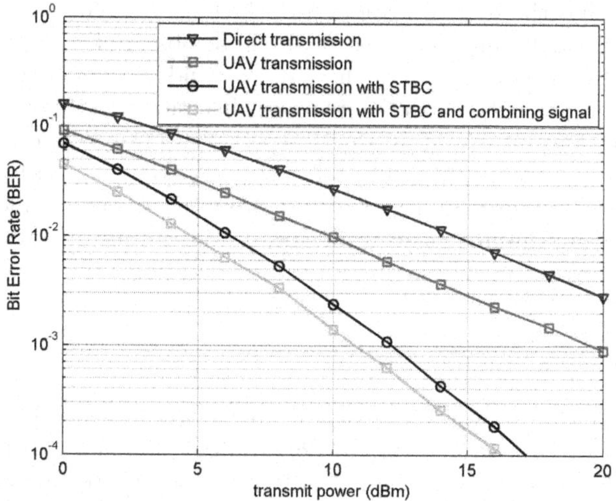

Fig. 8. The transmission performance with different transmit power.

system. As a result, the effect of the different location of the UAV (l) on vertical axis on the performance of system is quite small.

Finally, Fig. 8 shows the performance of system with the different transmit powers. From the figure, since the added UAV operates as a relay, the performance of the UAV transmission is improved significantly in comparison with the direct transmission, about 5 [dB]. It can be seen that the UAV transmission with STBC encoding also achieves much better performance than both direct transmission and UAV transmission. This is because that STBC cooperative diversity gain can be obtained in this case. In order to get higher performance, the signals in both phase 1 and phase 2 are combined at the receiver. As a result, the performance is also slightly improved for the UAV transmission with STBC and combining signal, about 1.5 [dB] in this scenario.

5 Conclusion

In this paper, a transmission method using UAV and distributed STBC encoding for multi-hop wireless relay networks is proposed to use in search and rescue operations. First, an UAV is considered to add to the hop with the minimum output signal-noise-ratio (SNR) and operates as a relay node to maintain the links between adjacency nodes in network, expand the transmission coverage area and improve the transmission performance. Next, in order to overcome the difficulty in assigning the STBC patterns to the distributed relays and also alleviate the complexity of system design and implementation, the original STBC pattern is modified while keeping the same cooperative diversity gain. Finally, an algorithm is proposed to find out the optimal location of the added UAV in the hop, where the UAV has the best contribution to the data transmission performance between the transmitter and the receiver. It can be observed from the simulation results that the optimal location of the UAV depends on not only the environment of real scenarios but also the cooperative diversity gain. In order to achieve the best performance of system in this scenario, the UAV should be located at $(0.2 \leq x \leq 0.3)$ according to horizontal axis and $(0.5 \leq l \leq 0.9)$ according to vertical axis. We can also confirm that the proposed method achieves the significant performance improvement while keeping the simple operation of system for UAV communications in search and rescue operations.

References

1. Valavanis, K.P., Vachtsevanos, G.J. (eds.): Handbook of Unmanned Aerial Vehicles. Springer, Dordrecht (2015). https://doi.org/10.1007/978-90-481-9707-1
2. Mozaffari, M., et al.: A tutorial on UAVs for wireless networks: applications, challenges, and open problems. IEEE Commun. Surv. Tutor. **21**(3), 2334–2360 (2019)
3. Onwuka, E., et al.: MANET: a reliable network in disaster areas. J. Res. Natl. Dev. (JORIND) **9**(2), 105–113 (2011)
4. Anjum, S.S., Noor, R.M., Anisi, M.H.: Review on MANET based communication for search and rescue operations. Wirel. Pers. Commun. **94**(1), 31–52 (2015). https://doi.org/10.1007/s11277-015-3155-y
5. Loo, J., Mauri, J.L., et al.: Mobile Ad Hoc Networks: Current Status and Future Trends. CRC Press, Boca Raton (2012)
6. Li, B., et al.: UAV communications for 5G and beyond: recent advances and future trends. IEEE Internet Things J. **6**(2), 2241–2263 (2019)
7. Waharte, S., Trigoni, N.: Supporting search and rescue operations with UAVs. In: 2010 International Conference on Emerging Security Technologies, September 2010. IEEE (2010)

8. Zeng, Y., et al.: Wireless communications with unmanned aerial vehicles: opportunities and challenges. IEEE Commun. Mag. **54**(5), 36–42 (2016)
9. Wu, Z., et al.: Performance evaluation of OFDM transmission in UAV wireless communication. In: 2005 Thirty-Seventh Southeastern Symposium on System Theory, March 2005. IEEE (2005)
10. Vahidi, V., et al.: Orthogonal frequency division multiplexing and channel models for payload communications of unmanned aerial systems. In: 2016 International Conference on Unmanned Aircraft Systems (ICUAS), June 2016. IEEE (2016)
11. Sendonaris, A., Erkip, E., Aazhang, B.: User cooperation diversity-part I: system description. IEEE Trans. Commun. **51**(11), 1927–1938 (2003)
12. Sendonaris, A., Erkip, E., Aazhang, B.: User cooperation diversity part II: system description. IEEE Trans. Commun. **51**(11), 1939–1948 (2003)
13. Alamouti, S.M.: A simple transmit diversity technique for wireless communications. IEEE J. Sel. Areas Commun. **16**(8), 1451–1458 (1998)
14. Harshan, J., Rajan, B.S.: Distributed space-time block codes for two-hop wireless relay networks. J. Commun. **5**(4), 282–296 (2010)
15. Bao, V.N.Q., Kong, H.Y.: A simple performance approximation for Multi-hop decode-and-forward relaying over Rayleigh fading channels. IEICE Trans. Commun. **E92-B**(11), 3524–3527 (2009)
16. Laneman, J.N., Tse, D.N.C., Wornell, G.W.: Cooperative diversity in wireless networks: efficient protocols and outage behavior. IEEE Trans. Inf. Theory **50**(12), 3062–3080 (2004)
17. Goldsmith, A.: Wireless Communications. Cambridge University Press, Cambridge (2005)
18. Obiedat, E.A., Cao, L.: Soft information relaying for distributed turbo product codes (SIR-DTPC). IEEE Signal Process. Lett. **17**(4), 363–366 (2010)
19. Dohler, M., Li, Y.: Cooperative Communications Hardware, Channel and PHY. Wiley, Hoboken (2010)
20. Atoev, S., Kwon, O., et al.: An efficient SC-FDM modulation technique for a UAV communication link. Electron. J. **7**(12), 352–370 (2018)
21. Mozaffari, M., Debbah, M., et al.: Unmanned aerial vehicle with underlaid device-to-device communications: performance and tradeoffs. IEEE Trans. Wirel. Commun. **15**(6), 3949–3963 (2016)

Hardware, Software, and Application Designs

Resolution-Improvement of Confocal Fluorescence Microscopy via Two Different Point Spread Functions

Xuanhoi Hoang, Vannhu Le, and MinhNghia Pham(✉)

Le Quy Don Technical University, Hanoi, Vietnam
levannhuktq@gmail.com, nghiapm2018@mta.edu.vn

Abstract. In this paper, we propose a new method to obtain the improvement of lateral axial resolution of confocal fluorescence microscopy. In this method, we employ two different beams to illuminate the sample: (1) the Gaussian beam; (2) the donut beam. Two different images are produced from these two illumination beams. A higher resolution image is generated by a multi-relationship between these two image. A set of simulation and experimental results are employed to compare the effectiveness of proposed method with the traditional confocal fluorescence microscopy. These results demonstrated that our method can be employed to achieve the resolution-enhancement of confocal fluorescence microscopy.

Keywords: High-resolution · Confocal fluorescence microscopy · Image processing

1 Introduction

Because of the diffraction limit, the maximum spatial resolution of tradition optical microscopy in the far field is about 0.61 λ/NA, in which λ is the illumination wavelength and *NA* is the numerical aperture [1–3]. Confocal fluorescence microscopy (CFM) provides the relatively high-resolution image that has low out-of-focus background-noise. Additionally, the CFM system can be used to achieve the resolution improvement by a factor of $\sqrt{2}$ [4, 5]. Therefore, the CFM is extensively employed to analyze of three-dimensional sample and becomes a powerful tool for investigating specimen in the life science. However, the resolution of the CFM only reaches to 200 nm under the recent experimental condition [6, 7].

Note that STED nanoscopy is based on the nonlinear excitation process as for breaking Abbe's diffraction limit and yet is limited by the high intensity radiation that may cause side effects like photo-bleaching or photo-toxicity due to non-linear saturated rationale [8, 9]. In addition, another non-negligible limitation is the problem of strong background signals mostly left by secondly-excited depletion beam. Requirement for special dyes of STED is another drawback. These disadvantages give rise to the booming development of other STED-like techniques with biological compatibility, such as

N.-S. Vo and V.-P. Hoang (Eds.): INISCOM 2020, LNICST 334, pp. 77–84, 2020.
https://doi.org/10.1007/978-3-030-63083-6_6

reversible saturable optical fluorescence transitions (RESOLFT) [10], charge state depletion (CSD) [11], ground state depletion (GSD) [12], fluorescence quenching microscopy (FQM) [13], excited state saturation (ESSat) [14] and fluorescence emission difference microscopy (FED) [15] and so on. Another super-resolution strategy, applying molecular energy state difference and multiply-beam illumination, as well as thousands of times of photon acquisitions, is concerning single-molecular localization imaging (SML), like stochastic optical reconstruction microscopy (STORM) [16, 17] and photo-activated localization microscopy (PALM) [18, 19]. Tens of thousands of wide-field illumination register fluorescent dyes while each illumination only excites sparsely-distributed molecules beyond diffraction limit. Digital reconstruction of these images will result in super-resolved structures. These SML techniques are limited by the long acquisition time and the special dyes. Structured illumination microscopy (SIM) is an alternative approach to break the diffraction barrier by a factor of 2, based on multiply-beam illumination [20, 21]. By virtue of periodically patterning the illumination light with different illumination angles and light polarizations, high frequency information could be extracted through post-processing. This technique, however, is costly due to expensive architecture in order to achieve video-rate acquisition speed. In addition, the resolution improvement is also limited although saturated SIM (SSIM) could further improve the spatial resolution at the price of high risk of photo-bleaching.

In this paper, we suggest a novel approach for the improvement of resolution of confocal fluorescence microscopy. In this method, we introduce two illumination beams. The first beam is the Gaussian beam. The second beam is the donut beam which are modulated by the 0–2 vortex phase mask. For obtaining the image with the higher resolution, we introduce a multiplying relationship between the two images. The proposed method is simple and performs easily.

2 Methodology

By using vector diffraction theory, the intensity distribution of incident light propagating through an objective lens can be calculated. The electric field near the focus imaging plane can be acquired explicitly by the formulae derived from the Debye integral as:

$$\vec{E}(r_2, \xi_2, z_2) = iC \iint_{\Omega} \sin(\varsigma) E_0 A(\varsigma, \xi) P \times e^{i\Delta a(\varsigma,\xi)} e^{ikn(z_2 \cos\varsigma + r_2 \sin\varsigma cons(\xi-\xi_2))} d\varsigma d\xi \quad (1)$$

where $\vec{E}(r_2, \xi_2, z_2)$ represents the electric field vector at the point (r_2, ξ_2, z_2) which is generated by cylindrical coordinates, C shows the normalized constant, E_0 represents the amplitude distribution of the input light beam, $A(\zeta, \xi)$ represents a 3×3 matrix related to the structure of the imaging lens and P presents Jones vector of the incident light beam. $\Delta a(\zeta, \xi)$ represents the parameter of phase delay using the phase mask.

When the sine condition is used with the object lens, $A(\zeta, \xi)$ can be represented by,

$$A(\varsigma, \xi) = \sqrt{\cos\varsigma} \begin{bmatrix} 1 + (\cos\varsigma - 1)\cos^2\varsigma & (\cos\varsigma - 1)\cos\xi\sin\xi & -\sin\varsigma\cos\xi \\ (\cos\varsigma - 1)\cos\xi\sin\xi & 1 + (\cos\varsigma - 1)\sin^2\xi & -\sin\varsigma\sin\xi \\ \sin\varsigma\cos\xi & \sin\varsigma\sin\xi & \cos\varsigma \end{bmatrix} \quad (2)$$

The point spread function (PSF), h, can be calculated through the intensity distribution as,

$$h = \left|\vec{E}\right|^2 \tag{3}$$

The image can be illuminated by the following equation

$$I = x * h \tag{4}$$

in which x denotes the observation-sample.

In this paper, we introduce one method based on the use of two kinds of PSFs to achieve the improvement of the resolution of the CFM. The two images can be captured by using the two PSFs. One image is acquired by employing the dot beam which is modulated by Gaussian beam. This image is called the dot image. Other image is achieved by using the donut beam that is modulated by the 0–2 vortex phase mask as depicted in Fig. 1. This image is called the donut image. The third image is generated by using the multiplying relationship between the dot image and the donut image as shown in Fig. 2. This image is called the multi-image. The proposed image can be presented by,

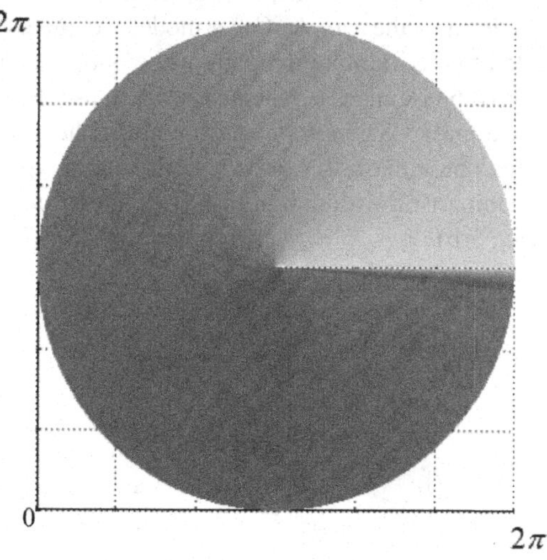

Fig. 1. The 0–2 vortex phase mask.

$$I = I_{Dot} \exp(\alpha./(I_{Donut} + \beta)) \tag{5}$$

We show the two feasible imaging schemes which can be used to perform the proposed method. The first model is shown in Fig. 3(a). It can be seen that two illumination paths are shown: the first path is Gaussian beam, the second path is donut one. The donut beam is modulated by 0–2π vortex phase mask that is generated by a spatial light modulator (SLM). Because two illumination paths are separated before the BS, we should

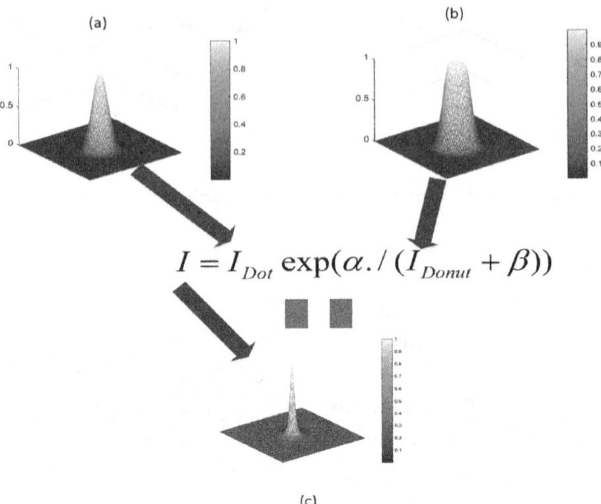

Fig. 2. The multi-theory. (a) the PSF is generated by the dot beam; (b) the PSF is generated by the donut beam; (c) the PSF is generated by using the multiplying relationship.

carefully adjust the two illumination beams, to protect that the two PSFs of these two beams have the same position on the sample. Other model is depicted in Fig. 3(b). In this model, a SLM is inserted to the illumination path of the optical system, which is used to switch different kinds of phase masks. When we want to use the dot beam, the SLM will generate the 0 phase mask. When we want to employ the donut beam, the SLM will produce 0–2π vortex phase mask. It can be seen that the setup of optical system in Fig. 3(b) is simpler in comparison with that in Fig. 3(a). Therefore, we use the setup of the optical system in Fig. 3(b).

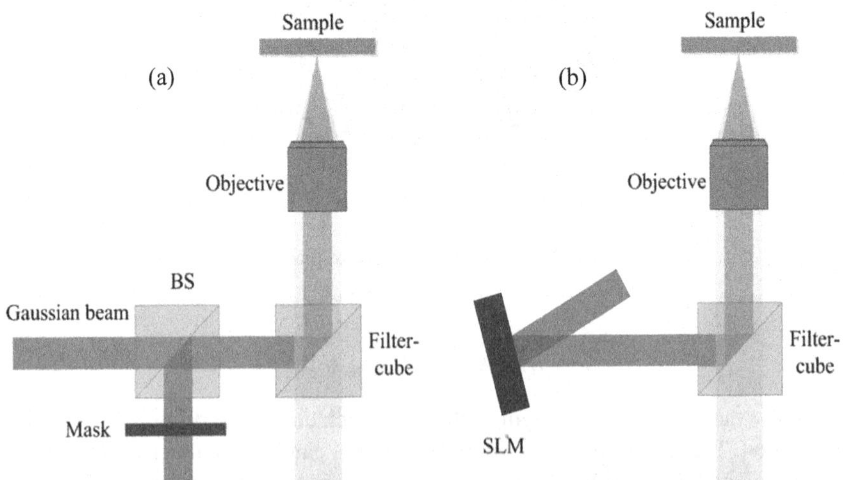

Fig. 3. Optical layouts via two illumination beams in CFM.

3 Simulation and Experimental Results

In order to highlight the proposed ability for the improvement of the resolution, we perform some simulation results. We use some initial parameters for simulation condition: $NA = 1.49$; the illumination laser, $\lambda = 640$ nm, is used to illuminate sample; the sample is placed in the medium for index refraction of 1.518. The first, we consider with the spokes image. The original image is illuminated in Fig. 4(a), its size is set to $6\lambda \times 6\lambda$. By using Eq. (4), Fig. 4(b) shows the confocal image, while the proposed image is illuminated in Fig. 4(c) with $\alpha = 5$ and $\beta = 1$. We draw additionally one blue circle one the confocal image and the proposed image. It can be seen that the outside of the blue circle, we can discern for the confocal image and the proposed image. However, the inside of the blue circle, the confocal image is not discerned, while some parts of the proposed image still discerns. This means that the effectiveness of our method for the improvement of the resolution can be achieved.

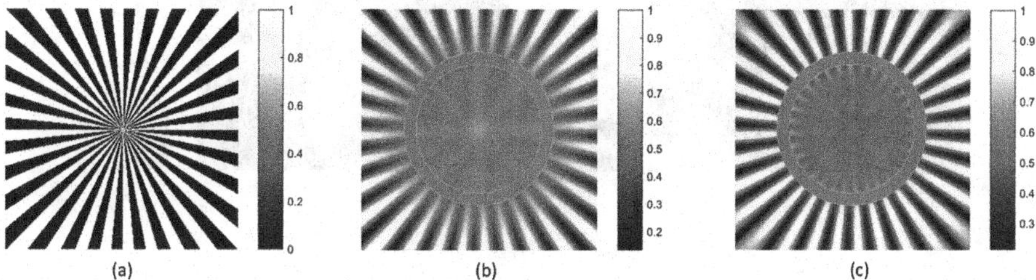

Fig. 4. (a) the original image; (b) the confocal image; (c) the proposed image (Color figure online)

Next, we simulate the imaging process with cell microtubules. The original image is illuminated in Fig. 5(a). The size of the original image is equal to $6\lambda \times 6\lambda$. The confocal image of the CFM is depicted in Fig. 5(b). Figure 5(c) indicates the proposed image. There are the two lines which nearly parallel at the positions (1) and (2) on the original image. It can be seen that the two lines of the confocal image which is shown at the positions (1) and (2) do not separate. While the two lines of the proposed image can be discerned.

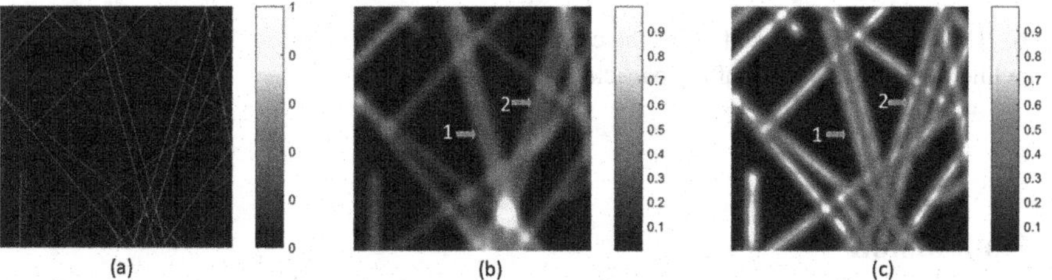

Fig. 5. (a) the original image; (b) the confocal image; (c) the proposed image.

The finally, in order to highlight the ability of the resolution-improvement of our method, 200 nm spherical fluorescence particles are used to perform experiment. The experimental result of CFM is represented in Fig. 6(a). Figure 6(b) depicts the experimental result of the proposed method. It can be seen that the size of spherical fluorescence particles of two images is the near same. However, the resolution-ability of the proposed image is better than that of the confocal image. This can be clearly shown by drawing a line though the center of the particle as depicted on the Fig. 6(a) and (b). This line is illuminated in the Fig. 6(c). Form Fig. 6(c), it can be clearly seen that the resolution by using the proposed method was remarkably improved. We use full-width at half-maximum (FWHM) to measure the ability of resolution. The FWHM of the particle on the proposed image is smaller 1.43 time that that of the particle on the confocal image. This result demonstrated that our method can be used to improve the resolution of the CFM.

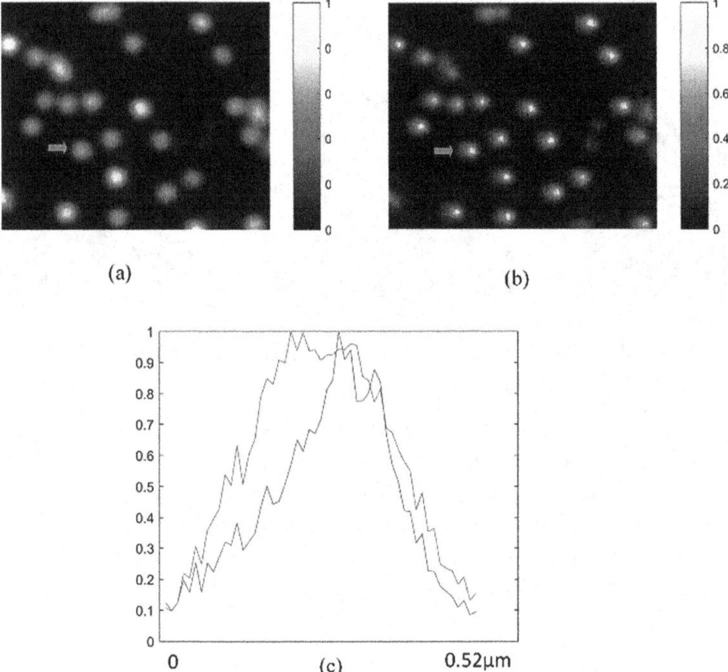

Fig. 6. The experimental result of (a) the confocal image; (b) the proposed image; (c) the image of one line through the center of the particle.

4 Conclusion

We have suggested one effective method using two PSFs to achieve the enhancement of the resolution of CFM. Two images are generated by two kinds of the PSFs. A novel relationship between these two images is built. The effectiveness of our method is demonstrated by some simulation results. The experiment with spherical fluorescence particles was shown. The experimental result illuminated that the proposed method can be employed to acquire the improvement of the resolution about 1.43 times. The proposed method can be used in many different fluorescence microscopies such as light-sheet fluorescence microscopy, wide-field fluorescence microscopy, etc.

Acknowledgment. This work is supported by Vietnam National Foundation for Science and Technology Development (NAFOSTED) under Grant number (103.03-2018.08).

References

1. Wan, C., et al.: Three-dimensional visible-light capsule enclosing perfect supersized darkness via antiresolution. Laser Photonic Rev. **5**, 743–749 (2014)
2. Pawley, J.: Handbook of Biological of Confocal Microscopy, 3rd edn. Springer, New York (2006)
3. Wilson, T.: Confocal Microscopy, vol. 426, pp. 1–64. Academic Press, London (1990)
4. Gu, M.: Principles of Three Dimensional Imaging in Confocal Microscopies. World Scientific, Singapore (1996)
5. Segawa, S., Kozawa, Y., Sato, S.: Resolution enhancement of confocal microscopy by subtraction method with vector beams. Opt. Lett. **39**(11), 3118–3121 (2014)
6. So, S., et al.: Overcoming diffraction limit: from microscopy to nanoscopy. Appl. Spectros. Rev. **53**(1) (2017)
7. Le, V., Wang, X., Kuang, C., Liu, X.: Resolution enhancement of confocal fluorescence microscopy via two illumination beams. Optics Lasers Eng. **122**, 8–13 (2019)
8. Hell, S.W., Wichmann, J.: Breaking the diffraction resolution limit by stimulated emission: stimulated-emission-depletion fluorescence microscopy. Opt. Lett. **19**, 780–782 (1994)
9. Gao, P., Prunsche, B., Zhou, L., Nienhaus, K., Nienhaus, G.U.: Background suppression in fluorescence nanoscopy with stimulated emission double depletion. Nature photonic. **11**, 163–169 (2017)
10. Wang, S., Chen, X., Chang, L., Xue, R., Duan, H., Sun, Y.: GMars-Q Enables long-term live-cell parallelized reversible saturable optical fluorescence transitions nanoscopy. ACS Nano **10**(10), 9136–9144 (2016)
11. Sharma, Reena., Singh, Manjot, Sharma, Rajesh: Recent advances in STED and RESOLFT super-resolution imaging techniques. Spectrochim. Acta Part A Mol. Biomol. Spectrosc. **231**(15), 117715 (2020)
12. Dixon, Rose E., Vivas, Oscar., Hannigan, Karen I., Dickson, Eamonn J.: Ground state depletion super-resolution imaging in mammalian cells. J. Vis. Exp. **129**, 56239 (2017)
13. Chen, X., Zou, C., Gong, Z., Dong, C., Guo, G., Sun, F.: Sub-diffraction optical manipulation of the chargestate of nitrogen vacancy center in diamond. arXiv:1410.4668
14. Trebbia, J.-B., Baby, R., Tamarat, P., Lounis, B.: 3D optical nanoscopy with excited state saturation at liquid helium temperatures. Opt. Express **27**(16), 23486–23496 (2019)
15. Li, Y., et al.: Image scanning fluorescence emission difference microscopy based on a detector array. J. Microsc. **266**, 288–297 (2017)
16. Rust, M.J., et al.: Sub-diffraction-limit imaging by stochastic optical reconstruction microscopy (STORM). Nat. Methods **3**, 793–796 (2006)
17. Huang, B., et al.: Three-dimensional super-resolution imaging by stochastic optical reconstruction microscopy. Science **319**, 810–813 (2008)
18. Betzig, E., et al.: Imaging intracellular fluorescent proteins at nanometer resolution. Science **313**, 1642–1645 (2006)

19. Hess, S.T., et al.: Ultra-high resolution imaging by fluorescence photoactivation localization microscopy. Biophys. J. **91**(11), 4258–4272 (2006)
20. Classen, A., et al.: Superresolution via structured illumination quantum correlation microscopy. Optica **4**(6), 480 (2017)
21. Gustafsson, M.G.L.: Nonlinear structured-illumination microscopy: wide-field fluorescence imaging with theorycally unlimited resolution. PNAS **102**(37), 13081–13086 (2005)

Estimations of Matching Layers Effects on Lens Antenna Characteristics

Phan Van Hung[1], Nguyen Quoc Dinh[1](✉), Hoang Dinh Thuyen[1], Nguyen Tuan Hung[1], Le Minh Thuy[2], Le Trong Trung[3], and Yoshihide Yamada[4]

[1] Le Quy Don Technical University, Hanoi, Vietnam
dinhnq@mta.edu.vn
[2] School of Electrical Engineering, Hanoi University of Science and Technology, Hanoi, Vietnam
[3] Telecommunications University, Nhatrang, Vietnam
[4] Malaysia-Japan International Institute of Technology UTM, Kuala Lumpur, Malaysia

Abstract. The dielectric lens antenna is a prime candidate for the mm-wave 5G communications system. The size and the radiation efficiency can be improved by using a high-density dielectric lens antenna. However, the dense dielectric material lenses can make some antenna properties deteriorate due to the reflections at the surface between the air and the dielectric. These unexpected effects can be solved by using a quarter-wavelength matching layer (ML). In this paper, the authors perform the study to estimate the influence of the ML on the antenna properties on specific dielectric materials. The results illustrate a marked improvement in gain, a significant reduction in the side-lobe level, and considerable changes in the electric field distribution on the plane with and without using the ML. Besides, the article also shows the abilities to minimize the antenna size when different types of dielectric materials are chosen while maintaining the antenna radiation characteristics.

Keywords: Lens antenna · Matching layer · Quarter-wavelength matching

1 Introduction

At the millimeter-wave frequency, the antenna element size becomes smaller ten times compared to the sub-6 GHz 5G base station antenna element. This frequency range allows selecting various antennas elements such as lens, array, and reflector antennas [1–4]. Reflector antennas with multiple feeds have high direction and the ability to create multi-beams. However, high cross-polarization and the side-lobe level if these antennas are affected by feeds [5].

In recent years, there has been an increased interest in using dielectric lens antennas among researchers and manufacturers [6–10]. This type of antenna is capable of producing multi-beams and broadband. With a unique structure, the lens antenna is not affected by the blockage of the reflectors and the feeds. Besides, the lens antenna allows electric waves from the front to penetrate backwards. Lens antennas have been studied

N.-S. Vo and V.-P. Hoang (Eds.): INISCOM 2020, LNICST 334, pp. 85–94, 2020.
https://doi.org/10.1007/978-3-030-63083-6_7

and applied to many kinds of broadband communication systems such as Ka-band [11], Q band [12], and V band [13], automotive cruise control radars and obstacle detection radars in W band [14, 15], in which lenses made of the dielectric materials, uniform size and shape are designed according to the system specifications.

In dielectric lens antenna applications, choosing the relative permittivity of the lens material is very important since its value has a major influence on the antenna radiation characteristics. Dielectric lens antennas made of dense dielectric materials like Mica, Alumina, and Silicon enhance the efficiency of transmitting energy through the lens and improve the front to back radiation ratio of the antenna. However, the dense dielectric material in the lens antenna causes reflection from the lens surface between the air and the dielectric layer of the lens. The reflection can significantly affect not only the input impedance of the power [16–18] but also the characteristics of lens antennas such as reduced gain, beam distortion, and increased side-lobe level. Therefore, the ML is applied to lessen these unwanted effects, thereby improving the efficiency of using the lens antenna in information systems [19, 20]. There are two fundamental methods to create a ML: using a quarter-wavelength matching layer and the reaction wall which are studied by Morita, S.B. Cohn in [21, 22].

However, up to now, the studies of effects of high-density dielectric materials used to make typical lenses such as Teflon, Mica, Alumina, and Silicon have not been conducted yet. In this paper, the authors perform a theoretical calculation of the power reflection coefficient and the simulation in the electromagnetic environment, thereby making accurate evaluations of the specific impact of each dielectric material type. The results serve as a basis for the selection of specific materials in the manufacture of lens antennas. The paper includes 4 parts: part 2 presents the theoretical calculations of the quarter-wavelength matching layer, constructing antenna models and simulation parameters. The results and the discussion are shown in part 3. The conclusion is presented in part 4.

2 Theoretical Matching Layers and Antenna Modeling

2.1 Theoretical Matching Layers

The relative effectiveness of the ML technique can be easily assessed by the power reflection coefficient at the surface between the air and the dielectric. In Fig. 1c, the ML is assigned between the air and the dielectric region: region 1 denotes the air, region 2 is the ML and region 3 denotes the dielectric. The power reflection coefficient at the surface between region i and region j is denoted as r_{ij}. Accordingly, r_{12} and r_{23} are the power reflection coefficients at the surface between Regions 1 and 2 and Regions 2 and 3, respectively. The power reflection coefficient is determined by the following equation [23, 24].

$$\mathrm{r_{ij}} = \left| \frac{\sqrt{\varepsilon_\mathrm{i}} - \sqrt{\varepsilon_\mathrm{j}}}{\sqrt{\varepsilon_\mathrm{i}} + \sqrt{\varepsilon_\mathrm{j}}} \right| \tag{1}$$

$$R = \frac{(r_{12} + r_{23})^2 - 4r_{12}r_{23}sin^2\alpha_2 d}{(1 + r_{12}r_{23})^2 - 4r_{12}r_{23}sin^2\alpha_2 d} \tag{2}$$

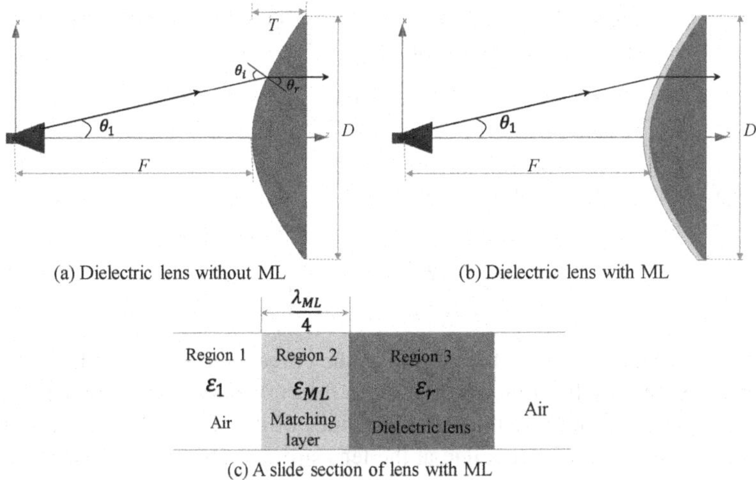

Fig. 1. Antenna modeling of a dielectric lens antenna

where $\alpha_2 = 2\pi/\lambda_{ML}$, λ_{ML} is the wavelength in the ML environment. If ε_{ML} is between the value of ε_1 and ε_r, then $r_{12}r_{23}$ is positive, thereby R changes and serves as a function of ML layer thickness d. Therefore, R reaches the minimum value when $d = \lambda_{ML}/4$ and equals zero when $r_{12} = r_{23}$, then ε_{ML} is the geometric mean of ε_1 and ε_r or $\varepsilon_{ML} = \sqrt{\varepsilon_1 \varepsilon_r}$.

The thickness of the ML layer is defined as follows:

$$d = \frac{\lambda_{ML}}{4} = \frac{\lambda}{4\,n_{ML}} \tag{3}$$

where λ is the wavelength in free space and n_{ML} is the refraction index of ML. Under these conditions, all power is transferred from the medium (1) into the environment (2), no power is reflected at the surface. Besides, assigning ML to a quarter wavelength with an appropriate dielectric constant also aims to coordinate the impedance at the surface of the two power lines.

2.2 Antenna Modeling

As can be seen in Fig. 1, the lens antenna structure consists of a conical horn antenna as a wide-angle power and a lens which has a simple structure of a curved surface and a flat surface and rotates around the oz axis. The conical horn antenna is designed to perform at 28 GHz and reaches a maximum gain of 15.15 dBi. The lenses are dense dielectric materials with a relative permittivity of $\varepsilon_r = 2.1$ (Teflon), $\varepsilon_r = 5.7$ (Mica), $\varepsilon_r = 9.2$ (Alumina) and $\varepsilon_r = 11.9$ (Silicon), respectively. These materials are commonly used in the design and manufacture of lens antennas. The lens surface curves towards the feed horn. The focal point of the lens is located at the center of the horn antenna, which is the interface between the waveguide and the flare of the conical horn. The curved surface of the lens is hyperbolic, and it is determined by Eq. (4) [25–27]. The diameter of the lens is D. F is a distance from the focal point to the lens vertex. The lens is located at the angles of $\theta_{1Max} = \pm 25.545^0$, where the electromagnetic field of the conical horn antenna is -10 dB compared to the maximum values.

$$r = \frac{(n-1)F}{ncos\theta_1 - 1} \quad (4)$$

The lens thickness at the center is calculated as:

$$T = \frac{1}{n+1}\left[\sqrt{F^2 + \frac{(n+1)D^2}{4(n-1)}} - F\right] \quad (5)$$

where r is the distance from the focal point to the curved surface of the lens. θ_1 is the angle created by the ray from the focal point to the curved surface of the lens and the optical axis in region 1, and n is the refractive index of the lens. θ_i and θ_r represent the incident angle and the refractive angle at the lens surface. These angles satisfy Snell's law.

$$n = \frac{sin\theta_i}{sin\theta_r} \quad (6)$$

The lens is covered with a quarter-wavelength matching layer whose structure is shown in Fig. 1b. In Fig. 1c, The ML has the relative permittivity (ε_{ML}) and the thickness ($d = \lambda_{ML}/4$), in which λ_{ML} is the wavelength inside the ML. The ML is set in region 2 (between region 1 and region 3). The rays are radiated from the conical horn antenna in region 1, passing through the quarter-wavelength matching layer to region 3, and then going through the other side of the lens forming the rays parallel to the oz axis.

2.3 Simulation Parameters

The simulation parameters, computer configurations, and software are shown in Table 1. The survey is conducted using the ANSYS HFSS electromagnetic field simulation software. A simultaneous setting of two adaptive cross approximation (ACA) and multi-pole multi-level fast method (MLFMM) are employed to analyze results to improve accuracy as well as save memory and calculation time.

Table 1. Simulation parameters

Computer specifications	CPU	Intel (R) 3.20 GHz
	RAM	32 GB
	Software	ANSYS HFSS
	Simulation method	MLFMM and ACA

(continued)

Table 1. (*continued*)

Dielectric lens material parameters	Lens diameter [mm]	D	100
	Distance from the origin to the lens vertex [mm]	F	100
	Teflon relative permittivity	ε_r	2.1
	Mica relative permittivity		5.7
	Alumina relative permittivity		9.2
	Silicon relative permittivity		11.9
	Air relative permittivity	ε_1	1
	ML relative permittivity	ε_{ML}	$\sqrt{\varepsilon_1\varepsilon_r}$
	Matching layer thickness [mm]	d	$\lambda_{ML}/4$
Survey frequency			28 GHz

3 Simulation Results

3.1 Power Reflection Coefficients

Figure 2 shows the effects of the thickness and relative permittivity of the ML on the power reflection coefficients of the lens antennas. Obviously, when the thickness of ML is equal to a quarter of the wavelength in ML medium, the power reflection coefficient is lowest. When the relative permittivity of the ML is equal to the geometric mean of the air and the dielectric substance of the lens and the thickness of ML is equal to a quarter of the wavelength in ML medium, the power reflection coefficient is zero, as illustrated in Fig. 2a. When the thickness of the ML layer changes, the power reflection coefficient at the interface of the air layer and the dielectric of the Silicon lens is always higher than

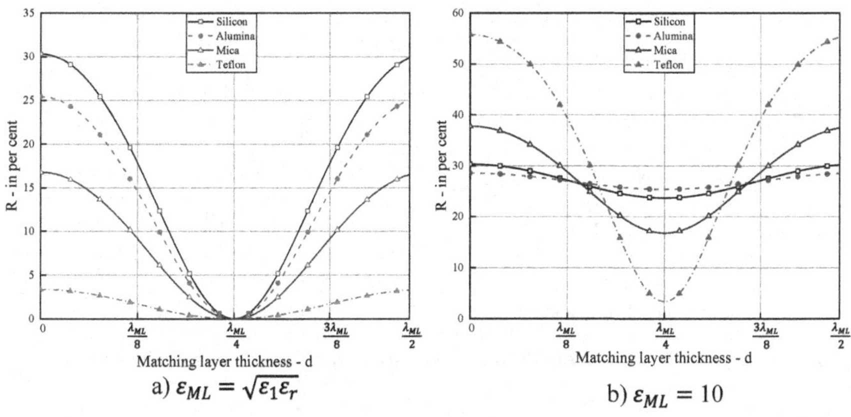

Fig. 2. Power reflection coefficients

that of other dielectric materials. The power reflection coefficient is even 30% higher when the thickness of the ML is zero or half the wavelength in the ML medium. In Fig. 2b, a relative permittivity value of the ML ($\varepsilon_{ML} = 10$) is randomly surveyed. The results show that the power reflection coefficient is minimized when $d = \lambda_{ML}/4$ and cannot eliminate the reflected energy at $\varepsilon_{ML} \neq \sqrt{\varepsilon_1 \varepsilon_r}$. The results show the accuracy of the theoretical calculation.

3.2 Radiation Patterns

The radiation pattern of the lens antenna with and without the ML using the different materials are presented in Fig. 3. It is easily observed that the antenna lens radiations with the ML is better than those of the ML-free antenna. Specifically, in the absence of the ML, the lens antenna has the highest gain of 27.07 dBi when Teflon is used as a dielectric, and it reaches the highest gain of 25.39 dBi when the lens dielectric is Mica. In the presence of the ML, the lens antenna has the highest gain of 27.82 dBi with the dielectric of Silicon, and 27.14 dBi with the dielectric of Mica. In Silicon dielectric, the maximum gain with the ML is 2.2 dB higher than the maximum gain without the ML, while in Teflon dielectric this figure is only 0.32 dB.

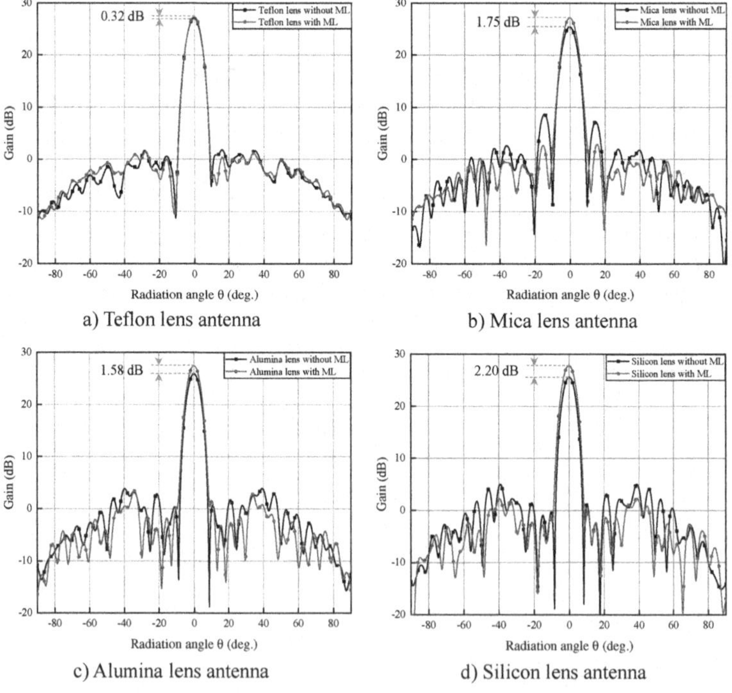

Fig. 3. Radiation patterns of different materials with and without matching layers.

3.3 Side-Lobe Levels

The changes in the gains and the side-lobe levels when using and not using the ML in the different dielectrics are shown in Fig. 4. As shown in the graph, when the ML is attached to the lens, the side-lobe levels of all four cases decrease. Mica reduces by 7.52 dB, from −16.69 dB to −24.21 dB, followed by Silicon dielectric and Alumina at 5 dB and 1.99 dB, respectively; whereas in the case of Teflon dielectric, there is a minor change in the side-lobe level (only 0.68 dB, from −25.21 dB to 25.89 dB).

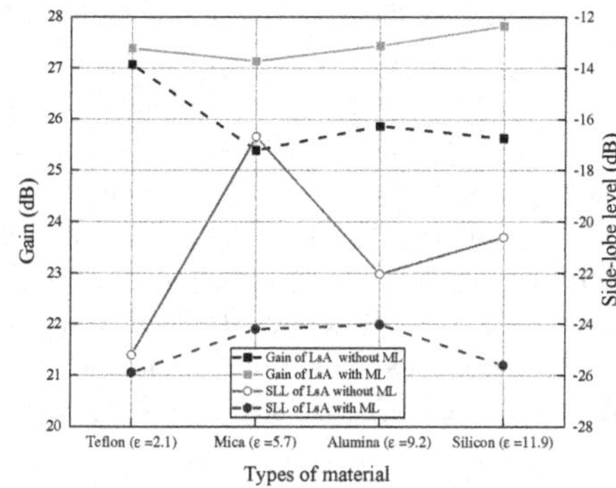

Fig. 4. The comparison of the gains and the side-lobe levels.

3.4 The Lens Thickness

The specific simulation results and calculations are presented in Table 2. Applying formula (5) and established simulation parameters, we can calculate the thickness of the lens. Accordingly, the lens with the Teflon dielectric is the thickest with 25.17 mm, while the lens with the Silicon dielectric is the thinnest with only 7.07 mm. The thicknesses of Alumina and Mica are 8.07 mm and 10.67 mm, respectively.

Table 2. Comparison results.

Types of material	lens without ML				Lens with ML			
	Gain [dBi]	SLL [dB]	HPBW [deg.]	Lens thickness [mm]	Gain [dBi]	SLL [dB]	HPBW [deg.]	Lens thickness [mm]
Teflon (ε_r = 2.1)	27.07	−25.21	7.42	22.94	27.39	−25.89	7.29	25.17
Mica (ε_r = 5.7)	25.39	−16.69	7.39	8.94	27.14	−24.21	7.16	10.67
Alumina (ε_r = 9.2)	25.86	−22.04	6.77	6.53	27.44	−24.03	6.84	8.07
Silicon (ε_r = 11.9)	25.62	−20.61	6.54	5.63	27.82	−25.61	6.84	7.07

3.5 Electric Field Distributions

Figure 5 shows the electric field distributions of the lens antennas with the dielectric Teflon and Silicon on the xz plane. From Fig. 5c and d, it is easily to see the differences in the electric field distributions between the Silicon antenna lens without the ML (5c) and with the ML (5d). As for the Silicon lens without the ML, the electric field distribution between the lens and the conical horn antenna is rough, even cancels out just before the lenses, and the spherical waveform of the feed horn is no longer clear. As for ML-coated Silicon lens, the field distribution is quite smooth and more uniform, and the planar waves behind the lens are shown more clearly. The first reason for this field distribution is that silicon has a dense dielectric structure; hence, the rays from the feed horn to the lens surface are reflected more. Another possible explanation is that the curvature of the inner surface of the lens is less than that of Teflon; therefore, the rays bounce back toward the feed horn more. Accordingly, the field distribution on the xz plane of the Teflon lens antenna undergoes a minor change on account of Teflon having a less dense dielectric structure and a higher curvature of the inner surface of the lens, thus the rays are reflected from the surface less and are deflected into the edge of the lens instead of being reflected toward the feed horn, as shown in Fig. 5a and b. This result shows the efficiency of using the ML for dense dielectric structure materials.

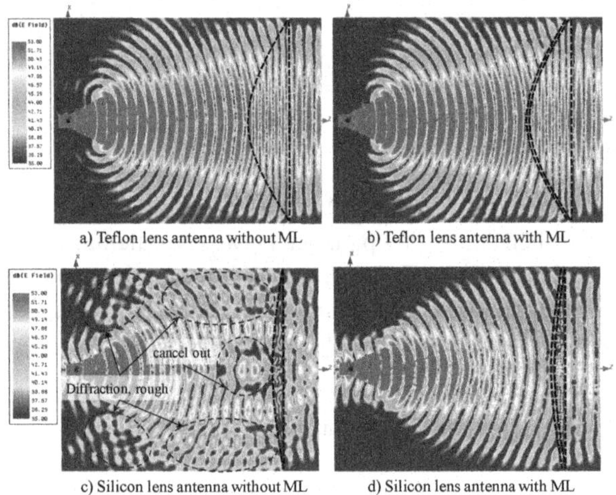

a) Teflon lens antenna without ML b) Teflon lens antenna with ML

c) Silicon lens antenna without ML d) Silicon lens antenna with ML

Fig. 5. Electric field distributions on the xz plane.

The above-mentioned results show the effectiveness of ML in improving the radiation ability of lens antennas, a significant reduction in the side-lobe level, and a more uniform distribution of the electric field. Besides, the results also show the corresponding sizes of the lenses. This will be helpful for researchers, designers, and manufacturers to select an appropriate lens antenna structure with a smaller size and better properties.

4 Conclusion

Through the simulation, analysis, and evaluation results, this paper has demonstrated the role of ML in improving the characteristics of a specific dense dielectric lens. The results show that covered with a ML, the Silicon dielectric can improve the maximum gain of 2.2 dB while the Mica dielectric can reduce the maximum side-lobe level of 7.52 dB. In addition, the Silicon dielectric lens is 18.1 mm thinner than the Teflon dielectric lens and achieves a higher gain. These results show the efficiency of using a ML for high-density dielectric lens antennas.

Acknowledgments. This research is funded by Vietnam National Foundation for Science and Technology Development (NAFOSTED) under grant number 102.04-2018.08.

References

1. Dinh, N.Q., Binh, N.T., Yamada, Y., Michishita, N.: Proof of the density tapering concept of an unequally spaced array by electric field distributions of electromagnetic simulations. J. Electromag. Waves Appl. **34**(5), 668–681 (2020). https://www.tandfonline.com/doi/full/10.1080/09205071.2020.1736185
2. Yamada, Y., et al.: Unequally element spacing array antenna with butler matrix feed for 5G mobile base station. In: 2nd International Conference on Telematics and Future Generation Networks (TAFGEN), pp. 72–76. Kuching, Malaysia (2018)
3. Quzwain, K., Yamada, Y., Kamardin, K., Rahman, N.H.A., Dinh, N.Q.: Reflector surface shaping method for a Cassegrain antenna, In: 6th International Conference on Space Science and Communication, pp. 1–5. Johor, Malaysia (2019)
4. Tajima, T., Yamada, Y.: Design of shaped dielectric lens antenna for wide angle beam steering. Electron. Commun. Japan Part 3, **89**(2), 1–12 (2006)
5. Hong, W., et al.: Multibeam antenna technologies for 5G wireless communications. IEEE Trans. Antennas Propag. **65**(12), 6231–6249 (2017)
6. Filipovic, D.F., Gearhart, S.S., Rebeiz, G.M.: Double-slot antennas on extended hemispherical and elliptical silicon dielectric lenses. IEEE Trans. Microw. Theory Tech. **41**(10), 1738–1749 (1993)
7. Costa, J.R., Fernandes, C.A., Godi, G., Sauleau, R., Le Coq, L., Legay, H.: Compact ka-band lens antennas for LEO satellites. IEEE Trans. Antennas Propag. **56**(5), 1251–1258 (2008)
8. Pasqualini, D., Maci, S.: High-frequency analysis of integrated dielectric lens antennas. IEEE Trans. Antennas Propag. **52**(3), 840–847 (2004)
9. Hung, P.V., Dinh, N.Q., Nguyen, T.V.D., Yamada, Y., Michishita, N., Islam, M.T.: Electromagnetic simulation method of a negative refractive index lens antenna. In: 2019 International Conference on Advanced Technologies for Communications (ATC), pp. 109–112. Hanoi, Vietnam (2019)
10. Nguyen, N.T., Sauleau, R., Ettore, M., Le Coq, L.: Focal array fed dielectric lenses: an attractive solution for beam reconfiguration at millimeter waves. IEEE Trans. Antennas Propag. **59**, 2152–2159 (2011)
11. Wu, X., Eleftheriades, G.V., Van Deventer Perkins, T.E.: Design and characterization of single- and multiple-beam mm-wave circularly polarized substrate lens antennas for wireless communications. IEEE Trans. Microw. Theory Tech. **49**, 431–441 (2001)
12. Barès, B., Sauleau, R.: Electrically small shaped integrated lens antennas: a study of feasibility in Q-band. IEEE Trans. Antennas Propag. **55**, 1038–1044 (2007)

13. Holzman, E.L.: A highly compact 60-GHz lens-corrected conical horn antenna. IEEE Antennas Wirel. Propag. Lett. **3**, 280–282 (2004)
14. Schoenlinner, B., Wu, X., Ebling, J.P., Eleftheriades, G.V., Rebeiz, G.M.: Wide-scan spherical-lens antennas for automotive radars. IEEE Trans. Microw. Theory Tech. **50**, 2166–2175 (2002)
15. Migliaccio, C., Dauvignac, J.Y., Brochier, L., Le Sonn, J.L., Pichot, C.: W-band high gain lens antenna for metrology and radar applications. Elect. Lett. **40**(22), 1394–1396 (2004)
16. Van der Vorst, M.J.M., et al.: Effect of internal reflections on the radiation properties and input impedance of integrated lens antennas – Comparison between theory and measurements. IEEE Trans. Microw. Theory Tech. **49**, 1118–1125 (2001)
17. Godi, G., Sauleau, R., Thouroude, D.: Performance of reduced size substrate lens antennas for millimeter-wave communications. IEEE Trans. Antennas Propag. **53**, 1278–1286 (2005)
18. Neto, A., Maci, S., De Maagt, P.J.I.: Reflections inside an elliptical dielectric lens antenna. IEE Proc. Microw. Antennas Propag. **145**(3), 243–247 (1998)
19. Nguyen, N.T., Sauleau, R., Martinez, P.: Very broadband extended hemispherical lenses: role of matching layers for bandwidth enlargement. IEEE Trans. Antennas Propag. **57**, 1907–1913 (2009)
20. Tokan F.: Matching layer design procedure for a novel broadband dielectric lens antenna, pp. 149–198 (2013). ISBN 978-0-9891305-3-0 ©2013 SDIWC (2013)
21. Morita, T., Cohn, S.B.: Simulated quarter-wave matching sheet: In: U.R.S.I. Meeting, Washington D. C. (1954)
22. Morita, T., Cohn, S.B.: Microwave lens matching layer by simulated quarter-wave transformer. IRE Trans. Antennas Propag. **4**(1), 33–39 (1956)
23. Jones, E.M.T., Cohn, S.B.: Surface matching of dielectric lens. J. Appl. Phys. **26**(4), 452–457 (1955)
24. Stratton, J.A.: Electromagnetic Theory. McGraw-Hill Book Company Inc., New York (1941)
25. Lo, Y.T., Lee, S.W.: Antenna Handbook, 2nd edn. Van Nostrand Rainhold Company, New York (1988)
26. Tajima, Y., Yamada, Y.: Design of shaped dielectric lens antenna for wide angle beam steering. Electron. Commun. Japan Part 3 **89**(2), 1–12(2006)
27. Sliver, S.: Microwave Antenna Theory and Design. McGraw-Hill Inc., New York (1949)

A 3-Stacked GaN HEMT Power Amplifier with Independently Biased Technique

Luong Duy Manh[1(✉)], Tran Thi Thu Huong[1], Bui Quoc Doanh[1], and Vo Quang Son[2]

[1] Le Quy Don Technical University, 236 Hoang Quoc Viet, Hanoi, Vietnam
duymanhcs2@mta.edu.vn
[2] University of Transport and Communications, No. 3 Cau Giay, Hanoi, Vietnam

Abstract. A design of 3-stacked GaN high-electron-mobility transistor radio-frequency power amplifier using independently biased technique is presented. The power amplifier operates at 1.6 GHz for wireless communications applications. By independently setting proper bias conditions, DC power consumption of the power amplifier can be reduced leading to efficiency enhancement without output power degradation. A performance comparison of the proposed power amplifier with a conventional 3-stacked power amplifier has been performed. The simulated results indicate that the proposed power amplifier offers superior efficiency over the conventional one.

Keywords: Power amplifier · GaN HEMT · Independently biased

1 Introduction

Traditional circuit configurations have been proposed to improve performance of power amplifier including Darlington [1,2], cascode [3,4] and hybrid [5] configurations. The Darlington configuration can handle very high current capability that is highly suitable for power amplifier design. The cascode configuration, however, can offer various advantages such as wide bandwidth and high gain. The hybrid configuration which combines a bipolar junction transistor (BJT) and a field effect transistor (FET) can offer low output impedance and low DC power consumption. Although these configurations have been proved to deliver various promising advantages, they still exhibit inherent drawbacks for high-frequency power amplifier design. One of the most critical drawbacks of these configurations is that they are not able to independently adjust bias condition of each transistor in the configuration. This reduces degrees of freedom in performance improvement. Recently, independently biased InGaP/GaAs HBT cascode [6] and independently biased 3-stacked InGaP/GaAs Heterojunction bipolar transistor (HBT) [7] configurations have been reported for RF power amplifier (PA) design. Theses proposed configurations can offer the possibility of independent control

This research is funded by Vietnam National Foundation for Science and Technology Development (NAFOSTED) under grant number 102.04-2018.14.

N.-S. Vo and V.-P. Hoang (Eds.): INISCOM 2020, LNICST 334, pp. 95–104, 2020.
https://doi.org/10.1007/978-3-030-63083-6_8

of bias condition. Nevertheless, they still exhibit low output power due to low power capability of a HBT device.

In addition to these configurations, an independently biased 3-stacked GaN high-electron-mobility transistor (HEMT) has been introduced in [8]. However, this structure was just in the form of a bare chip and the investigated results were performed for this type of chip but not for a power amplifier design. In this paper we practically implement a design of a PA based on the independently biased technique for a 3-stack GaN HEMT structure as presented in [8]. An investigation on how an independently biased approach in the 3-stacked GaN HEMT configuration can improve efficiency and output power is carried out. The rest of the paper is organized as follow: Sect. 2 will describe in details the proposed amplifier. Investigation of performance of the proposed amplifier will be conducted in Sect. 3. Then Sect. 4 will present experiment including experimental setup for large-signal measurement and measured results. Finally, Sect. 5 will conclude the paper.

2 Descriptions of the Proposed Amplifier

2.1 3-Stacked GaN HEMT MMIC Chip

Schematic of the 3-stacked GaN HEMT configuration is illustrated in Fig. 1a. Three GaN HEMT devices are connected to each other in a cascode-type topology. In addition to the conventional bias terminals including gate bias and drain bias terminals, two additional bias terminals are inserted to two floating points between the first and the second transistors and between the second and the third transistors. Owing to this approach, bias condition of each transistor in the proposed configuration can be independently controlled. In practice, the 3-stacked GaN HEMT structure is realized using a MMIC (Monolithic Microwave Integrated Circuit) technology from WIN Semiconductor Corp. foundry service [9]. The service includes following information: metal thickness = 4 μm, substrate thickness = 100 μm. The schematic and MMIC layout of this structure are described in Fig. 1. As can be seen in the figure, in the MMIC layout, two RF-bypass capacitors are realized using a metal-insulator-metal (MIM) technology while input and output terminals are ground-signal-ground (GSG) connections for chip evaluation using GSG probes in practice. Interconnects inside the MMIC chip are made using metal transmission lines. Three GaN HEMTs have the same size of 0.25 μm × 0.75 μm × 4 fingers. The total size of the MMIC chip is 674 μm × 1025 μm

2.2 Power Amplifier

Figure 2 shows descriptions of the proposed amplifier which is realized using the 3-stacked GaN HEMT topology as an active device. The MMIC chip is connected to the external components including input/output matching networks and biasing lines through gold bonding wires (BW) with a diameter of

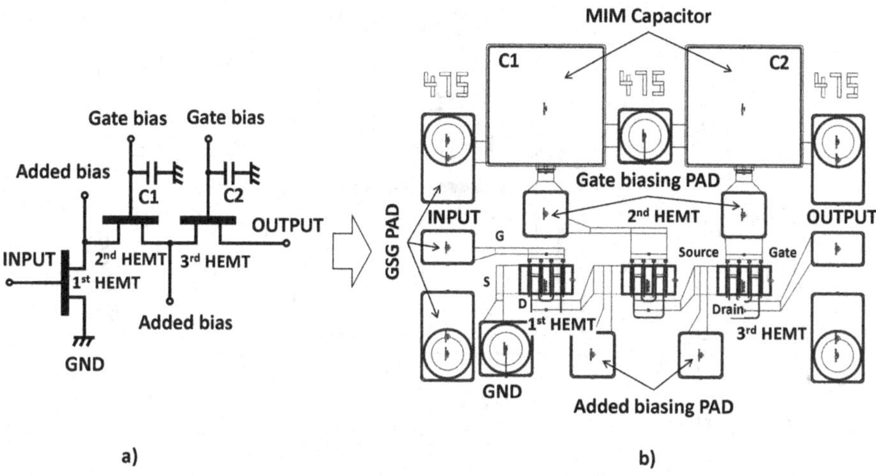

Fig. 1. Proposed 3-stacked GaN HEMT topology: a) Schematic and b) MMIC layout.

30 μm. Input matching network (IMN) and output matching network (OMN) are implemented using off-chip lumped components from Murata libraries [10]. Values of the IMN components are: L_1 : LQW18AN4N3C00 (4.3 nH), C_1 : GRM1555C1HR90WA01 (0.9 pF) and values of the OMN components are: L_2 : LQW18AN12NG00 (12 nH), C_2 : GRM1555C1HR30WA01 (0.3 pF). Gate bias of the first transistor and drain bias of the third transistor are implemented using Bias-Tees including a RF choke inductor and a block capacitor C_b. In addition, drain bias of the first and the second transistors are implemented using two quarter wavelength transmission lines. Here, it is noted that the IMN and OMN are designed to match source and load impedances of the MMIC chip to 50 Ω at only the fundamental frequency without using any harmonic termination techniques. This is a reason why the IMN and OMN just include a combination of a inductor and a capacitor as indicated in the figure. Optimum fundamental source and load impedances are found by using a simulated load/source pull method based on non-linear models of GaN HEMT provided by WIN Semiconductor Corp. Their values are found as follow: $Z_S = 79.3 + j138.6\,\Omega$, $Z_L = 134.7 + j174.1\,\Omega$ where Z_S and Z_L are the optimum source and load impedances, respectively.

3 Performance Evaluation

3.1 V_{d1} Variation

The most important feature of the proposed amplifier is the high degree of freedom in bias adjustment of each transistor leading to the PA performance improvement. In the proposed circuit topology as indicated in Fig. 1, two added

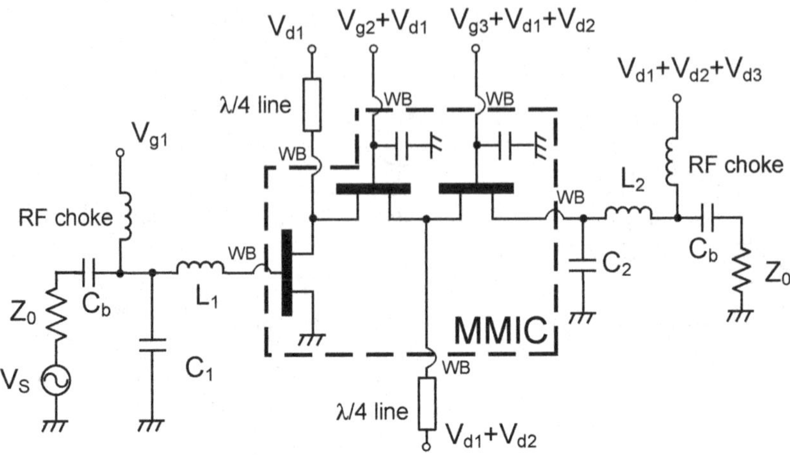

Fig. 2. Schematic.

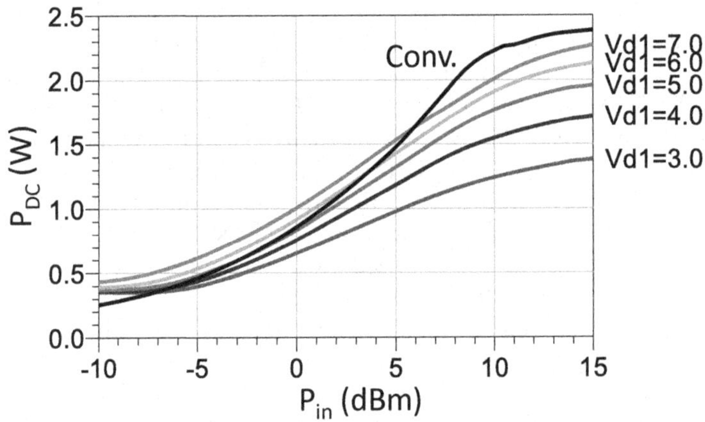

Fig. 3. DC power of the conventional and proposed 3-stacked PA vs. input power with variation of $V_{\mathbf{d1}}$. The black curve represents DC power of the conventional configuration.

bias terminals have been included in order to bias the drain terminals of the first ($V_{\mathbf{d1}}$) and the second ($V_{\mathbf{d2}}$) transistors. This means, by appropriately controlling these bias values ($V_{\mathbf{d1}}$ and $V_{\mathbf{d2}}$), PA performance such as power added efficiency (PAE) and output power can be significantly improved. Firstly, the variation of $V_{\mathbf{d1}}$ will be investigated. The main contribution of this independent bias control is to re-contribute the DC power consumption of the PA. This helps to increase efficiency while ensure sufficient output power. This re-contribution of the DC

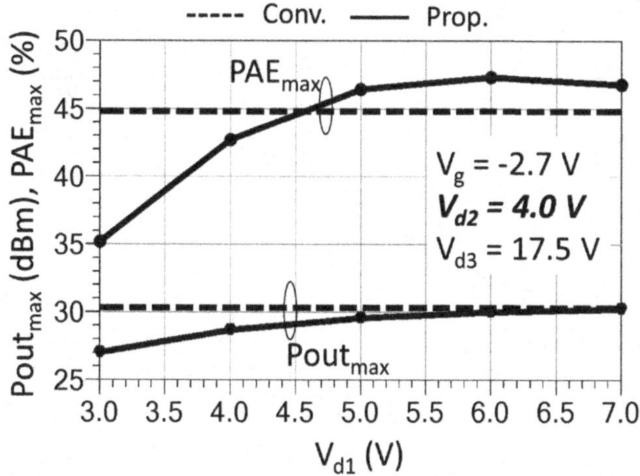

Fig. 4. Maximum PAE and Pout as a function of varied $V_{\mathbf{d1}}$. Performance comparison with a conventional structure is also shown.

power can be seen in Fig. 3. Decreasing $V_{\mathbf{d1}}$ while keeping $V_{\mathbf{d2}}$ and $V_{\mathbf{d}}$ results in lower DC power at high input power region. This means that maximum efficiency (PAE) can be enhanced since the maximum PAE occurs at the high input power region. This can be further understood because there is a re-contribution of quiescent drain currents among transistors leading to a re-contribution of DC power consumption when changing the bias values. On the other hand, a conventional 3-stacked configuration which has a similar topology with the proposed 3-stacked one as indicated in Fig. 1a but without using two added bias terminals is not able to make this re-contribution of the DC power. Here, it is noted that all three transistors of the proposed configuration are biased in a class-AB. All three gate bias voltages are set to −2.7 V and the third drain bias voltage ($V_{\mathbf{d3}}$) is kept at a constant value of 27.5 V. Similar bias conditions are made for both the conventional and proposed configurations in order to make a logical comparison between them. The effect of DC power reduction with respect to the change of $V_{\mathbf{d1}}$ on PA performance can be clearly seen in Fig. 4. As can be seen in the figure, when $V_{\mathbf{d1}}$ changes from 3.0 V to 7.0 V, both PAE and Pout can be significantly varied. When $V_{\mathbf{d1}}$ increases, $\mathbf{PAE_{max}}$ can be higher than that of the conventional one while $\mathbf{Pout_{max}}$ can still remain the same level as the conventional one. To make a superior trade-off between efficiency and output power compared with that of the conventional one, $V_{\mathbf{d1}}$ should be set to 5.5 V.

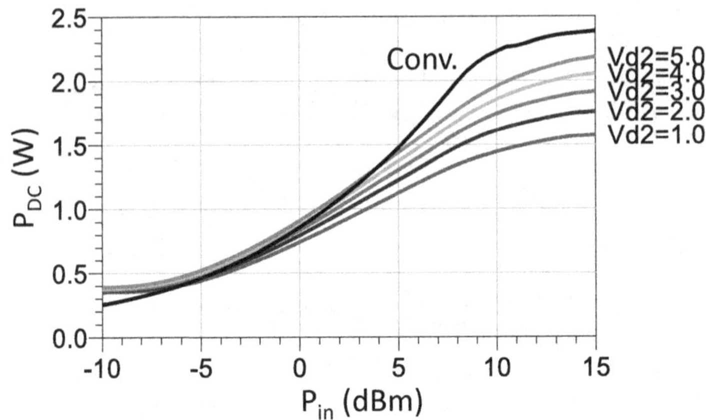

Fig. 5. DC power of the conventional and proposed 3-stacked PA vs. input power with variation of V_{d2}. The black curve represents DC power of the conventional configuration.

3.2 V_{d2} variation

After setting V_{d1}, V_{d2} then will be considered for performance improvement. Figure 5 illustrates the effect of V_{d2} variation on the DC power re-contribution inside the PA. It once gain can be seen that decreasing V_{d1} while keeping V_{d1} and V_d results in lower DC power at high input power region. This re-contribution of DC power also helps to increase PAE with an expense of low output power.

The dependence of maximum PAE and maximum Pout of the conventional and proposed 3-stacked GaN HEMT configurations with variation of V_{d2} is shown in Fig. 6. In this figure V_{d1} is kept at 5.5 V. The figure shows that when V_{d2} changes from 1 to 5 V, maximum PAE first increases and reaches maximum values at the middle region and drop at high values while maximum Pout increases with the increasing V_{d2}. According to this figure, to remain sufficient output power without degradation of efficiency over the conventional one, V_{d2} is chosen to be 4.0 V.

4 Experiment

4.1 PA Prototype

After considering optimum bias conditions for the proposed PA, a prototype of the PA has been fabricated as indicated in Fig. 7. The external Bias-Tees for biasing the gate of the first transistor and the drain of the third transistor are not shown in the figure. Female and male SMA connectors function as the input and output ports, respectively. The prototype was fabricated on a Megtron6 substrate from Panasonic with following parameters: dielectric constant = 3.7, substrate thickness = 0.75 mm, dissipation factor = 0.002 and copper thickness =

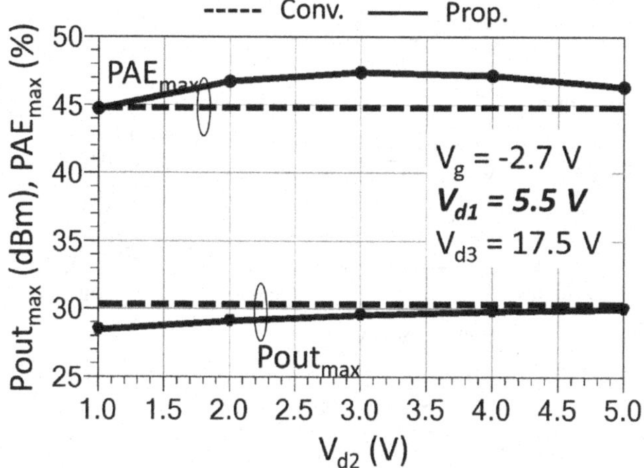

Fig. 6. Maximum PAE and Pout as a function of varied $V_{\mathbf{d2}}$. Performance comparison with a conventional structure is also shown.

Fig. 7. Fabricated PA prototype.

35 μm. In the measurement of the PA prototype, bias condition of each transistor which are found from the previous section are used. The bias conditions for the proposed configuration are as follow: $V_{\mathbf{g}} = -2.7\,\mathrm{V}$, $V_{\mathbf{d1}} = 5.5$ V, $V_{\mathbf{d2}} = 4.0\,\mathrm{V}$, $V_{\mathbf{d}} = 17.5\,\mathrm{V}$ while the bias conditions for the conventional one are as follow: $V_{\mathbf{g}} = -2.7\,\mathrm{V}$, $V_{\mathbf{d}} = 27\,\mathrm{V}$ which is equivalent to $V_{\mathbf{d1}} + V_{\mathbf{d2}} + V_{\mathbf{d3}}$.

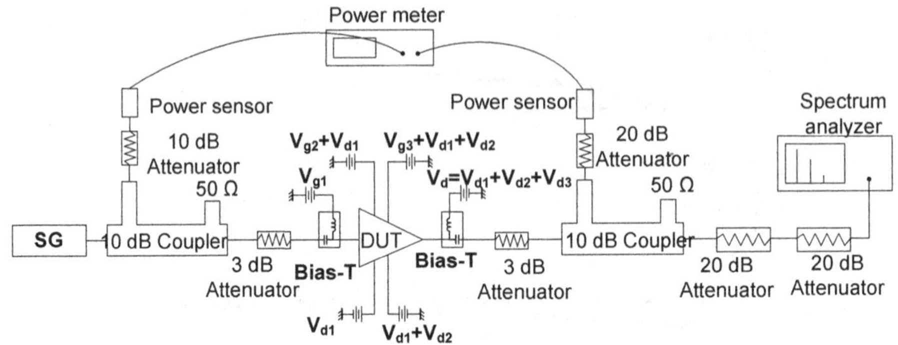

Fig. 8. Experimental setup.

4.2 Experimental Setup

The experimental setup for large-signal measurement of the PA prototype is described in Fig. 8. A microwave signal generator (SG) is employed to input RF signal to the PA at an operation frequency 1.6 GHz. Two Bias-Tee symbols are clearly visible in the figure. Two RF directional couplers are used to slit the input and output RF signals to make input and output power measurement. Power measurement is made by using power sensors combined with a power meter. All components for power measurement are carefully calibrated prior to make actual measurements. A spectrum analyzer (SA) is employed to check the output spectrum as well as to verify if the PA is unintentionally oscillated during the large-signal measurement.

4.3 Measured Results

The measured large-signal performance including PAE, output power and gain are shown in Fig. 9 which shows the measured and simulated PAE, Pout and gain of the designed PA. The PA is biased at the optimum bias condition which is found from the previous sections. The practival bias condition is given as follow: $V_{\mathbf{g}} = -2.7$ V, $V_{\mathbf{d1}} = 5.5$ V, $V_{\mathbf{d2}} = 4.0$ V, $V_{\mathbf{d}} = 17.5$ V. As can be seen in the figure, measured results agree well with the simulated one. This implies that the simulations predict well performance of the proposed 3-stacked Gan HEMT PA. There are some discrepancies between simulation and measurements caused by losses in the PA assembly process and losses in realistic lumped components.

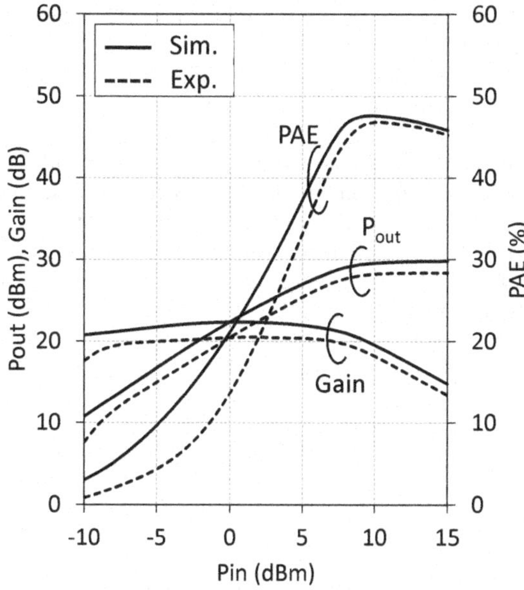

Fig. 9. Large-signal performance measurement vs. simulation.

5 Conclusion

This paper presents the design of the independently biased 3-stacked GaN HEMT power amplifier operating at 1.6 GHz for wireless communications. It has been shown that by independently adjusting the added bias terminals $V_{\mathbf{d1}}$ and $V_{\mathbf{d2}}$, efficiency of the proposed PA can be significantly improved due to the re-contribution mechanism of DC power consumption. When comparing with a conventional 3-stacked GaN HEMT amplifier, the proposed amplifier exhibits higher efficiency with similar output power level if $V_{\mathbf{d1}}$ and $V_{\mathbf{d2}}$ are properly controlled. The measured results agree well with the simulated ones validating the design method.

References

1. Armijo, C.T., et al.: A new wide-band Darlington amplifier. IEEE J. Solid-State Circ. **4**(24), 1105–1109 (1989)
2. Weng, S.H., et al.: Gain-bandwidth analysis of broadband Darlington amplifiers in HBT-HEMT process. IEEE Trans. Microw. Theory Tech. **11**(60), 3458–3473 (2012)
3. Ahy, R.M., et al.: 100-GHz high-gain InP MMIC cascode amplifier. IEEE J. Solid-State Circ. **10**(26), 1370–1378 (1991)
4. Sowlati, T., et al.: A 2.4-GHz 0.18-µm CMOS self-biased cascode power amplifier. IEEE Trans. Microw. Theory Tech. **8**(38), 1318–1324 (2003)

5. Roozbahani, R.G., et al.: BJT-BJT, FET-BJT, and FET-FET. IEEE Circ. Devices Mag. **6**(20), 17–22 (2004)
6. Manh, L.D., et al.: Power gain performance enhancement of independently biased heterojunction bipolar transistor cascode chip. Jpn. J. Appl. Phys. **04DF11**(54), 1–8 (2015)
7. Manh, L.D., et al.: Microwave characteristics of an independently biased 3-stack InGaP/GaAs HBT configuration. IEEE Trans. Circ. Syst. I, Reg. Papers **64**(5), 3487–3495 (2017)
8. Hoang, N.H., et al.: A novel Independently biased 3-stack GaN HEMT configuration for efficient design of microwave amplifiers. Appl. Sci. **7**(9), 1–16 (2019)
9. http://www.winfoundry.com/
10. https://www.murata.com/

Feasibility and Design Trade-Offs of Neural Network Accelerators Implemented on Reconfigurable Hardware

Quang-Kien Trinh[1], Quang-Manh Duong[1(✉)], Thi-Nga Dao[1], Van-Thanh Nguyen[2], and Hong-Phong Nguyen[1]

[1] Le Quy Don Technical University, Hanoi, Vietnam
{kien.trinh,manhdq}@lqdtu.edu.vn

[2] Posts and Telecommunications Institute of Technology, Hanoi, Vietnam

Abstract. In recent years, neural networks based algorithms have been widely applied in computer vision applications. FPGA technology emerges as a promising choice for hardware acceleration owing to high-performance and flexibility; energy-efficiency compared to CPU and GPU; fast development round. FPGA recently has gradually become a viable alternative to the GPU/CPU platform.

This work conducts a study on the practical implementation of neural network accelerators based-on reconfigurable hardware (FPGA). This systematically analyzes utilization-accuracy-performance trade-offs in the hardware implementations of neural networks using FPGAs and discusses the feasibility of applying those designs in reality.

We have developed a highly generic architecture for implementing a single neural network layer, which eventually permits further construct arbitrary networks. As a case study, we implemented a neural network accelerator on FPGA for MNIST and CIFAR-10 dataset. The major results indicate that the hardware design outperforms by at least 1500 times when the parallel coefficient p is 1 and maybe faster up to 20,000 times when that is 16 compared to the implementation on the software while the accuracy degradations in all cases are negligible, i.e., about 0.1% lower. Regarding resource utilization, modern FPGA undoubtedly can accommodate those designs, e.g., 2-layer design with p equals 4 for MNIST and CIFAR occupied 26% and 32% of LUT on Kintex-7 XC7K325T respectively.

Keywords: Neural network · FPGA accelerator · Data recognition

1 Introduction

The Development of the Neural Network is the Motivation to Improve Computing Capability on Different Platforms

In recent years, researches on the neural network have shown a significant advantage in machine learning over traditional algorithms based on handcrafted models. There

N.-S. Vo and V.-P. Hoang (Eds.): INISCOM 2020, LNICST 334, pp. 105–123, 2020.
https://doi.org/10.1007/978-3-030-63083-6_9

has been a growing interest in the study of neural networks, inspired by the nervous system in the human brain. Owing to the high accuracy and good performance, neural networks have been widely adopted in many applications such as image classification [1], face recognition [2], smart digital video surveillance [3], and speech recognition [4], etc. In general, neural network features a high fitting ability to a wide range of pattern recognition problems, which makes the neural network a promising candidate for many artificial intelligence applications.

Recent research on the neural network is showing great improvement over traditional algorithms, various neural network models, like Convolutional Neural Network (CNN), Recurrent Neural Network (RNN), have been proposed. CNN [5] improves the Top-5 image classification accuracy on ImageNet [6] dataset from 73.8% to 84.7% in 2012 and further helps improve object detection [7] with its outstanding ability in feature extraction. RNN [8] achieves state-of-the-art word error rate on speech recognition.

As neural network models become larger and deeper, numerous operations and data accesses are demanded in neural network-based implementations while higher accuracy typically demands more complex models. For example, Krizhevsky et al. [9] achieved 84.7% Top-5 accuracy in Take ImageNet Large-Scale Vision Recognition Challenge (ILSVRC) with a model including 5 convolution layers and 3 fully-connected layers, they get a recognition accuracy [10] of 95.1% surpassing human-level classification (94.9% [11]) with a 22-layer model and won the ILSVRC-2015 competition for achieving an accuracy of 96.4% with a model depth of 152 [12]. Such a model can take over 11.3 billion floating-point operations (GFLOPs) for the inference procedure, and even more for training.

The operations in CNNs are computationally intensive with over billion operations per input image [13], thus requiring high-performance hardware platforms. The rapidly changing field of deep learning makes it even more difficult for a generic accelerator to match for a wide range of neural network algorithms. In this context, there is a timely need to reform the mapping strategy of neural networks to the hardware platform and to support modular and scalable hardware customization for specific applications without losing design flexibility. Choosing an appropriate computing platform for neural network-based applications is extremely essential.

FPGA, GPU, and ASIC are the widely-applied selections in addition to using the traditional CPU usage for accelerators available in the market today. For FPGAs, recently there have been major efforts from technology leaders to better integrate FPGA accelerators. There is also a growing number of GPU and ASIC accelerator solutions offered commercially, such as NVIDIA GPU and IBM PowerEN processor with edge network accelerators.

Application-Specific Standard Processor (ASSP) Based Approaches for Neural Network Accelerators. Neural networks are implemented on CPU and GPU platforms, i.e., currently widely adopted ASSPs; however, they are not efficient either in terms of implementation speed (CPU) or energy consumption (GPU) [14]. Indeed, a typical CNN architecture has multiple convolutional layers that extract features from the input data, followed by classification layers. This essentially requires massive parallel calculations. General-purpose processors (CPUs) rely on a few processing elements and operate sequentially, hence they are not efficient for CNN implementation and can hardly meet the performance requirement. In contrast to CPU, GPUs can offer Giga to Tetra FLOPs [15] per second's computing speed due to their single-instruction-multiple-data (SIMD) architecture and high clock frequency, therefore they are good choices for high-performance neural network applications. However, the power consumption of typical GPUs is exceedingly high - for an NVIDIA Tesla K40 GPU, the thermal design power (TDP) is 235 W [16], thus GPUs are not suitable for embedded systems such as mobile devices, robots, etc., which are mostly powered by batteries and low power consumption becomes essential to them. Besides, both CPU and GPU have the disadvantage of poor integration capability, neither the CPU nor the GPU is specifically designed for neural network calculations so they are not optimized for neural networks, resulting in poor energy efficiency, especially in the real-time applications that require large bandwidth.

Application-Specific IC (ASIC), which is rigorously optimized for neural networks, could solve both poor performance and high energy consumption of CPU and GPU [17]. This hardware solution undoubtedly is superior to any other platform when performing calculations on the same neural network. Nonetheless, the ASIC design cycle is relatively long due to high complexity and very costly for low volume production. More important, ASIC is non-hardware-reconfigurable technology, thus, no single ASIC platform could meet the rapid improvements and the diversity of problems on the neural network application. Therefore, the implementation of ASIC for neural network accelerators, in reality, needs to be carefully considered.

Reconfigurable Hardware-Based Approaches for Neural Network Accelerators
As a balancing approach among the mentioned ASSP platforms and ASIC, along with distinct features, FPGAs present as promising platforms for the hardware acceleration of CNNs [18]. FPGA-based neural network accelerators have become increasingly popular thanks to their high reconfigurability, fast turn-around time (compared to ASICs), high-performance, and better energy efficiency (compared to GPUs) [19].

In a particular study, Marco Bettoni *et al.* [20] implemented a CNN design on FPGA and obtained results showing that the proposed implementation is as efficient as a general-purpose 16-core CPU, and almost 15 times faster than an SoC GPU for mobile application. Research by Eriko Nurvitadhi *et al.* [21, 22] implemented on BNN and RNN networks showed that in comparison to 14-nm ASIC, GPU, and multi-core CPU, FPGA provides superior performance/watt over CPU and GPU because FPGA's on-chip BRAMs, hard DSPs, and reconfigurable fabric allow for efficiently extracting fine-grained parallelisms. Moreover, newer FPGAs with more DSPs, on-chip BRAMs, integrated hard accelerators IP cores, and higher frequency have the potential to narrow the FPGA-ASIC efficiency gap.

Nonetheless, those prior works are targeted at either high accuracy or high performance for specific architecture [20–22] and ignore the intrinsical trade-offs between resource utilization, performance, accuracy, and network architecture. Therefore, the scalability and the integrability of the neural network design has not fully explored and studied.

In this work, we developed fundamental and highly generic building blocks that allow constructing virtually any neural networks. These base components allow us to systematically study the feasibility of using FPGAs as an accelerator for neural network applications. The design trade-offs on aspects including network architecture, resource utilization, accuracy, and performance for a wide range of devices to understand the real power and limitation of the FPGAs as the reconfigurable platform for neural network implementation. These assessments will be the basis for the application of FPGAs as hardware accelerators for practical neural network applications.

The main contributions of this work are summarized as follows

- A high-performance generic design of neural network accelerator combining software (for parameters training) with the powerful capability of hardware computation (on matrix additions and multiplications). In particular, we analyze the design by theoretically deriving performance metrics including the memory size and processing latency of the FPGA-based neural network accelerator. To support the design analysis, a numerical format selection method based on trained parameter values domain on two considered datasets.
- A methodology is proposed on how to optimize parallelism strategy with different parallel coefficients for each layer to achieve high throughput.
- An in-depth discussion on the design tradeoffs between resource utilization, model accuracy, and performance of the image classification models with different parameters including numerical formats, parallel coefficients, and network architectures through the practical accelerators (for the most representative datasets (MNIST and CIFAR-10).
- On-board demonstrations of FPGA implementation using single or multilayer neural networks and CNN that achieve mostly the same accuracy as the software implementation.

The remaining of this paper is organized as follows: Sect. 2 introduces the basic background of neural networks. Section 3 presents the results of data recognition performed on the software. Section 4 proposes a generic design for the data recognition problem on the hardware and describes our FPGA-based implementation details upon this proposed design. Section 5 concludes the paper.

2 Background

2.1 Neural Network

Generally, a fully-connected neural network consists of three consecutive layers: input, hidden, and output layers as shown in Fig. 1. First, input features (e.g., image data) are collected and fed into the input layer. Then, input features are fully connected to hidden layers that learn the underlying patterns of input data. Finally, hidden features are progressively propagated to the output layer which provides predicted discrete labels (for classification models) or continuous values (for regression models).

This work considers the case study of image recognition tasks on MNIST [23] and CIFAR-10 datasets [24]. In the case of the MNIST set, we aim to construct a neural network-based classifier that can understand the handwritten digits. Specifically, the classification model should output the most likely digit among 10 possible single digits with a given input image of 28×28 pixels.

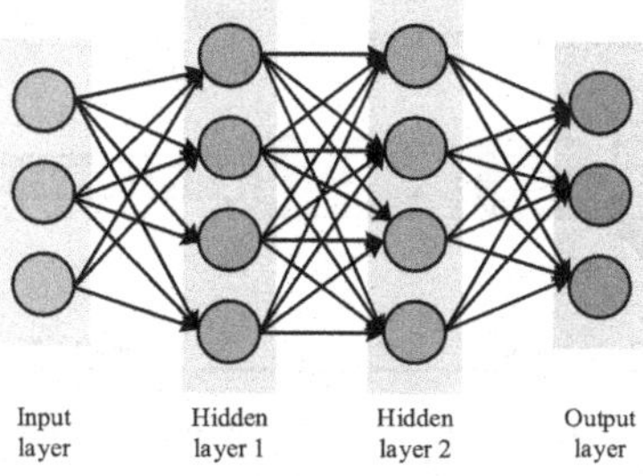

Fig. 1. The basic structure of the neural networks.

Fully-connected neural networks have shown promising performance on image classification. However, with large and high-resolution image inputs, the fully-connected neural networks suffer from a complex network architecture, which requires a large memory size to store training parameters and high-performance computing units. Therefore, a more effective network architecture should be designed to overcome the drawback of fully-connected neural networks, and convolutional neural networks (CNN) were invented for achieving superhuman performance on complex visual tasks.

2.2 Convolutional Neural Network

Emerged from the study of the brain's visual cortex, CNNs have been widely applied in image recognition. Typically, CNNs are composed of three types of layers: convolutional layers, pooling layers, and fully-connected layers. Multiple convolutional and pooling layers are stacked one after another followed by a series of fully-connected layers. Each neuron in the convolutional layer corresponds to learn a specific pattern of a limited area by only connecting to features related to that area. The pooling layer then simply performs downsampling on activation units of the previous convolutional layer for further reducing the number of training parameters. Finally, the fully-connected layers conduct the same duties found in traditional neural networks and produce class scores from the extracted features provided by the convolutional and pooling layers.

A CNN model consists of two components: the feature extraction part and the classification part. The convolution layers and pooling layers perform feature extraction. For example, given an image, the convolution layer detects features such as two eyes, long ears, four legs, a short tail, and so on. The fully connected layers then act as a classifier on top of these features and assign a probability for the input image being a dog.

In our study, the popular CIFAR-10 dataset was selected as the case study to evaluate the implementation of the image classification task on the hardware, using the CNNs. The 32×32 pixel RGB images in the CIFAR-10 dataset are sent to the feature extraction layer to filter out the most basic characteristics of the object.

As shown in Fig. 2, the featured extraction layer consists of 7 component layers and the output of the feature extraction layer will be transformed into a one-dimensional vector, which will be the input of the fully connected layer. This input through a multi-layer perceptual algorithm is used to calculate the probability and draw conclusions: the input data belongs to which of the 10 labels of the CIFAR-10 dataset.

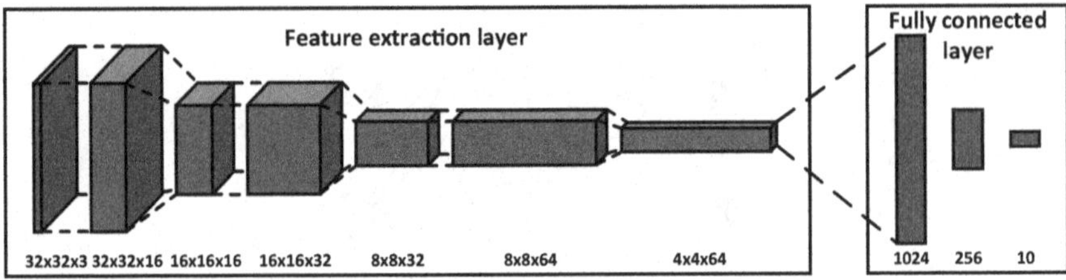

Fig. 2. Block diagram of the image recognition model on the CIFAR-10 dataset.

3 Performance Evaluation of Neural Network-Based Classifier on a Software Tool

Although this work will eventually focus on hardware implementation, implementing the neural network-based classification model on a software platform is needed for the parameter learning process and architectural optimization. Parameter training should be conducted using a software tool since this phase is generally performed only once using the offline training data, we can perform parameter training on any powerful computing units. Also, this process runs highly sophisticated learning algorithms and complex activation functions that are not efficient for implementation on the FPGA. Then, the inference phase, which requires much less computational resources, can be conducted on the FPGA board for the real-time data. Therefore, to compare the neural network-based classifier performance between the software platform and an FPGA board, this section constructs an optimal neural network-based image classification model and evaluates the neural network-based classifier performance on a software platform using a variety of network hyper-parameters.

3.1 Software Platform

There are many software-based approaches for modeling the neural network. Among those, Python is the most popular and widely-used programming language for evaluating neural network-based models. Most data scientists and machine learning developers (57% [25]) are currently using a variety of Python-based libraries such as TensorFlow, Keras, Theano, Scikit-learn, PyTorch [26]. In this work, we conduct modeling neural networks using the programming language Python and TensorFlow library running on the Ubuntu 16.04 64 bit operating system (Intel Core i5 5200U, RAM 12 GB, SSD 256 GB) as a software platform to conduct model training and evaluation of the entire image recognition result.

3.2 Analysis of the Recognition Accuracy

Using the above-mentioned software platform, we first conduct the study on the impacts of design parameters on the accuracy of the model. We consider the following four cases to calculate the accuracy: 1 layer, 2 layers, 3 layers, and 4 layers. In these cases, the parameters to be adjusted include the learning rate, epoch, and batch size. The learning rate shows the degree of adjustment of the weight matrix value W after each learning to reduce the value of the loss function. The greater the learning rate, the faster the loss function decreases. Epoch is the number of times a model is learned in the training session. Batch size is the amount of data to be included in a training session. The image recognition results on the two sets of MNIST and CIFAR-10 databases with the presented software settings are shown in the following figure (Fig. 3).

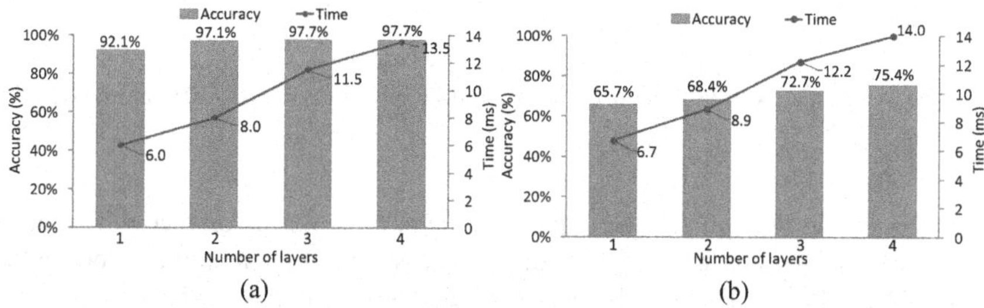

Fig. 3. Software-based recognition accuracy for the (a) MNIST dataset and the (b) CIFAR-10 dataset corresponding to different numbers of layers.

Based on the results obtained on the graph, it can be seen that image recognition accuracy in the MNIST dataset is relatively high (at least 92.1%) compared to object classification accuracy in the CIFAR-10 dataset (maximum up to 75.4%). The higher prediction accuracy on the MNIST dataset than CIFAR-10 is expected since CIFAR-10 images are undoubtedly more complex than MNIST ones. Also, the average time to recognize MNIST images is relatively lower (9.75 ms) compared to CIFAR-10 images (10.45 ms). This can be explained by the neural network structure for CIFAR-10 image recognition much larger than that of MNIST, therefore the average inference time for CIFAR-10 should be longer than that for MNIST pictures.

3.3 Analysis of the Number Format

Most software tools for machine learning techniques by default support floating-point arithmetic operations and floating-point training parameters, which achieves mostly absolute calculation accuracy. However, implementing floating-point arithmetic operations on hardware is inherently complicated and area-inefficient. Therefore, we need to look for an alternative way to implement a neural network-based classification model on hardware. First, we analyze the range domain of input data and weight matrix elements values extracted from the software implementation. Then, an appropriate number format for parameters and unit values of the neural network is selected. For both MNIST and CIFAR-10 datasets, as shown in Fig. 4, the weight matrix elements values are fundamentally concentrated in the range $(-0.25 \div 0.25)$.

Based on the statistical analysis, an 8-bit fixed-point for numerical representation, accuracy can be up to 2^{-6} (i.e., using 6 bits for representing the fraction) could be enough for representing the value domain of the parameters we have calculated. Compared to the floating-point number (single precision), an 8-bit fixed-point number would drastically reduce the design complexity and resource usage. Nonetheless, to evaluate the proposed

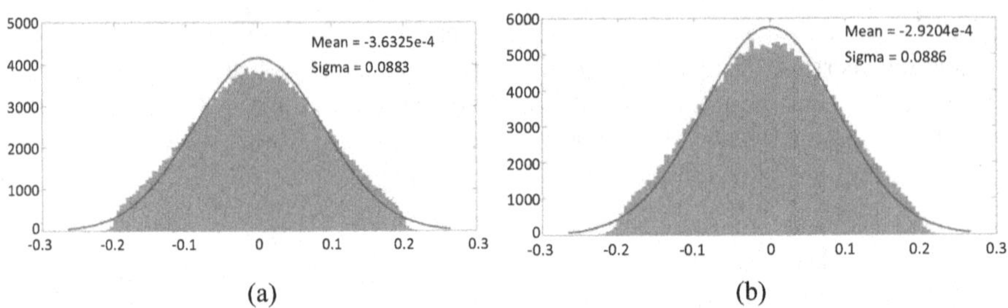

Fig. 4. The weight matrix value domain of the (a) MNIST and the (b) CIFAR-10 image recognition model.

numerical domain selection and the impacts of the number format on recognition accuracy, we still need to conduct the performance assessment of the neural network on the actual hardware design. This will be presented and discussed in Sect. 5.2.

4 Design of Neural Network Accelerator on FPGA

In this section, we first introduce a standard and fully-parameterized hardware architecture design for a single neural network layer as shown in Fig. 5. Then, this fundamental design can be used to construct the whole complex neural network with an arbitrary number of hidden layers. The generic hyperparameters of a single layer are n_i and n_{i+1}, where n_i is the number of input units and n_{i+1} is the number of output units at layer i (or the number of input units at the layer i + 1), respectively. For each layer, there are multiple processing units including multiplier-accumulator (MAC), adder, and activation and memory buffer for input, output, and training parameters. Note that the higher number of hidden layers results in more resource utilization on the hardware. To reduce the computational complexity of hardware architecture, the training parameters including weights and biases matrices are learned using the neural network software tool and

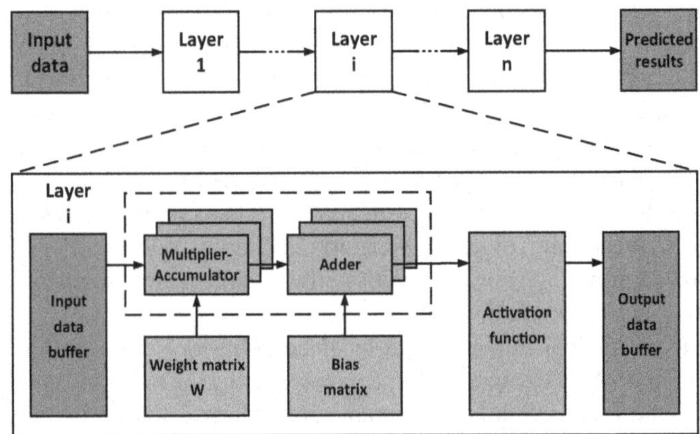

Fig. 5. Block diagram of neural network accelerator implemented on reconfigurable hardware.

are restored in the memory of the hardware. The detailed design of these processing elements and memory buffer is presented in the following subsections.

4.1 Design of Processing Units

Multiplier-Accumulators (MACs)

Assume that each MAC unit corresponds to multiplication between two binary numbers: an input feature and a weight value. During a clock cycle, an input feature and a column of weight matrix are multiplied in parallel using n_{i+1} MACs. To complete multiplication between the input vector and the weight matrix, n_i clock cycles corresponding to n_i input features are required.

Taking inputs from the input data buffer and the weight matrix, MACs are the main processing element used to perform multiplication between input features X of $1 \times n_i$ and the weighted matrix W of $n_i \times n_{i+1}$. The MAC output, called vector c_k, is calculated as below.

$$c_k = \sum_{j=1}^{n_i} x_j w_{jk},\ 1 \leq k \leq n_{i+1} \tag{1}$$

Where c_k is the kth element of the output vector, x_j is the jth element of the input vector, and w_{jk} is the weight element at column j and row k. Thus, the number of cumulative adders used is n_{i+1} and the number of multiplier-accumulators used is n_{i+1} (Fig. 6).

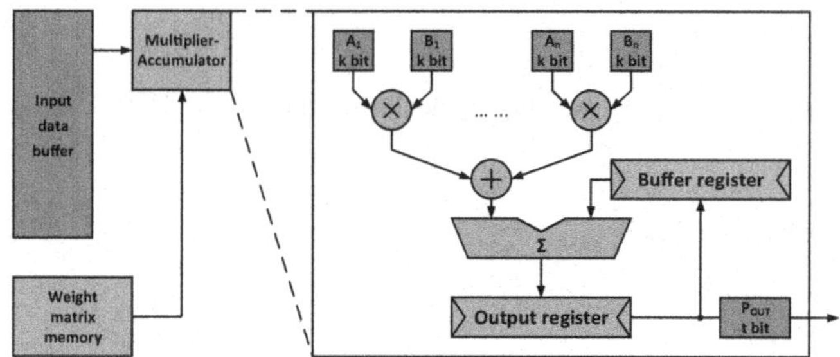

Fig. 6. Hardware design of the multiplier-accumulator.

Parallel Coefficient

To reduce the number of clock cycles needed for matrix multiplication, it is possible to read j input units and j columns of the weight matrix at the same time. Then the required number of clock periods for the matrix multiplication can be reduced by a factor of j; however, the number of MACs in a clock cycle will increase accordingly by $p = n_i/j$ times. Herein, the value of parameter p is called a parallel coefficient. The number of multipliers, accumulators, and clock cycles are estimated as follows:

$$N^{MAC} = pn_{i+1};\ N^{add} = pn_{i+1} + 1;\ N_i^{clocks} = \frac{n_i}{p} \tag{2}$$

where N^{MAC} is the number of multipliers, N^{add} is the number of adders, and N_i^{clocks} is the number of clocks.

Activation Function

Among the most commonly used activation functions for neural networks, there are some options, including Sigmod, Tanh, ReLU, or leak ReLU as shown in Fig. 7. From the hardware design point of view, we essentially selected the Rectified Linear Unit (ReLU) function because of its simplicity and feasibility for hardware implementation. As we can see in Fig. 7, the ReLU activation function is a piecewise linear function that outputs the input directly in case of the positive input value and returns zero otherwise. Using the ReLU function can accelerate the training process thanks to the fast calculation of the loss function's gradient concerning parameters. ReLU is also proven to be good enough for achieving an adequate level of accuracy for different neural network problems [27–29].

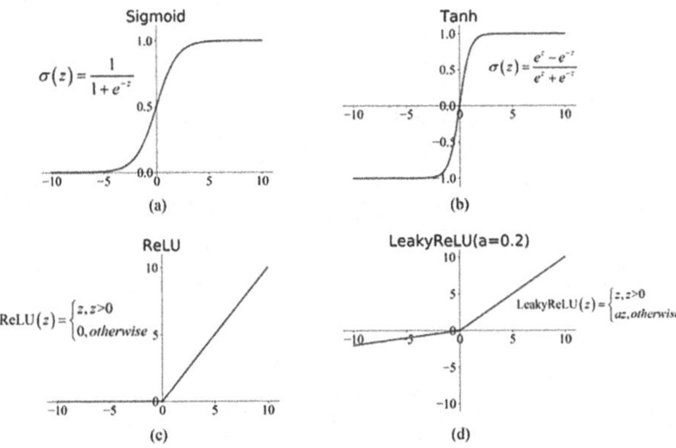

Fig. 7. Common activation functions: (a) Sigmoid, (b) Tanh, (c) ReLU, and (d) LeakyReLU.

4.2 Design of Data Buffer

Input Data Buffer

The input data buffer has memory cells arranged in rows and operates under the FIFO mechanism. The FIFO width is equal to the number of bits for input representation multiplied by the parallel coefficient p. This permits p elements that can be read or written at the same time by issuing a FIFO read and write, respectively. Regardless of the value of the parallel coefficient, the total memory required for the ith layer with n_i elements, each represented by k bits, is equal to

$$Mem_i^{Input} = kn_i(bits) \tag{3}$$

Weight Matrix Memory.

Recall that the weight matrix is optimized during the training process implemented by the software tool since parameter learning consumes a lot of hardware resources. The values of the weight matrix can be represented by the k bits binary number. Similar to the organization of the input buffer memory, the data width of the weight memory has to be matched the designed parallel coefficient. The size of the weight matrix W is $n_i \times n_{i+1}$. To represent the address for the n_i registers we need to take up to the following hardware resources:

$$Mem_i^{Weight} = kn_in_{i+1}(bits) \tag{4}$$

Considering the analysis of number format in Subsect. 3.3, which shows that trained weight values are real numbers with a limited value range, we design the format of weight values as follows. Those values in the actual hardware design if remapped to convenient fixed-point representation values, in turn, can be treated as the equally scaled-down of the integer values. This can be done first by multiplied by a scale-up factor m (m is a non-negative number) and then is rounded to a signed integer number. Therefore, the actual hardware multiplier eventually is the just an integer multiplier, which is much simpler than the real-number multiplier as can be shown in Fig. 8.

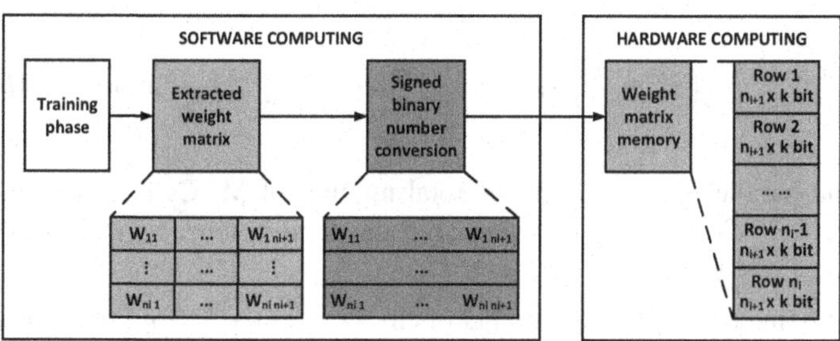

Fig. 8. Hardware design of the weight matrix.

Bias Vector Memory

The bias matrix extracted from the training phase of the classification model is a vector with n_{i+1} elements. Similarly to weight values, each bias-element needs k bits to represent. The resource occupied by the hardware for the bias vector is estimated as below:

$$Mem_i^{Bias} = kn_{i+1}(bits) \tag{5}$$

Output Data Buffer

The output data buffer with the FIFO mechanism is designed to store the results of the multiplier-accumulator (vector of n_{i+1}). To overcome the overflow phenomenon when performing the scalar product between n_i-element vectors (each element occupies k bits), the output result should be represented by $(\log_2 n_i + 2k)(bits)$. Theoretically, the number of bits for an output unit is at least 2 times higher than that for an input unit, which can cause the memory shortage especially in case of a large number of hidden layers. Therefore, we can reduce the number of bits occupied by each output value by

m bits. More specifically, before storing in the buffer, the output value is divided by 2^{m}. Dividing the real-number output value by 2^{m} can be simply implemented in the case of binary numbers by removing the m lowest bits of the binary data. The output data buffer consists of n_{i+1} registers for n_{i+1} output features and each register is represented by $\left(\log_2 n_i + 2k - m\right)(bits)$ and requires $add_{b\_out} = \log_2 n_i(bits)$ to specify a register address. Finally, the amount of memory resources on the hardware for the output data buffer at layer $i + 1$ equals:

$$Mem_i^{Output} = n_{i+1}\left(\log_2 n_i + 2k - m\right)(bits) \quad (6)$$

4.3 Hardware Utilization and Processing Latency

We have derived the resource utilization on hardware for the hidden layer $i + 1$ including input data buffer, weight matrix, bias vector, output data buffer. If there are L consecutive hidden layers, the number of matrix multiplication blocks are $L + 1$. Then, the total amount of memory occupied is estimated as below:

$$Mem^{total} = \sum_{i=1}^{L+1}\left(Mem_i^{Input} + Mem_i^{Weight} + Mem_i^{Bias}\right) + Mem_{l+1}^{Output}(bits) \quad (7)$$

Given the parallel coefficient p, the total number of MACs is equal to $N^{MAC} = \sum_{i=1}^{L+1} N_i^{MAC} = p\sum_{i=1}^{L+1} n_{i+1}$ and similarly, the total number of adders is $N^{Add} = \sum_{i=1}^{L+1} N_i^{Add} = p\sum_{i=1}^{L+1} n_{i+1}$.

The processing latency (or the number of clock cycles) N^{clocks} for the neural network with parallel coefficient, p is calculated as:

$$N^{clocks} = \sum_{i=1}^{L+1} N_i^{clocks} = \sum_{i=1}^{L+1} \frac{n_i}{p} \quad (8)$$

We can infer from hardware consumption and processing delay that the neural network-based hardware architecture requires hardware resource that is linearly proportional to the parallel coefficient. Meanwhile, the processing delay is linearly reduced by a factor of p. The selection of parallel coefficients should be considered based on the FPGA memory, the number of given processing units, and the required delay of a specific application. The actual performance metrics and resource utilization will be presented and discussed in the subsequent section.

5 Performance Evaluation of Neural Network Accelerator on FPGA

5.1 Experimental Setup

In this subsection, we introduce two considered datasets and describe the neural network architecture for each data. Then, the performance metrics and network parameters are also presented. Based on the generic model described in the previous section, we have implemented the hardware models targeted for MNIST and CIFAR-10 datasets. Those models are fully described using synthesizable VHDL optimized for FPGA. Those case studies will be further used for evaluating the performance and other design aspects.

For the MNIST dataset, each input image will be converted to a 1 × 784 binary matrix. Input matrix and the network parameters (i.e., weights and biases values) obtained from the training phase are fed into the Xilinx Vivado Simulator [30] to collect performance metrics including computational speed and recognition accuracy (Fig. 9).

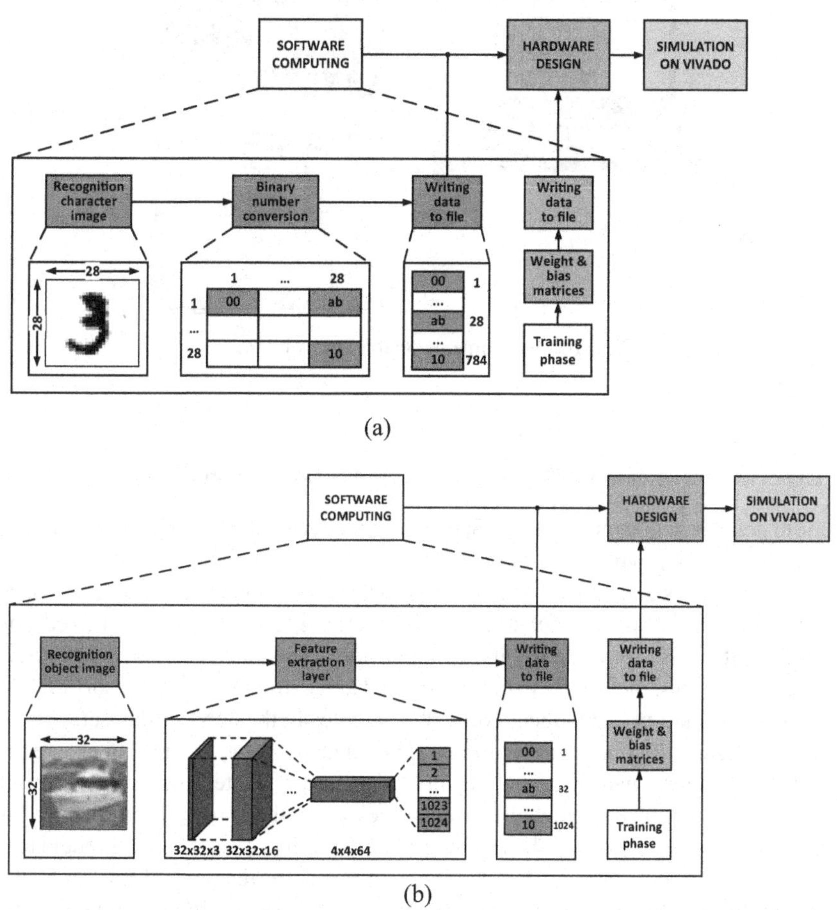

Fig. 9. Design model of image recognition on hardware for the (a) MNIST dataset and the (b) CIFAR-10 dataset.

In cases of the CIFAR-10 dataset, images are larger and more complex than those in the MNIST database. To temporarily simplify the hardware design, we only implement fully-connected layers on the hardware while feature extraction layers are pre-processed. Note that the implementation of those layers follows the classical image classifier and does not cost much in cases of the small filter kernel [31]. In the first fully-connected layer, the number of input units is 1024, the number of output features is 256; in the second fully-connected layer, the number of output neurons is 10. Therefore, the size of the weight matrix for the first layer is 1024 × 256, for the second layer is 256 × 10; the size of the bias vectors for the first layer is 256 × 1, for the second layer is 10 × 1.

To examine the performance of the hardware design of the neural network accelerator, we have collected the performance of the image classification model on a real FPGA board as can be seen in Fig. 10.

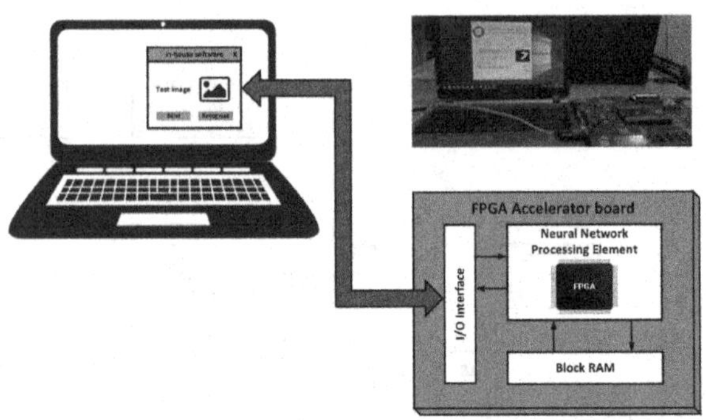

Fig. 10. Experiments on the FPGA board.

5.2 Experimental Results of Neural Network Accelerator on FPGA

Our entire generic design presented in Sect. 4 has been described by a hardware description language (HDL), where the number format and parallel coefficient are considered as design parameters and can be set to desired values. The hardware architecture of NN is evaluated on the Vivado Simulator. The parameters are converted into a fixed-point number format (with 2, 4, 8, or 16 bits) by an in-house software on C ++. The featured parameters of the built neural networks are trained and extracted using TensorFlow. The performance evaluation is conducted on 1,000 samples in the MNIST dataset.

After designing the image recognition model on the hardware, experiments are conducted to evaluate the hardware architecture. First, the extracted image data from the text files are put into the designed block and processed by the hardware simulator Xilinx Vivado. The classified label which is the output of the image classification model is then compared with the true label and the number of correctly recognized images will be recorded in the counter. When the last image in the test set is classified, the classification accuracy is obtained.

The Impacts of the Number Formats and Network Architectures

In this subsection, we focus on analyzing the dependence of recognition accuracy on two main parameters: the number of hidden layers and number format to represent the values of the weight matrix. The results of MNIST image recognition accuracy with different parameters are shown in Table 1. It can be seen that the classification accuracy depends more on the number format than the number of hidden layers. When the number of bits used to represent the weight matrix is too small, the recognition accuracy is very low (e.g., 16% accuracy for 2-bit number format). Meanwhile, the accuracy sharply improves if the number of bits changes from 2 to 4 or more. However, there is no significant improvement in cases of using more than 4 bits for each training parameter. E.g., the accuracy converges to 89% and 96.9% with 1 and 2 hidden layers, respectively. From Table 1, we observe that the 8-bit format can be used to save hardware resources while ensuring accurate performance on image classification.

Table 1. Accuracy of MNIST image recognition implemented on the hardware with different hidden layers and number formats for 1,000 samples.

Number of hidden layers	2-bit	4-bit	8-bit	16-bit
1	16.8%	88.6%	89%	89%
2	16%	96.8%	96.9%	96.9%

Hardware Utilization on Different Chips and with Different Parallel Coefficients
Recall that we have derived the hardware resources required for MACs, adders, and memory in Eq. (2). The number of MACs, the number of adders, and memory size increase proportionally to the parallel coefficient and the number of hidden layers. Indeed, using a higher parallel coefficient p results in short prediction time but a significant increase in the system resources demand. At the same time, multilayer neural networks can produce higher accuracy while demanding more computing resources. Therefore, it is necessary to study the relationship between these two factors (recognition time and resource demand) for a better selection of neural network architectures in reality.

To understand the feasibility of FPGA for the neural network application, we first implemented and compared the designs on some representative FPGA devices from Xilinx. Then, with the HDL designed and fully logical verified, we have implemented on the actual FPGAs. The main results are presented in Fig. 11.

In this work, we have chosen some representative and active FPGA families from Xilin [32], including the low-cost (Artix-7), the best price/performance (Kintex-7), the performance-optimized (Virtex-7) solutions with different resource capabilities. In terms of resources, except for the Artix XC7A100T FPGA, where all DSP is fully utilized, the remaining FPGA devices are considered large enough for accommodating 2-layer neural networks in either CIFAR or MNIST. The actual resource utilization of slice registers, LUT, RAM only accounts for a small proportion of the total availability (e.g., less than 15% LUT for Kintex 7 XC7K325T, or less than 5% LUT for Virtex 7 XC7V980). This is the strong validation for the feasibility of the implementation of the more sophisticated recognition and classification engines (e.g., up to 5 hidden layers with more complex activation functions) on the next generation FPGA. The other high-end devices such as Ultra-scale [33] and Ultra-scale plus [34] with their extremely large logic and computing resource are essentially capable not only for neural networks but complete AI system implementations. In terms of performance, the achievable clock frequencies are technically dependent mainly on the latency of MAC (i.e., FPGA DSP macro). Therefore, the reported clock frequencies range from 153 MHz (Artix 7) to ~ 200 MHz (Virtex 7).

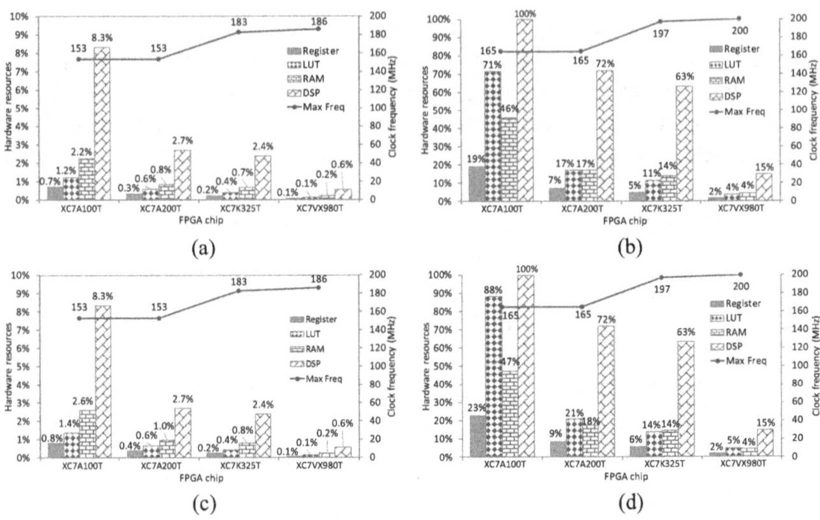

Fig. 11. Hardware utilization and corresponding maximum clock frequency (a) 1-layer MNIST, (b) 2-layer MNIST, (c) 1-layer CIFAR, and (d) 2-layer CIFAR implementations using an 8-bit fix-point number format targeted for different Xilinx FPGA devices.

Furthermore, we evaluated the impacts of the parallel coefficient on the resource utilization and bandwidth of the hardware architecture. Figure 12 presents the resource utilization for the case of MNIST implementation on the Xilinx Kintex-7 FPGA series XC7K325T.

Based on the results obtained, we found that the image recognition speed also increases almost proportionally to the parallel coefficient p. Specifically, when the parallel computing is not applied (i.e., $p = 1$), the implemented network can process 250,000 images image recognition per second, and when this coefficient increases to 8, the image recognition speed reaches 1,762,000 images per second for the MNIST dataset. Similarly, for the CIFAR-10 dataset, the recognition speed is 153,000 and 1,084,000 images per second, respectively. The practical performance hence increases by more than 7X in either implementation when changing p from 1 to 8. This is explained by a slight degradation in maximum clock frequencies when the design becomes larger. There is an inevitable trade-off between increasing image recognition speed and the cost in logical resource, and this is explicitly shown by the dependency of the growth rate of the resource utilization and the parallelism level.

Performance Comparison Between FPGA-Based Neural Network Accelerator and Neural Network Using the Software Tool

From the simulation results of the software design and the implementation results of the design on the hardware, we can see that the achievable accuracy of the image recognition on the hardware as good as achievable accuracy by software. Meanwhile, hardware implementation is more beneficial in terms of detection speed than that on software.

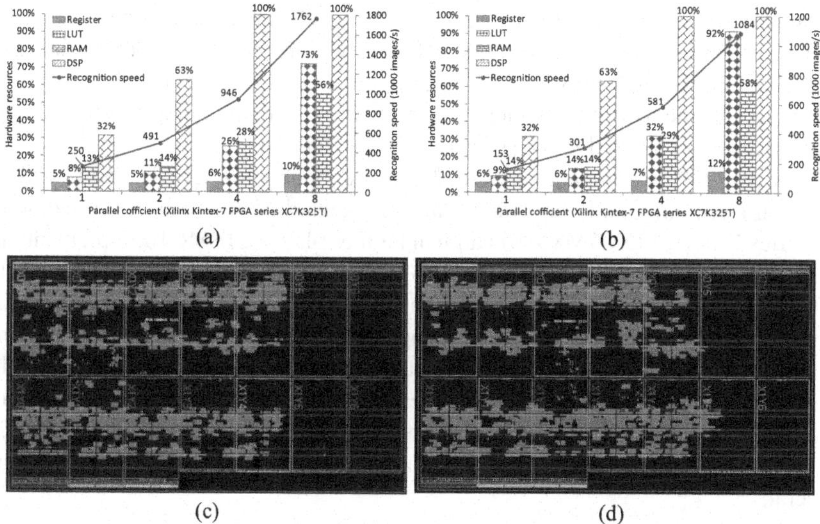

Fig. 12. Resource utilization and recognition speed on Xilinx Kintex-7 FPGA series XC7K325T of 2-layer neural network for (a) MNIST, (b) CIFAR implementations with different parallel coefficients and images depicting physical layout on the FPGA chip for (c) MNIST, (d) CIFAR implementations when parallel coefficient equals 4.

Full comparative figures between image recognition on hardware and software are given in Tables 2. From this table, we observe that the time to recognize images when performing on hardware is faster than image recognition on software at least 1500 times when the parallel coefficient equals 1 (6000 μs vs 4 μs, using one hidden layer for recognition), and maybe faster up to 20,000 times when the parallel coefficient is 16 (8000 μs vs 0.4 μs with two hidden layers).

Table 2. Comparison of software and hardware performance in MNIST image recognition.

Platform	Software		Hardware									
Number of layers	1	2	1	1	1	1	1	2	2	2	2	2
$p^{(*)}$	N/A	N/A	1	2	4	8	16	1	2	4	8	16
$T^{(**)}(\mu s)$	6000	8000	4	2	1.1	0.6	0.3	5.3	2.7	1.4	0.8	0.4

$^{(*)}$ Parallel coefficient, $^{(**)}$ Recognition time

6 Conclusion

In this work, we have proposed a generic design for neuromorphic computing upon one layer, that allows us to construct any other sophisticated neural networks. Along with the generic design, a systematic study has been conducted on the resource utilization, performance, and accuracy of the neural network models and their dependencies on the design hyper-parameters. From the statistical study, we have practically proven that using a fixed-point number for hardware implementation could greatly reduce the complexity and resource for the hardware implementation while still maintaining mostly the same level of accuracy compared to the software implementation.

As a case study, we implemented the hardware models for MNIST and CIFAR-10 datasets on a reconfigurable hardware platform. Regarding the resource utilization, a Xilin Virtex 7 device (XC7VX980) can handle the 2-layer CIFAR-10 implementation with spending less than 5% of LUT and 15% in DSP. Furthermore, at iso-accuracy, the FPGA-based neural network implementations are notably faster in recognition speed. If no parallel computing is considered, the proposed hardware accelerator is 1,500 times quicker than the baseline software implementation and could reach 20,000 with a higher degree of parallelism. Though our initial design in this work is limited for the neural networks, the impressive results proved the potential of reconfigurable devices and FPGA as the flexible and powerful platform for neuromorphic computing and AI applications in general.

Acknowledgment. This research is funded by the Vietnam National Foundation for Science and Technology Development (NAFOSTED) under grant number 102.01-2018.310.

References

1. Krizhevsky, A., Sutskever, I., Hinton, G.E.: ImageNet classification with deep convolutional neural networks. In: NIPS 25, pp. 1106–1114. Curran Associates, Inc. (2012)
2. Sun, Y., Wang, X., Tang, X.: Deep learning face representation by joint identification-verification. In: Neural Information Processing Systems, pp. 1988–1996 (2014)
3. Ji, S., Xu, W.: 3D convolutional neural networks for automatic human action recognition. Pattern Anal. Mach. Intell. **35**(1), 221–231 (2013)
4. Abdel-Hamid, O.: Convolutional neural networks for speech recognition. In: Audio Speech & Language Processing (2014)
5. https://github.com/Xilinx/chaidnn. Accessed 31 Mar 2020
6. https://www.xilinx.com/support/documentation/whitepapers/wp504-accel-dNeuralnetworks.pdf. Accessed 31 Mar 2020
7. http://www.deephi.com/technology/dnndk. Accessed 31 Mar 2020
8. Abadi, M., et al.: Tensorflow: Large-scale ML on heterogeneous distributed systems. arXiv preprint arXiv:1603.04467 (2016)
9. Krizhevsky, A., Sutskever, I., Hinton, G.E.: ImageNet classification with deep convolutional neural networks. In: Neural Information Processing Systems, pp. 1097–1105 (2012)
10. He, K., Zhang, X., Ren, S., Sun, J.: Delving deep into rectifiers: surpassing human-level performance on ImageNet classification. In: Proceedings of the IEEE International Conference on Computer Vision, pp. 1026–1034 (2015a)
11. Russakovsky, O., et al.: Imagenet large scale visual recognition challenge. Int. J. Comput. Vis. **115**(3), 211–252 (2015)
12. He, K., Zhang, X., Ren, S., Sun, J.: Deep residual learning for image recognition. In: Proceedings of the IEEE Conference on Computer Vision and Pattern Recognition, pp. 770–778 (2016)
13. Szegedy, C., et al.: Going deeper with convolutions. In: CVPR, pp. 1–9 (2015)

14. Guo, K., Zeng, S., Yu, J., Wang, Y., Yang, H.: A survey of FPGA-based neural network accelerator (2017). arXiv:1712.08934v3
15. Liang, S., Yin, S., Liu, L., Luk, W., Wei, S.: FP-BNN: binarized neural network on FPGA. Neurocomputing **275**, 1072–1086 (2017). Accessed 18 Oct 2017. https://doi.org/10.1016/j.neucom.2017.09.046
16. NVIDIA, Tesla K40 GPU Active Accelerator, NVIDIA (2013)
17. Chen, Y.-H., et al.: Eyeriss: an energy-efficient reconfigurable accelerator for deep convolutional neural networks. IEEE Int. Solid-State Circ. Conf. (ISSCC) (2016)
18. Ovtcharov, K., Ruwase, O., Kim, J.-Y., Fowers, J., Strauss, K., Chung, E.S.: Accelerating deep CNNs using specialized hardware. In: Microsoft Research Whitepaper, vol. 2, no. 11 (2015)
19. Putnam, A., et al.: A reconfigurable fabric for accelerating large-scale datacenter services. In: International Symposium on Computer Architecture (ISCA), p. 1324 (2014)
20. Bettoni, M., Urgese, G., Kobayashi, Y., Macii, E., Acquaviva, A.: A convolutional neural network fully implemented on FPGA for embedded platforms. In: 2017 New Generation of CAS (NGCAS), Genova, pp. 49–52 (2017). https://doi.org/10.1109/ngcas.2017.16
21. Nurvitadhi, E., Sheffield, D., Sim, J., Mishra, A., Venkatesh, G., Marr, D.: Accelerating binarized neural networks: comparison of FPGA, CPU, GPU, and ASIC. In: 2016 International Conference on Field-Programmable Technology (FPT), Xi'an, pp. 77–84 (2016). https://doi.org/10.1109/fpt.2016.7929192
22. Nurvitadhi, E., Sim, J., Sheffield, D., Mishra, A., Krishnan, S., Marr, D.: Accelerating recurrent neural networks in analytics servers: comparison of FPGA, CPU, GPU, and ASIC. In: 2016 26th International Conference on Field Programmable Logic and Applications (FPL), Lausanne, pp. 1–4 (2016). https://doi.org/10.1109/fpl.2016.7577314
23. http://yann.lecun.com/exdb/mnist/. Accessed 31 Mar 2020
24. Krizhevsky, A.: CIFAR-10 AND CIFAR-100 DATASETS (2009). https://www.cs.toronto.edu/~kriz/cifar.html
25. https://becominghuman.ai/best-languages-for-machine-learning-in-2020-6034732dd24. Accessed 31 Mar 2020
26. https://opensource.com/article/18/5/top-8-open-source-ai-technologies-machine-learning. Accessed 31 Mar 2020
27. Feng, J., He, X., Teng, Q., Ren, C., Chen, H., Li, Y.: Reconstruction of porous media from extremely limited information using conditional generative adversarial networks. Phys. Rev. E. **100**, 033308 (2019). https://doi.org/10.1103/physreve.100.033308
28. Hu, J., Shen, L., Sun, G.: Squeeze-and-excitation networks. 2018 IEEE/CVF Conference on Computer Vision and Pattern Recognition (2018)
29. Krizhevsky, A., Sutskever, I., Hinton, G.E.: ImageNet classification with deep convolutional neural networks. Commun. ACM (2017)
30. https://www.xilinx.com/support/documentation/sw_manuals/xilinx2018_1/ug937-vivado-design-suite-simulation-tutorial.pdf. Accessed 06 Jun 2020
31. Aurelien Gron.: Hands-On Machine Learning with Scikit-Learn and TensorFlow: Concepts, Tools, and Techniques to Build Intelligent Systems, 1st edn. O'Reilly Media (2017)
32. https://www.xilinx.com/support/documentation/selection-guides/7-series-product-selection-guide.pdf. Accessed 06 Jun 2020
33. https://www.xilinx.com/support/documentation/selection-guides/ultrascale-plus-fpga-product-selection-guide.pdf. Accessed 06 Jun 2020
34. https://www.xilinx.com/support/documentation/data_sheets/ds890-ultrascale-overview.pdf. Accessed 06 Jun 2020

Information Processing and Data Analysis

Adaptive Essential Matrix Based Stereo Visual Odometry with Joint Forward-Backward Translation Estimation

Huu Hung Nguyen[1(✉)], Quang Thi Nguyen[1], Cong Manh Tran[1], and Dong-Seong Kim[2]

[1] Le Quy Don Technical University, 236 Hoang Quoc Viet, Hanoi, Vietnam
{hungnh.isi,thinq.isi}@lqdtu.edu.vn, manhtc@gmail.com

[2] Kumoh National Institute of Technology, Gumi-si, South Korea
dskim@kumoh.ac.kr

Abstract. Visual Odometry is widely used for recovering the trajectory of a vehicle in an autonomous navigation system. In this paper, we present an adaptive stereo visual odometry that separately estimates the rotation and translation. The basic framework of VISO2 is used here for feature extraction and matching due to its feature repeatability and real-time speed on standard CPU. The rotation is accurately obtained from the essential matrix of every two consecutive frames in order to avoid the affection of the stereo calibration uncertainty. With the estimated rotation, translation is rapidly calculated and refined by our proposed linear system with non-iterative refinement without the requirement of any ground truth data. The further improvement of the translation by joint backward and forward estimation is also presented in the same framework of the proposed linear system. The experimental results evaluated on the KITTI dataset demonstrate around 30% accuracy enhancement of the proposed scheme over the traditional visual odometry pipeline without much increase in the system overload.

Keywords: Visual odometry · Essential matrix · Non-iterative translation estimation · Forward-backward translation estimation

1 Introduction

Visual Odometry(VO) [1] is one of the important parts in robotics research, especially for autonomous navigation. With the unavailability of GPS signals such as indoor extra-terrestrial and in space, image-based localization becomes necessary. Specifically, a single movement between previous and current images is estimated by resolving geometric constraints. Subsequently, the full camera trajectory is finally recovered via the accumulation of these movements. Recently, the

N.-S. Vo and V.-P. Hoang (Eds.): INISCOM 2020, LNICST 334, pp. 127–137, 2020.
https://doi.org/10.1007/978-3-030-63083-6_10

survey [2] classified VO in different approaches such as monocular/stereo camera-based, geometric/learning-based and feature/appearance-based. The feature-based VO pipeline has a long history and has been detailed in Nister's [1] work. Scaramuzza and Fraundorfer conducted a comprehensive review of feature-based VO [3,4]. Accordingly, relative ego-motion between the two frames was obtained by three following major approaches

- **2D-to-2D**: Motion is estimated only from 2D feature correspondences.
- **3D-to-3D**: Motion is estimated only from 3D feature correspondences.
- **3D-to-2D**: Motion is estimated from 3D features in one frame and their corresponding 2D features in other.

The first approach, 2D-to-2D,is a methodology that recovers the rotation and translation direction from the essential matrix computed from 2D feature correspondences using the epipolar constraint. Nister proposed an efficient implementation [5] for the minimal case solution with five 2D correspondences and that has become the standard for 2D-to-2D motion estimation due to the efficiency in the presence of outliers. The second approach, 3D-to-3D, computes the camera motion by determining the aligning transformation of the two 3D feature sets that minimizes the Euclidean distance between the two sets of 3D features. As shown in [6], the minimal case solution involves three 3D-to-3D non-collinear correspondences, which can be used for robust estimation in the presence of outliers. 3D-to-2D method is well-known as perspective from n points (PnP). The pose is obtained via iteratively minimizing the summation of projection error between the projected points of 3D features and corresponding 2D observations. The minimal case involving 3D-to-2D correspondences in [7] is called perspective from three points ($P3P$). With $n \geqslant 6$, there is a simple and straightforward solution for PnP by solving the direct linear transformation (DLT) [8]. The conventional framework VISO2 [9] applied PnP approach with a robust against outliers. They adopted PnP using 3 randomly drawn correspondences into a RANSAC scheme, by first estimating (R, t) for 50 times independently. All inliers of the winning iteration were then used for refining the parameters, yielding the final transformation (R, t).

Since VO works by estimating the camera path incrementally (pose after pose), then over time, the errors introduced by each frame-to-frame motion accumulate. This generates a drift of the estimated trajectory from the real path. For some applications, it is utmost important to keep drift as small as possible, which can be done through local optimization over the last m camera poses. This approach called sliding window bundle adjustment or windowed bundle adjustment has been used in several works. However, it takes additional computational time because of being an iterative method. So it only is used as the final step for refining or executed at some special location. For real-time applications, reducing computational cost is important so proposing the fast and accurate frame to frame VO is still an active research area.

Note that, 3D-to-3D and 3D-to-2D approaches require triangulation of 3D points from 2-D image correspondences which are determined by intersecting back-projected rays from 2D image correspondences of at least two image frames.

In perfect conditions, these rays would intersect in a single 3D point. However, they never intersect because of image noise, camera model and calibration errors as well as feature matching uncertainty. Therefore, the point at a minimal distance in the least-squares sense from all intersecting rays can be taken as an estimate of the 3D point position. As pointed out by Nister [1], the 2D-to-2D method and 3D-to-2D method are evaluated to be more accurate than 3D-to-3D methods because 3D-to-3D minimizes feature position error whereas 3D-to-2D approach minimizes re-projection error or 2D-to-2D approach minimizes the Sampson error. In the case of using the 2D-to-2D approach, we do not need triangulation to calculate the rotation and scalable translation. However, we need to use triangulation for computing absolute translation. There is an easy way to obtain scale from relative distances between any combination of two 3D points or exploiting the trifocal constraint between 3 view matches of 2D features or iteratively minimizing re-projection error with known rotation for features on pairs of stereo images.

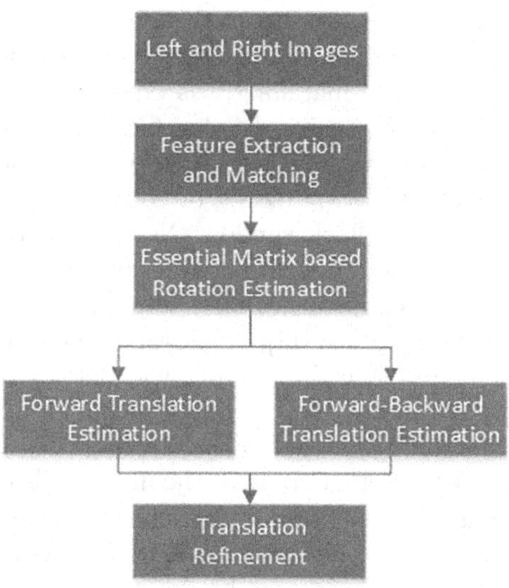

Fig. 1. Our proposed algorithm. Utilizing the feature extraction and matching component of VISO2 library and replacing the pose estimation part based on PnP by our proposed adaptive stereo VO with essential matrix based rotation estimation and join forward-backward translation estimation

In this paper, we propose adaptive stereo camera which estimates the transformation of two consecutive frames separately. The proposed algorithm is depicted in Fig. 1. Here, rotation is extracted from essential matrix estimated by using five point algorithm with preemptive RANSAC and translation is computed by solving a novel linear equation system modified from the re-projection equation. The proposed approach has following benefits:

- Accurate rotation from essential matrix estimation that avoids uncertainty of stereo calibration.
- Translation is estimated directly without iterative optimization for both initiation and final refinement joint forward and backward approaches.

The paper is organized as follows. Section 2 briefly provides a description of feature extraction and matching. Section 3 describes the proposed visual odometry approach based on the essential matrix estimation and non-iterative translation calculation. Section 4 presents the results of KITTI dataset in comparison to VISO2. Section 5 provides the concluding marks.

2 Feature Extraction

The input of our visual odometry is feature correspondences between four images in the previous and current stereo camera frames. We took advantage of feature detector and matching (active matching) proposed by Chli [9] because of its feature repeatability and speed. It was employed as open-source by Geiger in the VISO2 library [10]. In particular, firstly, 5×5 blob and corner masks were used to filter the input image to extract four feature classes: blob min/max, corner min /max. Additionally, feature matching was done by comparing the 11×11 block windows of horizontal and vertical Sobel filter responses to give two features using the sum of absolute differences (SAD) error metric. To speed-up, the matching process, sum over a sparse set of 16 locations was used instead of being summed over the whole block window. Note that, the matching process was done on the same class (blob max/min, corner max/min) to reduce the computational cost without reducing feature matching quality. Some of the outliers were rejected by circular matching [11], suggesting that each feature needs to be matched between left and right images of two consecutive frames, requiring four matches per feature. Finally, the bucketing technique is used to divide the corresponding features into 50×50 grids and selecting only a limited number of features in each bucket. This step guarantees the uniform distribution of selected features along the z-axis, the roll axis of the vehicle. It means that both close and far features are used for pose estimation resulting in high accuracy of vehicle trajectory. This feature detector and matching have been used for several visual odometry approaches such as [12] and [13] that achieved good performance on the KITTI dataset.

3 Proposed Pose Estimation

Single movement between previous and current frames is separated into two parts. Firstly, the five-point algorithm is performed to obtain the essential matrix and then rotation is extracted. Secondly, the translation initially is estimated by one point RANSAC by close features and finally is obtained by forward-backward refinement. Different from conventional approaches using iterative optimization such as [13], we proposed a closed-form linear equation for both initial and final estimation.

3.1 Rotation Estimation

The geometric relation between two consecutive frames of a calibrated camera is described by the so-called essential matrix E which contains the camera motion parameters up to a unknown scale factor for translation. It is represented as following form:

$$E = T^{\times} R \tag{1}$$

where skew matrix $T^{\times}$ is rewritten in detail as follow

$$T^{\times} = \begin{bmatrix} 0 & -t_z & t_y \\ t_z & 0 & -t_x \\ -t_y & t_x & 0 \end{bmatrix} \tag{2}$$

Each correspondence of two image frames satisfies the epipolar constraint

$$p^T E q = 0 \tag{3}$$

Where a 2D feature, p in previous frame corresponds to another 2D feature, q, in current frame. Essential matrix E has two additional properties

$$det(E) = 0 \tag{4}$$

And

$$2EE^T E - tr(EE^T)E = 0 \tag{5}$$

Note that, essential matrix E is 3×3 matrix with 8 unknown variables with an un-observable scale can be solved by five point equations by search solution of root of tenth degree polynomial proposed by Nister [5]. Each of the five point correspondences gives rise to a constraint (3). It can be also rewritten in a linear equation formular as follows

$$\hat{a}\hat{E} = 0 \tag{6}$$

where

$$\hat{a} = [p_1 q_1, q_1 q_2, p_1, p_2 q_1, p_2 q_2, p_2, q_1, q_2, 1] \tag{7}$$

and

$$\hat{E} = [E_{11}, E_{12}, E_{13}, E_{21}, E_{22}, E_{23}, E_{31}, E_{32}, E_{33}]^T \tag{8}$$

Stacking the constraints of five-point correspondences gives the linear equation (6) and by solving the system the parameters of E can be computed. Equations (6), (4), and (5) are extended to 10 cubic constraints, and then to a ten-degree polynomial. As a result, a maximum of 10 essential matrix solutions was obtained for any five-point set. The solution yielding the highest number of inliers was selected as a set representative. This five-point algorithm is applied in conjunction with preemptive RANSAC. A number of five-point sets are randomly taken from the total set of features. The five-point algorithm is applied to taken subsets and generate a number of hypotheses. The hypothesis with the best preemptive scoring which has the largest set of inliers is chosen as the final solution. This five-point algorithm may not always converge to a global minimum but can offer superior performance in the rotation because of some reasons:

- Essential matrix is estimated from a closed-form tenth degree polynomial.
- Five-point is a minimal set for essential estimation so the affection of outlier is small.
- Monocular method is not affected by imperfect calibration between left and right image of stereo camera.

Therefore, only left or right camera is used for rotation estimation.

3.2 Translation Estimation

The relative orientation between the previous and current frames was obtained by the algorithm described above. Here, we propose a joint forward-backward translation estimation. Firstly, an initial translation was estimated by one point RANSAC. Secondly, it is further improved by joint forward-backward non-iterative translation estimation with the rotation estimated previously.

Consider the projection equation from a 3D point feature from current frame to previous frame.

$$\begin{pmatrix} u_p \\ v_p \\ 1 \end{pmatrix} = \begin{pmatrix} f & 0 & u_c \\ 0 & f & v_c \\ 0 & 0 & 1 \end{pmatrix} \left[\left(R_{3\times3} \; t_{3\times1} \right) \begin{pmatrix} x \\ y \\ z \\ 1 \end{pmatrix} - \begin{pmatrix} s \\ 0 \\ 0 \end{pmatrix} \right] \tag{9}$$

with:

- Homogenous image coordiate $(u_p, v_p, 1)^T$ in left or right frame of preivious frame.
- Focal length f.
- Rotation R and translation t from current frame to previous frame.
- 3D point (x, y, z) in current left frame.
- Value s is equal to 0 for left frame or baseline for right frame.

This projection equation is re-written detail as follows

$$\begin{cases} u_p & = f\dfrac{(x_{Rot} + t_x + s)}{z_{Rot} + t_z} + u_c \\ v_p & = f\dfrac{(y_{Rot} + t_y)}{z_{Rot} + t_z} + u_c \end{cases} \tag{10}$$

where

$$\begin{pmatrix} x_{Rot} \\ y_{Rot} \\ z_{Rot} \end{pmatrix} = R_{3\times3} \begin{pmatrix} x \\ y \\ z \end{pmatrix} \tag{11}$$

$$\begin{pmatrix} -1 & 0 & \dfrac{u-u_c}{f} \\ 0 & -1 & \dfrac{v-v_c}{f} \end{pmatrix} \begin{pmatrix} t_x \\ t_y \\ t_z \end{pmatrix} = \begin{pmatrix} x_{Rot} + s - z_{Rot}\dfrac{(u-u_c)}{f} \\ y_{Rot} - z_{Rot}\dfrac{(v-v_c)}{f} \end{pmatrix} \tag{12}$$

A 3D feature of current frame is projected to both left and right of previous frame with $s = 0$ and $s = baseline$, respectively. So from Eq. (12), we can form a linear system of 4 equations of 3 unknown variables t_x, t_y, t_z as follows

$$A \begin{pmatrix} t_x \\ t_y \\ t_z \end{pmatrix} = B \tag{13}$$

They are known to be solution of Pseudo Inverse method

$$\begin{pmatrix} t_x \\ t_y \\ t_z \end{pmatrix} = (A^T A)^{-1} A^T B \tag{14}$$

In idea case without feature noise, translation is successfully obtained by Eq. (14) using only one feature correspondence. However, in the real situation, feature noise is unavoidable, using one feature does not guarantee the success of estimation. To obtain higher accuracy of translation estimation, we wrap this algorithm into the RANSAC scheme, 100 samples of closest 3D features are used to estimate candidate translations. The best one producing the largest of number inliers is considered as the final solution. These inliers are further used for the refinement step. Different from conventional methods that minimize the re-projection error iteratively, our proposed refinement quickly estimates absolute translation by solving a linear system. In particular, all n inliers are plug into Eq. (13) to create

$$\begin{bmatrix} A_1 \\ A_2 \\ . \\ . \\ . \\ A_n \end{bmatrix} \begin{pmatrix} t_x \\ t_y \\ t_z \end{pmatrix} = \begin{bmatrix} B_1 \\ B_2 \\ . \\ . \\ . \\ B_n \end{bmatrix} \tag{15}$$

Similar to above, Pseudo Inverse method is re-used to refine the initial estimation.

The above paragraph describes for backward estimation. To improve translation accuracy, both forward t_f and backward t_b translations are estimated by using same Eq. (15). The final solution is obtained by

$$t_{final} = 0.5(t_b - R^{-1} t_f) \tag{16}$$

where R is rotation estimated from previous section using the essential matrix.

4 Experimental Results

The proposed approach is evaluated on the KITTI dataset in comparison to its performance against the traditional VO pipeline, VISO2, proposed in [10]. The KITTI dataset consists of different traffic scenarios that are widely used

Table 1. Performance evaluation on KITTI Dataset

Sec Num	VISO2			Backward			Join For-Backward		
	t_e (%)	r_e (deg/ 100 m)	t_{abs} (m)	t_e (%)	r_e (deg/ 100 m)	t_{abs} (m)	t_e (%)	r_e (deg/ 100 m)	t_{abs} (m)
1	2.46	1.18	86.0	1.28	0.41	25.5	1.22	0.46	18.7
2	4.41	1.01	188.3	4.40	0.56	121.1	3.30	0.38	85.7
3	2.19	0.81	140.7	1.19	0.36	59.0	1.11	0.36	20.9
4	2.54	1.20	32.6	2.57	0.32	14.9	2.43	0.36	13.1
5	1.02	0.87	4.2	2.45	0.32	10.2	2.29	0.33	9.0
6	2.07	1.12	46.5	1.42	0.40	18.9	1.41	0.40	15.4
7	1.31	0.92	8.9	2.31	0.42	17.8	1.98	0.33	9.4
8	2.30	1.77	21.2	1.76	1.00	14.8	1.66	0.99	13.1
9	2.74	1.33	35.1	1.68	0.41	16.9	1.51	0.41	13.9
10	2.76	1.15	79.3	1.80	0.29	17.8	2.04	0.34	15.2
11	1.63	1.12	25.8	1.23	0.53	18.8	1.44	0.65	20.6
Avg	2.43	1.11	-	1.60	0.41	-	1.49	0.42	-

for evaluating autonomous driving algorithms. The dataset also accommodates challenging aspects such as different lighting, shadow conditions, and dynamic moving objects. In order to evaluate the performance of the VO approaches, RMSEs of measuring rotation/translation errors are computed from all possible sub-sequences of lengths (100, 200...800 m) as described in [14]. There is an evalution tool on their web page.

The detail error of 11 sections of dataset are shown in Table 1. For each approach, the Table displays the average rotation error r_e in degree/100 m, average translation in percentage (%) t_e and absolute error t_a in (m) of final frame compared to the ground-truth. The results of VISO2 library is named VISO2 and our proposed VO are named 'Backward' as well as 'Joint For-Backward', respectively, for two cases. This table indicates that the proposed approach achieves lower error for both translation and rotation, in general. Specifically, the rotation error of our approach using essential matrix is 0.46 deg/100 m while that of VISO2 is 1.1 deg/100 m; our translation error is 1.6% while that of VISO2 is 2.4%. That indicates around 30% translation error enhancement. The error of translation is further reduced to 1.49% by joint forward and backward translation estimation.That mean, the joint forward-backward estimation improve the translation accuracy 7%.

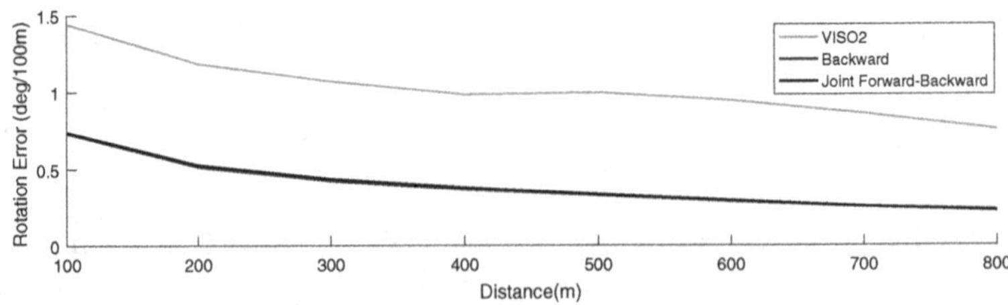

Fig. 2. Average rotation error along travel distance (Color figure online)

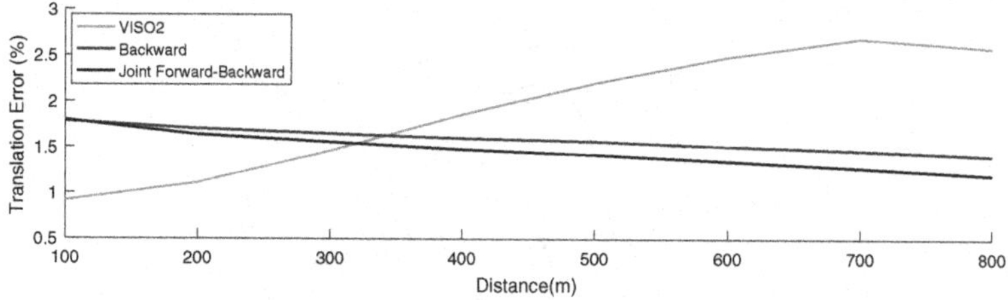

Fig. 3. Average translation error along travel distance

We also measure the transformation error along with the distance of travel from 100, 200,..., 800 m. The change of translation and rotation errors are shown in Fig. 2 and Fig. 3, respectively. For all travel distances, both rotation and translation errors of proposed approaches are smaller than those of VISO2. Specifically, Our translation error of backward estimation gradually reduces from 1.8% at 100 m to 1.5% at 800 m while the translation error of VISO2 increases monotonically from 1.0% at 100 m to 2.5% at 800 m. The change of joint forward-backward translation estimation in black is a little bit lower than that of backward estimation in blue. Consider the change of rotation error along the path length, our rotation errors are similar because we only focus on translation optimization. They gradually reduce from around 0.78 °/100 m at 100 m to around 0.23 °/100 m at 800 m. Similarly, The error of VISO2 reduces from 1.4 °/m at 100 m and 0.8 °/100 m at 800 m. However, at every travel distance 100, 200,..., 800 m. The error of the proposed approaches slower than that of VISO2.

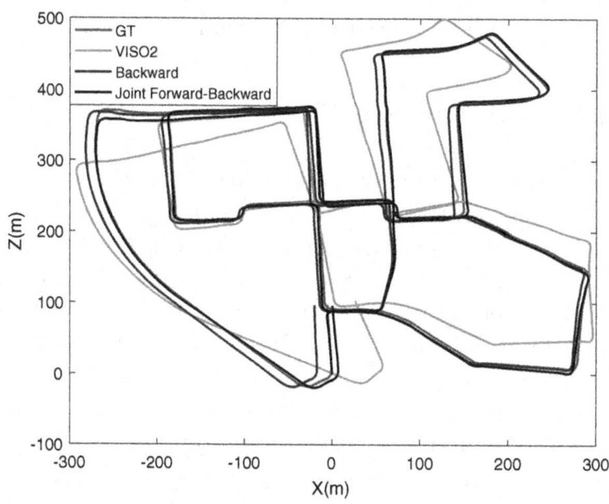

Fig. 4. Trajectory of Sect. 1 for three approaches compare to the ground-truth.

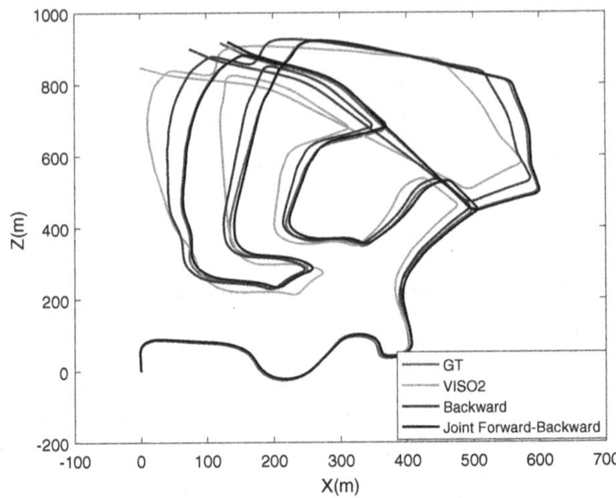

Fig. 5. Trajectory of Sect. 3 for three approaches compare to the ground-truth.

The accuracy improvement of our method compared to conventional approach VISO2 is confirmed by visualizing several camera trajectories in Sect. 1 and Sect. 3 in Fig. 4 and Fig. 5, respectively. It is clear that camera tracks of our approaches closer to the ground-truth than those of VISO2.

This evaluation has been done on an Intel Core i5-2400S CPU running at 2.5 Hz. The average single thread run-time per image for joint forward-backward translation estimation is 80 ms in total with 70 ms for rotation estimation and 10 ms for both forward and backward translation estimation. That means only forward or backward translation estimation spends on 5 ms.

5 Conclusion and Furture Work

An adaptive stereo visual odometry based on the essential matrix is presented by introducing the non-iterative translation estimation. The joint forward and backward translation estimation is proved to enhance performance. Compared to conventional methods that used *PnP* such as VISO2 library, the proposed method relies on the essential matrix estimation and the direct translation estimation by solving a linear system. The effectiveness of the proposed approach is verified by evaluating the errors in terms of translation and rotation on the KITTI dataset. The experimental result indicates that our approach achieves 1.6% and 1.49% with-out and with joint forward-backward translation estimation, respectively. In the future, we plan to widen the scope of applications utilizing the non-iterative translation estimation and refinement rotation estimation.

References

1. Nistér, D., Oleg ,N., James, B.: Visual odometry. In: Proceedings of the 2004 IEEE Computer Society Conference on Computer Vision and Pattern Recognition, 2004. CVPR 2004, vol. 1, pp. I-I. IEEE (2004)
2. Poddar, S., Kottath, R., Karar, V.: Motion estimation made easy: evolution and trends in visual odometry. In: Hassaballah, M., Hosny, K. (eds.) Recent Advances in Computer Vision. Studies in Computational Intelligence, vol. 804, pp. 305–331. Springer, Cham (2019). https://doi.org/10.1007/978-3-030-03000-1_13
3. Scaramuzza, D., Friedrich, F.: Visual odometry [tutorial]. IEEE Robot. Autom. Mag. **18**(4), 80–92 (2011)
4. Friedrich, F., Scaramuzza, D.: Visual odometry: Part II: matching, robustness, optimization, and applications. IEEE Robot. Autom. Mag. **19**(2), 78–90 (2012)
5. Nistér, D.: An efficient solution to the five-point relative pose problem. IEEE Trans. Pattern Anal. Mach. Intell. **26**(6), 756–770 (2004)
6. Arun, K.S., Huang, T.S., Blostein, S.D.: Least-squares fitting of two 3D point sets. IEEE Trans. Pattern Anal. Mach. Intell. **5**, 698–700 (1987)
7. Fischler, Martin A., Bolles, Robert C.: Random sample consensus: a paradigm for model fitting with applications to image analysis and automated cartography. Commun. ACM **24**(6), 381–395 (1981)
8. Longuet-Higgins, H.C.: A computer algorithm for reconstructing a scene from two projections. In: Fischler, M.A., Firschein, O. (eds.) Readings in Computer Vision: Issues, Problems, Principles, and Paradigms, pp. 61–62. Morgan Kaufmann Publishers Inc., San Francisco, CA, USA (1987)
9. Chli, M., Davison, A.J.: Active matching. In: Forsyth, D., Torr, P., Zisserman, A. (eds.) European Conference on Computer Vision – ECCV 2008. Lecture Notes in Computer Science, vol. 5302, pp. 72–85. Springer, Berlin, Heidelberg (2008). https://doi.org/10.1007/978-3-540-88682-2_7
10. Kitt, B., Geiger, A., Lategahn, H.: Visual odometry based on stereo image sequences with RANSAC-based outlier rejection scheme. In: 2010 IEEE on Intelligent Vehicles Symposium (IV), pp. 486–492. IEEE (2010)
11. Geiger, A., Ziegler, J., Stiller, C.: Stereoscan: dense 3D reconstruction in real-time. In: 2011 IEEE Intelligent Vehicles Symposium (IV). IEEE (2011)
12. Cvišić, I., Petrović, I.: Stereo odometry based on careful feature selection and tracking. In: 2015 European Conference on Mobile Robots (ECMR), pp. 1–6. IEEE (2015)
13. Nguyen, H.H., Lee, S.: Orthogonality index based optimal feature selection for visual Odometry. IEEE Access, p. 7 (2019)
14. Geiger, A., Lenz, P., Urtasun, R.: Are we ready for autonomous driving? The KITTI vision benchmark suite. In: 2012 IEEE Conference on Computer Vision and Pattern Recognition (CVPR). IEEE (2012)

A Modified Localization Technique for Pinpointing a Gunshot Event Using Acoustic Signals

Thin Cong Tran[1], My Ngoc Bui[1](✉), and Hoang Huy Nguyen[2]

[1] Institute of Electronics, Military Institute of Science and Technology, Hanoi, Vietnam
thin.vdt@outlook.com

[2] Department of Electromagnetic Compatibility, Faculty of Radio-Electronic Engineering, Le Quy Don Technical University, Hanoi, Vietnam

Abstract. This paper proposes a method for localizing a gunshot event using four acoustic sensor nodes mounted at the four corners of a rectangular working area. Each of these nodes involves three sensors to acquire acoustic signals of any gunshot inside the working area. The approach analyzes individual signals received by the nodes to identify sound events using false alarm probability and determine their emission directions exploiting a minimum mean square error estimator and the time difference of arrival of the events. The gunshot location is the quadrilateral center of four crossing points resulting from pairs of adjacent event emission directions. For evaluating the proposed method, a signal including ten real gunshots recorded by a nearby acoustic sensor is delayed and attenuated according to a theoretical wave propagation model to create various signal patterns, which simulates signals received by the installed sensor nodes. Furthermore, the Gaussian noise is added to the simulated signals to emulate the influence of wave propagation environment. This article also implements some mechanisms to compute the time difference of arrival for comparison. They are comprised of the first crossing of threshold and signal, maximum amplitude, Akaike's Information Criterion, and the cross-correlation function. Hence, one of them can be selected for a real application. Experimental results show that the proposed method achieves high accuracy of gunshot localization.

Keywords: Gunshot detection · Gunshot localization · Event detection · Signal detection · Signal processing

1 Introduction

Nowadays firearm violent crime frequently happens. However, relevant people would not be arrested easily if shooting incidents were not located quickly. Therefore, an automatic method of gunshot localization is really helpful in supporting the police to chase and catch criminals immediately.

A firing gun usually emits acoustic muzzle blast waves propagating spherically outwards with a speed of sound in the air [1]. Many papers utilized these signals to

N.-S. Vo and V.-P. Hoang (Eds.): INISCOM 2020, LNICST 334, pp. 138–149, 2020.
https://doi.org/10.1007/978-3-030-63083-6_11

localize gunshot using sound source localization techniques [2–6]. There are two common localization approaches, which use energy and time difference of arrival (TDOA) [7]. Although the energy-based localization technique can bring about high location accuracy, it is challenging to have a real wave energy propagation model as used in Ref. [7–9]. In contrast, the TDOA-based method is extremely simple because it can localize a sound source without a complex propagation model whose parameters depend on the environment as the energy-based one requires in Ref. [4, 7, 10–13]. Thus, this paper exploits the TDOA-based technique and acoustic signals to localize a gunshot event.

The novel point of the proposed method is that a sound source location is returned by minimizing mean square errors of a known TDOA map and a set of measured TDOA instead of using a direct solution [7, 11–16]. The direct solution is challenging due to complexity of mathematical equations and unavoidable TDOA errors could lead to no solution of those equations. For example, hyperbolic curves [13, 16] do not cross each other at the same point, thus resulting in no gunshot location. As a result, a minimum mean square error (MMSE) estimator is exploited to address the problem in this paper. First, the working area where the firearm situation is monitored is divided into small cells. Depending on acoustic sensor distribution and a specific sound speed in the air, a TDOA map is calculated for all the vertexes of the cells. Next, a MMSE estimator searches all over the cell vertexes for the most appropriate location whose TDOA least differs from measured TDOA resulting from acoustic signals.

Aside from implementing a MMSE estimator, this paper employs a signal detection technique based on false alarm probability [17] for detecting sound events in an acoustic signal. This procedure involves several signal processing stages. The first step estimates the noise level of signal. Next, detection thresholds are computed using the estimated noise level and a given false alarm probability; thus, they can adaptively vary with noise fluctuation. Afterwards, all the adjacent samples that exceed the thresholds are grouped into sound events. The last step eliminates false sound events with the predefined minimum values of amplitude and duration of true events. Hence, TDOA can be directly extracted from true sound events.

Additionally, to choose a suitable approach to the TDOA computation in real applications, the paper compares the effectiveness of various TDOA techniques while evaluating the proposed method.

For evaluating the proposed method, gunshot signals must be available. The next section therefore offers a way to simulate acoustic signals emitted by a gunshot.

2 Signal Simulation

Figure 1 (a) shows a rectangular working area $M_1M_2M_3M_4$ with the size of d x h ($d = h = 500$ m) and a coordinate system Oxy that is established by the rectangle's edges and vertexes. We arrange sensor nodes at the corners of the rectangle to receive signals propagating from any sound source S (x_S, y_S) in the area. A node is comprised of three acoustic sensors M_{ij} ($i = 1, 2, 3, 4$ and $j = 1, 2, 3$) and they are equivalently arranged around a circle with center M_i and a radius of $r = 5$ cm as illustrated in Fig. 1 (b). Hence, the coordinate of sensor M_{ij} is given by:

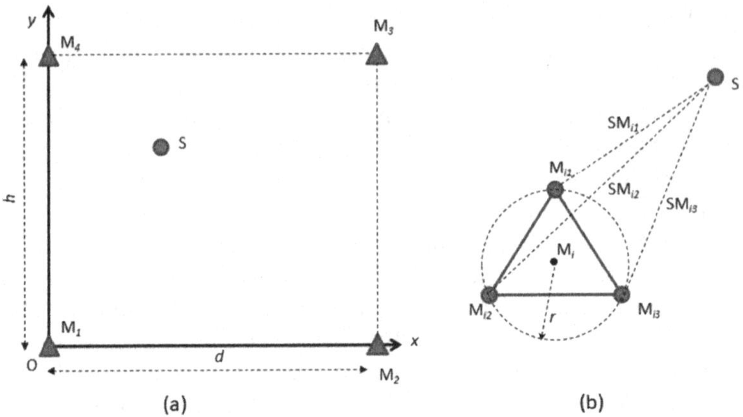

Fig. 1. Sensor arrangement: (a) working area, (b) a sensor node

$$x_{ij} = x_{M_i} - r\sin\left(\frac{2\pi(j-1)}{3}\right), \quad y_{ij} = y_{M_i} + r\cos\left(\frac{2\pi(j-1)}{3}\right) \tag{1}$$

where (x_{Mi}, y_{Mi}) is coordinate of M$_i$ and (x_{ij}, y_{ij}) is coordinate of M$_{ij}$ ($i = 1, 2, 3, 4$ and $j = 1, 2, 3$).

To simulate acoustic signals received by sensors, we exploit a wave propagation model [7, 8] to delay and attenuate a signal loaded from a record of gunshot signals emitted by an AK-47 rifle. As a result, a simulated signal of sensor M$_{ij}$ is given by:

$$z_{ij}(t) = G\frac{z_0\left(t - t_{ij} + \varepsilon_{ij}\right)}{SM_{ij}^{\alpha}} + n_{ij}, \quad t_{ij} = \frac{SM_{ij}}{C} \tag{2}$$

where t is time, $z_{ij}(t)$ and $z_0(t - t_{ij} + \varepsilon_{ij})$ are simulated and recorded signals respectively, G and α ($1 \le \alpha \le 2$) are ratios related to a gain of acquisition device and wave attenuation, t_{ij} is flight time of wave from source S to sensor M$_{ij}$, C is sound speed, ε_{ij} and n_{ij} are added Gaussian noise with means $\mu_\varepsilon = 0$, $\mu_n = 0$ and standard deviations $\sigma_\varepsilon \ge 0$, $\sigma_n \ge 0$, SM_{ij} is the wave propagation distance from source S to sensors M$_{ij}$, which is calculated as follows:

$$SM_{ij} = \sqrt{\left(x_S - x_{ij}\right)^2 + \left(y_S - y_{ij}\right)^2} \tag{3}$$

The two noises ε_{ij} and n_{ij} represent the influence of wave propagation environment such as reflection, diffraction and attenuation on both the arrival time and amplitude of signal, thus making simulated signals become roughly similar to real signals. Figure 2 depicts a gunshot signal source and a simulated signal, in which they are digitized by a sampling frequency of 8 kHz. It can be seen that the simulated signal is lagged behind the signal source, its amplitude and signal-to-noise ratio as well are smaller than the signal source's.

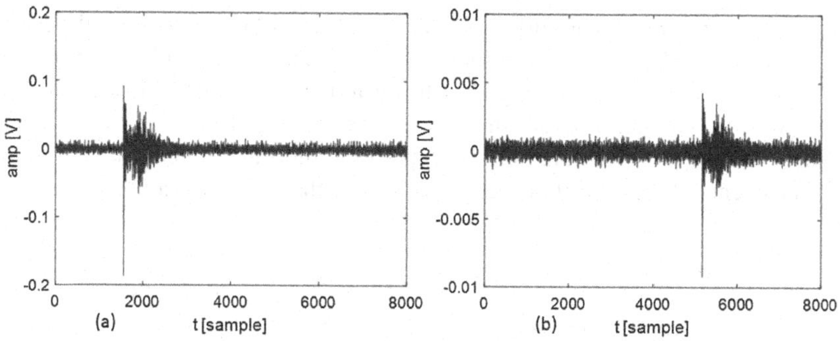

Fig. 2. a) Signal source, (b) simulated signal

3 Methodology

The proposed gunshot localization method comprises two main stages: event detection and source localization as illustrated in Fig. 3.

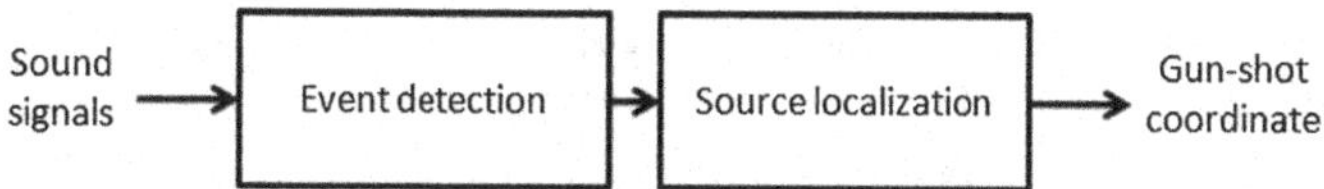

Fig. 3. The overall diagram of gunshot localization

3.1 Event Detection

A gunshot event is defined as a burst in an acoustic signal. To detect it, we exploit a Neyman-Pearson theorem of signal detection probability [17] to calculate a threshold using the following expression:

$$L(z) = \frac{p(z|Hp_1)}{p(z|Hp_0)} > \gamma, \quad p_{FA} = \int_{\{z:L(z)>\gamma\}} p(z|Hp_0) \tag{4}$$

where $L(z)$ is the likelihood ratio, Hp_0 is the signal absence hypothesis, Hp_1 is the signal presence hypothesis, z is an observed set, $p(z)$ is the probability density function, P_{FA} is the false alarm probability, and γ is a threshold. In our application, Hp_0 and Hp_1 are hypotheses of noise and gunshot event respectively and the likelihood ratio $L(z)$ is computed according to the signal amplitude.

Figure 4 illustrates a typical gunshot event with primary characteristic parameters: amplitude, duration, flight time. This event is separated from background noise by a positive threshold (the upper red dash line) or a negative threshold (the lower red dash line). The algorithm of event detection based on Eq. (4) is shown in Fig. 5. The entire process is composed of five phases. The first step estimates the noise background of an input acoustic signal to provide for the second step calculating the positive and negative thresholds. These detection thresholds are adaptive, which vary with the noise level.

The third step compares signal samples with the thresholds. Samples above the positive threshold or below the negative threshold are considered as candidate samples of a gunshot event. Gunshot events are found in the forth step by grouping adjacent candidate samples. Because the detection thresholds are resulting from a false alarm probability, there exist false events. Therefore, the last step (event filtering) is necessary to eliminate false sound events. A true gunshot event must satisfy the following condition:

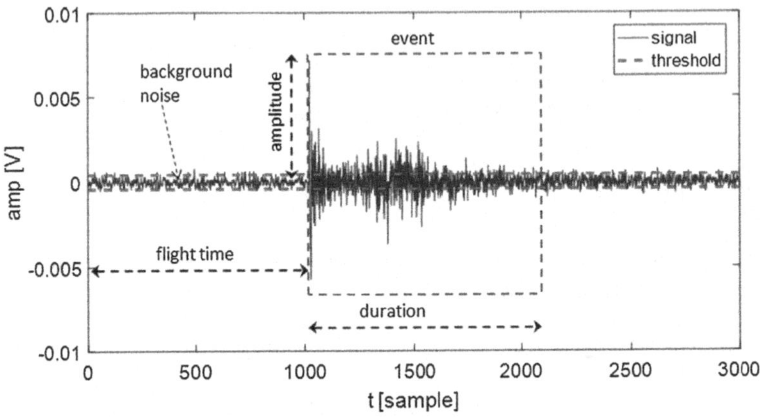

Fig. 4. A typical gunshot event

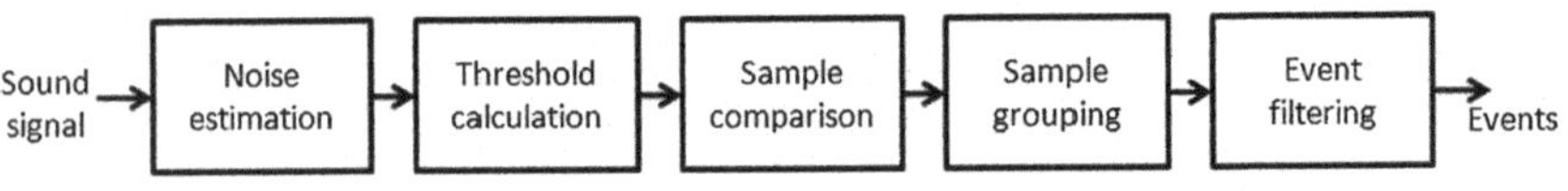

Fig. 5. Gunshot event detection in a sound signal

$$A \geq A_m \wedge N \geq N_m \tag{5}$$

where A and N are the average amplitude and duration of event, respectively, A_m and N_m are their minimum values.

3.2 Source Localization

The working area $M_1M_2M_3M_4$ is divided into cells with the size of $\Delta d \times \Delta h$ as illustrated in Fig. 6. For every vertex H_k (x_k, y_k) of cells, we compute TDOA as follows:

$$\Delta t_{ki} = \frac{\begin{bmatrix} H_kM_{i2} - H_kM_{i1} \\ H_kM_{i3} - H_kM_{i1} \end{bmatrix}}{C}, \quad H_kM_{ij} = \sqrt{\left(x_k - x_{Mij}\right)^2 + \left(y_k - y_{Mij}\right)^2} \tag{6}$$

where Δt_{ki} is an array of TDOA accounted for a vertex H_k and measuring points of a sensor node M_i, H_kM_{ij} is the distance between H_k and M_{ij} $(i = 1, 2, 3, 4; j = 1, 2, 3)$, C is sound speed.

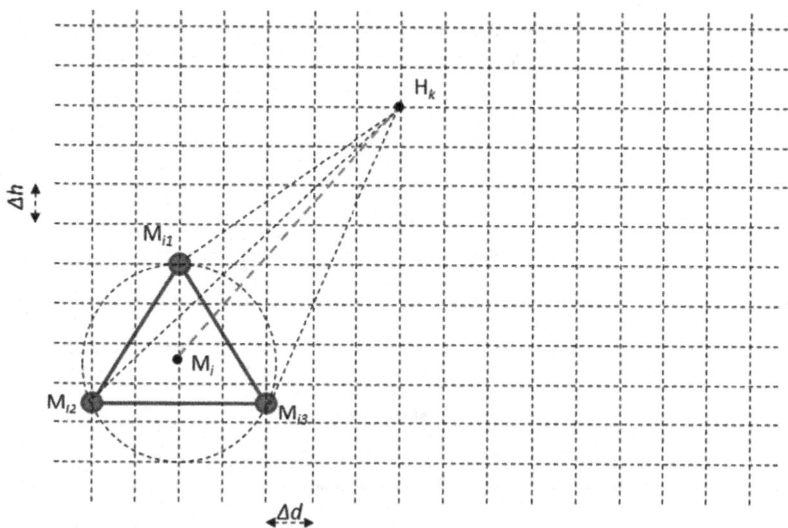

Fig. 6. Determination of gunshot emission direction using a sensor node M_i

It can be seen that TDOA depends on synchronization of signal acquisition between sensor channels. In other words, we hardly synchronize signals of faraway nodes due to complex hardware and software (we need external synchronous devices such as wireless transceivers, Ethernet cables, etc.) while we can simply do this for near sensors without external elements. Accordingly, we consider TDOA between sensors of individual nodes to improve TDOA accuracy as presented by Eq. (6).

Figure 7 shows determining the sound emission direction of a sensor node M_i ($i = 1, 2, 3, 4$). Sound events resulting from sensors M_{ij} ($j = 1, 2, 3$) are conducted to the event grouping. This block picks and groups neighboring events, relying on their arrival time via the following expression:

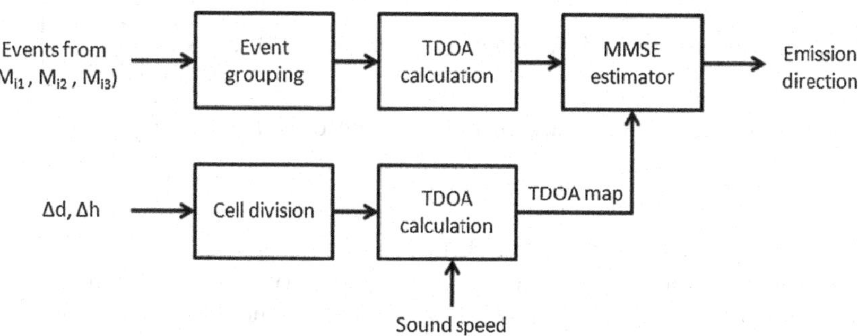

Fig. 7. Determination of sound emission direction

$$|t_2 - t_1| \leq \Delta_m \wedge |t_3 - t_1| \leq \Delta_m, \quad \Delta_m = \frac{M_3M_{12} - M_3M_{11}}{C} \tag{7}$$

where t_j ($j = 1, 2, 3$) is flight time of an event detected in a signal of the jth sensor, Δ_m is a possible maximum TDOA between neighboring events, which is obtained if the gunshot source is assumed to be at the node M_3 and Δ_m is TDOA accounted for sensors of the node M_1. Five event groups from the three sensors M_{1j} ($j = 1, 2, 3$) are indexed as in Fig. 8.

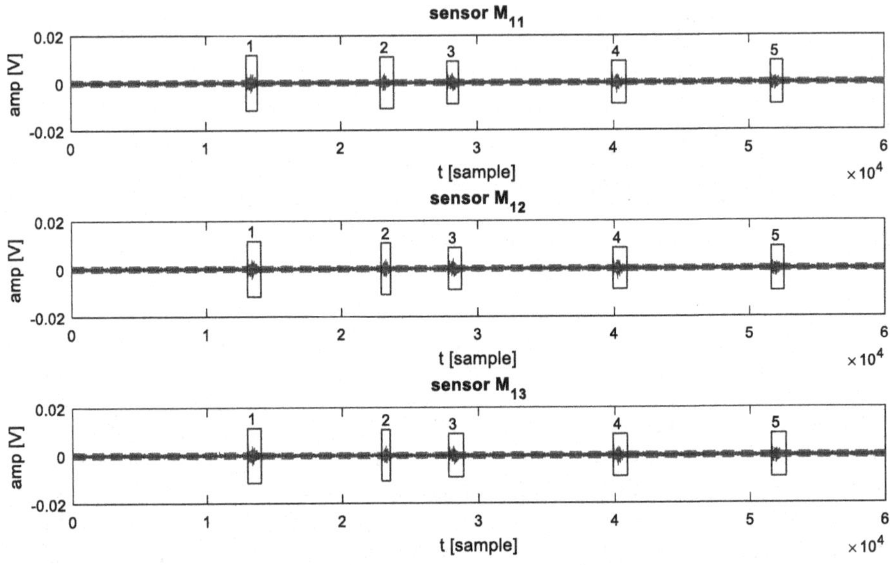

Fig. 8. Grouped events resulting from sensors M_{1j} ($j = 1, 2, 3$)

Based on the TDOA map in terms of cell division and the TDOA of events, we search for a satisfactory vertex H_i (x_i, y_i) against a sensor node M_i ($i = 1, 2, 3,4$) via a MMSE estimator. The MMSE estimator is constructed as follows:

$$\mathrm{H}_i \equiv \arg\min_{\mathrm{H}_k} MSE, \quad MSE = (\Delta - \Delta t_{ki})(\Delta - \Delta t_{ki})^T, \quad \Delta = \begin{bmatrix} t_2 - t_1 \\ t_3 - t_1 \end{bmatrix} \tag{8}$$

Afterwards, a sound emission direction for a sensor node is given by:

$$v_i = \left(x_i - x_{M_i}, y_i - y_{M_i}\right) \tag{9}$$

where v_i is a direction vector, $i = 1, 2, 3, 4$.

With pairs of vectors (v_1, v_2), (v_2, v_3), (v_3, v_4), and (v_4, v_1), we determine crosses S_1, S_2, S_3, and S_4, respectively. Finally, we achieve a source location S_e which is a quadrilateral center of $S_1S_2S_3S_4$, as illustrated in Fig. 9.

$$x_{S_e} = \frac{1}{4}\sum_{i=1}^{4} x_{S_i}, y_{S_e} = \frac{1}{4}\sum_{i=1}^{4} y_{S_i} \tag{10}$$

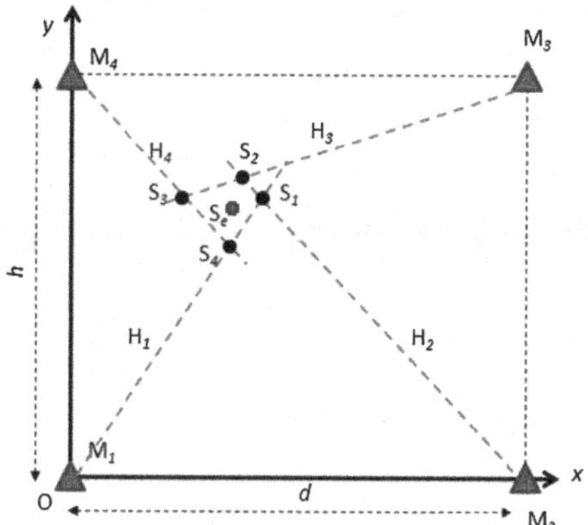

Fig. 9. Gunshot location estimation

4 Experimental Results

To evaluate the proposed method, we initially recorded signals emitted by ten real gunshots using only one nearby sensor as illustrated in Fig. 10. We subsequently simulated signals at sensors M_{ij} ($i = 1, 2, 3, 4; j = 1, 2, 3$) for the four gunshot source locations P_1 (200, 350), P_2 (400, 390), P_3 (340, 150), and P_4 (142, 266) based on Eq. (2) in which $G = 100$, $\alpha = 1$, $C = 340$ m/s (a specific sound speed in the air at 20 °C). We examined two cases: no noise when $\sigma_\varepsilon = 0$, $\sigma_n = 0$ and added noise when $\sigma_\varepsilon = 5$ ms (this results in a statistical location error $\sigma_d = \text{C} \times \sigma_\varepsilon = 1.7$ m), $\sigma_n = 0.3\sigma$ (σ is standard deviation of background noise of the recorded signal). Because the added noises randomly vary with the normal probability density function, we can create diverse signal patterns to trial the proposed method. Besides, we practically chose parameters $A_m = 1.5\sigma$, $N_m = 200$ samples to remove unwanted events using Eq. (5). Finally, the selected cell size is $\Delta d = \Delta h = 2$ m for dividing the working area into cells. Both the simulation and evaluation were implemented in Matlab software.

For a simulation (a gunshot at a location), our approach turned out an estimated location corresponding to a gunshot location as shown in Fig. 11. Additionally, we also implemented four common mechanisms determining TDOA in a trial. The first technique determines the arrival time of signal through the first crossing of threshold and signal (CRS); the second one defines the arrival time of a signal as the position of maximum amplitude (PAK); the third one estimates the arrival time using Akaike's Information Criterion (AIC) [18–21]; the last one directly exploits the cross-correlation function to compute TDOA between two signals (CCR) [4, 22]. Relying on the comparison between their location error averages of the ten gunshots, thus specifying which TDOA determination technique is appropriate for localizing gunshots.

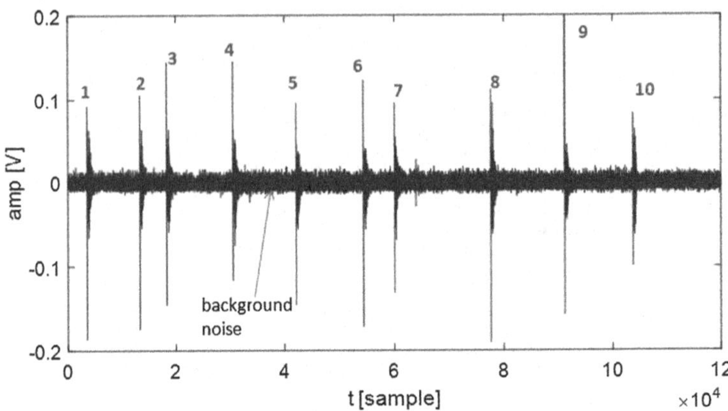

Fig. 10. An acoustic signal containing ten real gunshot events (**1, 2, …, 10**)

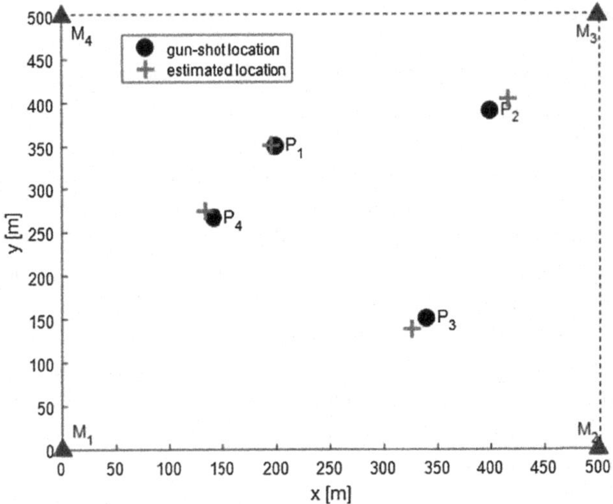

Fig. 11. Gunshots and estimated locations

Tables 1 and 2 show errors of gunshot locations returned by various TDOA techniques in two cases (no noise and added noise). Those are averages of distance difference of the ten gunshots and estimated locations as follows:

$$\delta = \frac{1}{10}\sum_{m=1}^{10}\sqrt{(x_P - x_{em})^2 + (y_P - y_{em})^2} \quad , \quad \delta_r = 100\text{x}\frac{\delta}{\max(d, h)} \tag{11}$$

where (x_P, y_P) is a coordinate of locations P_1, P_2, P_3, and P_4, (x_{em}, y_{em}) is the estimated location coordinate of the mth gunshot ($m = 1, 2, \ldots, 10$), and δ is the average location error, δ_r is a proportion of δ to distance between two sensor nodes.

Table 1. The average location error δ (m) and relative error δ_r (%) using various TDOA techniques without noise

Location	CRS	PAK	AIC	CCR
P1	1.9 (0.4)	1.0 (0.2)	1.0 (0.2)	1.0 (0.2)
P2	10.0 (2.0)	9.3 (1.9)	9.3 (1.9)	9.3 (1.9)
P3	4.9 (1.0)	4.6 (0.9)	4.6 (0.9)	4.6 (0.9)
P4	13.2 (2.7)	13.6 (2.7)	13.7 (2.7)	13.6 (2.7)
Mean	7.5 (1.5)	7.1 (1.4)	7.2 (1.4)	7.1 (1.4)

Table 2. The average location error δ (m) and relative error δ_r (%) using various TDOA techniques with noise adding

Location	CRS	PAK	AIC	CCR
P1	68.1 (13.6)	1.0 (0.2)	2.7 (0.5)	1.0 (0.2)
P2	11.3 (2.3)	9.3 (1.9)	9.1 (1.8)	9.3 (1.9)
P3	23.3 (4.7)	5.5 (1.1)	6.8 (1.4)	4.6 (0.9)
P4	31.3 (6.3)	13.6 (2.7)	13.0 (2.6)	13.6 (2.7)
Mean	33.5 (6.7)	7.4 (1.5)	7.9 (1.6)	7.1 (1.4)

It can be seen that if we do not add noise to acoustic signals, all the TDOA techniques turn out similar location accuracy (location error is roughly 7.5 m, corresponding to a relative error of 1.5%). Conversely, their effectiveness is different in the case of noise adding. The CRS method demonstrates the biggest error with an average location error of 33.5 m (relative error 6.7%). This is resulting from the noise presence in acoustic signals. Because the noise fluctuation distorts these signals, the first crossing points of the signals and their detection thresholds are not true arrival times. Although the AIC method (the average location error $\delta = 7.9$ m, relative error $\delta_r = 1.6\%$) improves the location accuracy compared with the CRS method, its error is still bigger than the PAK and CCR methods. The PAK and CCR methods bring about the highest location accuracies in which the CCR method (location error $\delta = 7.1$ m, $\delta_r = 1.4\%$) is slightly better than the PAK one (location error $\delta = 7.4$ m, $\delta_r = 1.5\%$). Obviously, the noise adding does not much influence on the performance of PAK, AIC and CCR methods. In other words, the AIC and CCR methods necessitate more computation than the CRS and PAK ones because they are comprised of extensive computing regarding Akaike's Information Criterion and the cross-correlation function. Hence, we claim that the PAK method is appropriate for localizing gunshots in our evaluation.

5 Conclusions

The paper introduces a modified method localizing a gunshot event based on the time difference of arrival of acoustic signals. Twelve acoustic sensors are arranged in four nodes (each of them comprises three sensors). Sound events are detected in individual signals of nodes through a false alarm probability. Then, the direction of a gunshot event is found with the assistance of a minimum mean square error estimator using the time difference of arrival between acoustic signals acquired by sensors of a node. The sound source location is the center of tetrahedron which is created by four crossing points of four pairs of adjacent emission directions from sensor nodes to the sound event. To evaluate the proposed method, the paper manipulates a signal record of real gunshots to simulate acoustic signals received by sensors relying on a wave propagation model. Furthermore, the article determines the time difference of arrival in various mechanisms to select the best one. The result evaluation for four gunshot locations inside a rectangular working area with size 500 m $\times$ 500 m reveals that the proposed gunshot localization method can turn out high location accuracy and its performance is not much affected by noise (the location errors are 7.1 m (1.4%) and 7.4 m (1.5%) for noise absence and noise adding respectively.)

References

1. Maher, R.C.: Acoustical Characterization of Gunshots. 5
2. Nimmy, P., Nair, K.R., Murali, R., Rajesh, K.R., Nimmy, M., Vishnu, S.: Analysis of acoustic signatures of small firearms for Gun Shot localization. In: 2016 IEEE 13th International Conference on Signal Processing (ICSP). pp. 228–231. IEEE, Chengdu, China (2016). https://doi.org/10.1109/ICSP.2016.7877829
3. Lopez-Morillas, J., Canadas-Quesada, F.J., Vera-Candeas, P., Ruiz-Reyes, N., Mata-Campos, R., Montiel-Zafra, V.: Gunshot detection and localization based on Non-negative Matrix Factorization and SRP-hat. In: 2016 IEEE Sensor Array and Multichannel Signal Processing Workshop (SAM). pp. 1–5. IEEE, Rio de Janerio, Brazil (2016). https://doi.org/10.1109/SAM.2016.7569648
4. Jain, P., Orhan, A., Wright, O.: Point-Source Localization of a Gunshot Group 9, 18-551, Fall 2009. 26
5. Akman, Ç., Sönmez, T., Özuğur, Ö., Başlı, A.B., Leblebicioğlu, M.K.: Sensor fusion, sensitivity analysis and calibration in shooter localization systems. Sens. Actuators, A **271**, 66–75 (2018). https://doi.org/10.1016/j.sna.2017.12.042
6. Bandi, A.K., Rizkalla, M., Salama, P.: A novel approach for the detection of gunshot events using sound source localization techniques. In: 2012 IEEE 55th International Midwest Symposium on Circuits and Systems (MWSCAS). pp. 494–497. IEEE, Boise, ID, USA (2012). https://doi.org/10.1109/MWSCAS.2012.6292065
7. Cobos, M., Antonacci, F., Alexandridis, A., Mouchtaris, A., Lee, B.: A survey of sound source localization methods in wireless acoustic sensor networks. Wireless Commun. Mob. Comput. **2017**, 1–24 (2017). https://doi.org/10.1155/2017/3956282
8. Deng, F., Guan, S., Yue, X., Gu, X., Chen, J., Lv, J., Li, J.: Energy-based sound source localization with low power consumption in wireless sensor networks. IEEE Trans. Ind. Electron. **64**, 4894–4902 (2017). https://doi.org/10.1109/TIE.2017.2652394

9. Dranka, E., Coelho, R.F.: Robust maximum likelihood acoustic energy based source localization in correlated noisy sensing environments. IEEE J. Sel. Top. Signal Process. **9**, 259–267 (2015). https://doi.org/10.1109/JSTSP.2014.2385657
10. Alameda-Pineda, X., Horaud, R.: A geometric approach to sound source localization from time-delay estimates. IEEE/ACM Trans. Audio Speech Lang. Process. **22**, 1082–1095 (2014). https://doi.org/10.1109/TASLP.2014.2317989
11. Gillette, M.D., Silverman, H.F.: A linear closed-form algorithm for source localization from time-differences of arrival. IEEE Signal Process. Lett. **15**, 1–4 (2008). https://doi.org/10.1109/LSP.2007.910324
12. Kundu, T.: Acoustic source localization. Ultrasonics **54**, 25–38 (2014). https://doi.org/10.1016/j.ultras.2013.06.009
13. Jin, B., Xiaosu, X., Zhang, T.: Robust time-difference-of-arrival (TDOA) localization using weighted least squares with cone tangent plane constraint. Sensors **18**, 778 (2018). https://doi.org/10.3390/s18030778
14. Gaber, M., Elnady, T., Elsabbagh, A.: Sound Source Localization in 360 Degrees Using a Circular Microphone Array. Conference Proceedings. 8 (2018)
15. Tiete, J., Domínguez, F., Silva, B., Segers, L., Steenhaut, K., Touhafi, A.: SoundCompass: a distributed mems microphone array-based sensor for sound source localization. Sensors **14**, 1918–1949 (2014). https://doi.org/10.3390/s140201918
16. Cobos, M., Marti, A., Lopez, J.J.: A modified SRP-PHAT functional for robust real-time sound source localization with scalable spatial sampling. IEEE Signal Process. Lett. **18**, 71–74 (2011). https://doi.org/10.1109/LSP.2010.2091502
17. Steven, M.: Kay: Fundamentals of Statistical Signal Processing. Prentice Hall PTR. (1998)
18. Zhou, Z., Cheng, R., Rui, Y., Zhou, J., Wang, H.: An improved automatic picking method for arrival time of acoustic emission signals. IEEE Access. **7**, 75568–75576 (2019). https://doi.org/10.1109/ACCESS.2019.2921650
19. Li, C., Huang, L., Duric, N., Zhang, H., Rowe, C.: An improved automatic time-of-flight picker for medical ultrasound tomography. Ultrasonics **49**, 61–72 (2009). https://doi.org/10.1016/j.ultras.2008.05.005
20. Shang, X., Li, X., Morales-Esteban, A., Dong, L.: Enhancing micro-seismic P-phase arrival picking: EMD-cosine function-based denoising with an application to the AIC picker. J. Appl. Geophys. **150**, 325–337 (2018). https://doi.org/10.1016/j.jappgeo.2017.09.012
21. Pinal Moctezuma, F., Delgado Prieto, M., Romeral Martinez, L.: Performance analysis of acoustic emission hit detection methods using time features. IEEE Access. **7**, 71119–71130 (2019). https://doi.org/10.1109/ACCESS.2019.2919224
22. Pavlidi, D., Griffin, A., Puigt, M., Mouchtaris, A.: Real-time multiple sound source localization and counting using a circular microphone array. IEEE Trans. Audio Speech Lang. Process. **21**, 2193–2206 (2013). https://doi.org/10.1109/TASL.2013.2272524

Table Structure Recognition in Scanned Images Using a Clustering Method

Nam Van Nguyen[1(✉)], Hanh Vu[2], Arthur Zucker[3], Younes Belkada[3], Hai Van Do[1], Doanh Ngoc- Nguyen[1], Thanh Tuan Nguyen Le[1], and Dong Van Hoang[1]

[1] Thuyloi University, 175 Tay Son, Dong Da, Hanoi, Vietnam
nvnam@tlu.edu.vn

[2] Viettel CyberSpace Center, 41st floor, Keangnam Landmark 72, Hanoi, Vietnam

[3] Sorbonne University, Polytech Sorbonne 75005, Paris, France

Abstract. Optical Character Recognition (OCR) for scanned paper invoices is very challenging due to the variability of 19 invoice layouts, different information fields, large data tables, and low scanning quality. In this case, table structure recognition is a critical task in which all rows, columns, and cells must be accurately positioned and extracted. Existing methods such as DeepDeSRT, TableNet only dealt with high-quality born-digital images (e.g., PDF) with low noise and apparent table structure. This paper proposes an efficient method called CluSTi (Clustering method for recognition of the Structure of Tables in invoice scanned Images). The contributions of CluSTi are three-fold. Firstly, it removes heavy noises in the table images using a clustering algorithm. Secondly, it extracts all text boxes using state-of-the-art text recognition. Thirdly, based on the horizontal and vertical clustering algorithm with optimized parameters, CluSTi groups the text boxes into their correct rows and columns, respectively. The method was evaluated on three datasets: i) 397 public scanned images; ii) 193 PDF document images from ICDAR 2013 competition dataset; and iii) 281 PDF document images from ICDAR 2019's numeric tables. The evaluation results showed that CluSTi achieved an F_1-*score* of 87.5%, 98.5%, and 94.5%, respectively. Our method also outperformed DeepDeSRT with an F_1-*score* of 91.44% on only 34 images from the ICDAR 2013 competition dataset. To the best of our knowledge, CluSTi is the first method to tackle the table structure recognition problem on scanned images.

Keywords: Table structure recognition · Object recognition · Clustering method

1 Introduction

Our paper aimed to recognize the table's structure from scanned images of invoices. Data tables are the main content of the documents, especially invoices.

N.-S. Vo and V.-P. Hoang (Eds.): INISCOM 2020, LNICST 334, pp. 150–162, 2020.
https://doi.org/10.1007/978-3-030-63083-6_12

Direct application of OCR techniques to the whole data table has been impossible since the recognized texts do not follow precisely the original table structure, which led to the recognition results on large images for most OCR techniques were not highly accurate, especially for tables with many data items [11]. Therefore, the table structure in scanned images has to be recognized so that each table cell can be correctly located and individually processed using OCR techniques. However, this has never been a trivial task because of the different shapes, sizes, and colors of the cell separators. In addition, canned invoice images are typically noisy, which can make these separations become blurred or even lost. Besides, cells must be aligned to rows and columns, this alignment nevertheless can easily be biased in specific images due to the noise. Most existing table structure recognition methods dealt with relatively clean table images, such as PDF document images [3,10,15–17,22], the recognition results however were not highly accurate due to the table complexity. Hence, those methods would not efficient to apply to noisy scanned invoice images.

In this paper, we proposed CluSTi, an efficient approach for table structure recognition in scanned invoice images, which is mainly based on clustering algorithms. CluSTi considers an invoice table as a set of text boxes, which are sorted by certain vertical and horizontal orders. CluSTi firstly uses Character Region Awareness for Text Detection (CRAFT), a semantic segmentation method, for text boxes detection in an image [2]. Given the coordinates of the text boxes, CluSTi then recognizes the correct cell row and column using Density-Based Spatial Clustering of Applications with Noise (DBSCAN) clustering algorithm [5]. Our method was evaluated with 397 public scanned table images, 193 document images from ICDAR 2013 competition, and 281 document images from ICDAR 2019 competition. The achieved $\mathbf{F_1}$**-score** were 87.5%, 98.5%, and 94.5%, respectively, which outperformed the accuracy of existing methods.

The rest of the paper is organized as follows. Section 2 presents in detail about our CluSTi method. In Sect. 3, we evaluated the results of our method using three public datasets. Finally, Sect. 4 concludes on our work and the perspectives for the future.

2 Methods

Existing table structure recognition methods can be divided into two groups: *top-down* and *bottom-up* methods. Herein, we proposed an efficient *bottom-up* approach for table structure recognition named CluSTi.

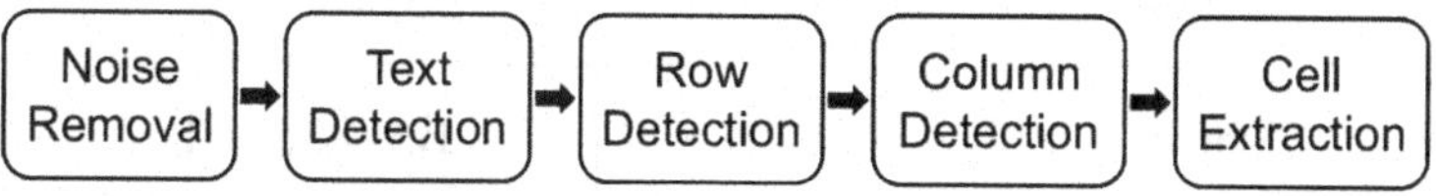

Fig. 1. Block diagram of CluSTi recognition process.

Basically, we applied and optimized a clustering technique at every steps of our recognition process. As shown in Fig. 1, CluSTi includes the following steps: i) Firstly, in the Noise Removal, we applied the DBSCAN clustering technique to clean the table images; ii) In the Text Detection, textual table elements are extracted from the images based on an object recognition deep learning model; iii) In the Row Detection, textual elements are horizontally regrouped using the above clustering technique of which the parameters was optimized; iv) Similarly, in the Column Detection, textual elements are vertically clustered with optimized parameters; v) Finally, in the Cell Reconstruction, the whole table structure are reconstructed cell by cell. The whole CluSTi process is demonstrated in Fig. 2. After Row detection step, all of the text boxes in the same rows are marked with the number on the top of the boxes, denoting the sequence number (i.e., 0, 1, 2, etc.) of their correct rows. Next, after Column detection step, all of the text boxes are labelled with the number on the top of the boxes, denoting the correct sequence number of their corresponding columns. The empty cells are also filled with blank text boxes. Then, after Cell reconstruction step, all of text boxes with the same sequence number of row and column are merged together to form the entire cells.

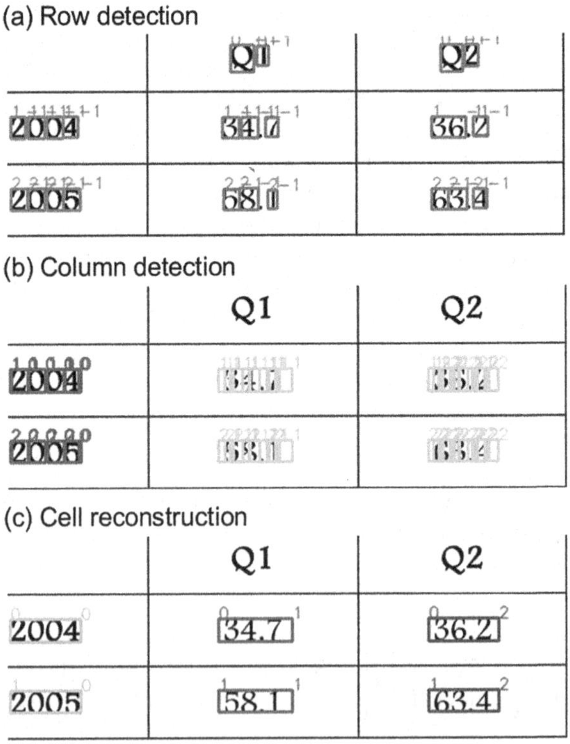

(a) Row detection

	Q1	Q2
2004	34.7	36.2
2005	58.1	63.4

(b) Column detection

	Q1	Q2
2004	34.7	36.2
2005	58.1	63.4

(c) Cell reconstruction

	Q1	Q2
2004	34.7	36.2
2005	58.1	63.4

Fig. 2. Overall description of CluSTi recognition process. (a) Row detection. (b) Column detection. (c) Cell reconstruction.

2.1 Noise Removal

In scanned images, noise is defined as the image's elements which can bias the text recognition [6]. From our available scanned images, we observed that the characters are the clusters containing a high number of neighboring pixels. In contrast, noise is the disjointed clusters of several pixels. Therefore, to remove noise from a scanned image, we relied on DBSCAN, which can segment the low and high density clusters. DBSCAN, which was introduced by Ester *et al.* (1996), groups the data points and their closest neighbors into clusters, and marks the lonely points into low-density region as the outliers [5]. The input parameters of DBSCAN include ϵ and *min_samples*. ϵ (eps) corresponds to the upper limit of the distance between two neighbors in a cluster, and *min_samples* corresponds the minimum number of points in a cluster. DBSCAN starts by choosing a random point, then it checks a nearest point (i.e., neighbor) in a circle of radius ϵ. The neighbors found are added into the group and the process continues with these new members of the group. If there is no more neighbors are found, and the number of group's members is greater or equal to the *min_samples*, then the group becomes a cluster. Otherwise, the group's points are marked as the outliers. DBSCAN is therefore suitable for clustering the texts since the characters are written one after the other. Moreover, this method can also be used to remove noise (or outliers) in the text images [5,24].

After applying DBSCAN with ϵ equals to 1 (i.e., pixel) and *min_samples* equals to the number of pixels of the smallest character in that specific language (e.g., Japanese), most of the noise is marked as the outliers and removed, as presented in Fig. 3.

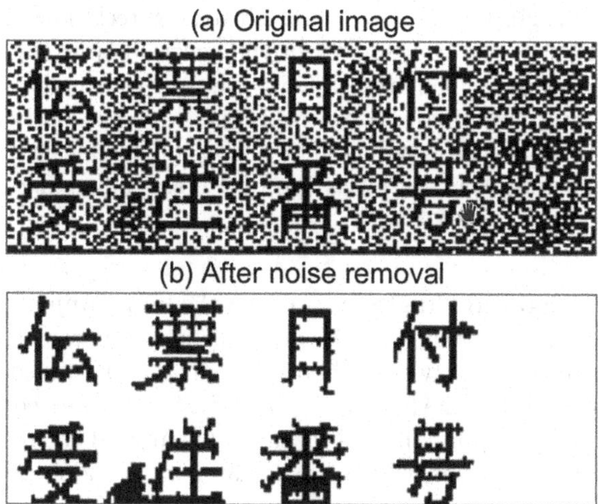

Fig. 3. An example image after applying DBSCAN for noise removal.

2.2 Text Detection

Recently, many deep learning scene text detectors have been proposed, and developed their applications in various fields [2,4,7–9,13,25]. Efficient methods have usually been inherited from object detection and semantic segmentation models such as Faster Region-based Convolutional Neural Network (Faster R-CNN) [18], Single Shot Multibox Detector (SSD) [12] and FCN [14].

CRAFT has been the best among current text detection methods thanks to its convolutional neural networks yielding the region score and affinity score [2]. Specifically, CRAFT detects character regions and links them to a text instance. This method is thus efficient for detecting any character including tiny, extremely long, curved, rotated and arbitrarily shaped characters. By applying CRAFT on the noiseless table images, we aimed to detect the texts as much as possible. Hence, CRAFT is configured to detect only character regions without linkage between them. We also fine-tuned the CRAFT's magnifier and bounding box parameters so that the small characters can be recognized, and there is limited white space in the character bounding regions.

2.3 Row Detection

In the previous step, we bounded every textual elements in the image with distinct rectangle boxes. Based on the coordinates of these character bounding regions, we then grouped them into their corresponding rows, and determined the number of rows in the table images using the following horizontal clustering algorithm.

Horizontal Clustering. The horizontal clustering technique is described in **Algorithm 1**. Firstly, the coordinates of the centroids (i.e., (x_c, y_c)) of every detected text boxes are calculated. Then, they are normalized according to the x-axis. Finally, the normalized centroids (i.e., (x_n, y_n)) are clustered using DBSCAN with optimized parameters. The output of the horizontal clustering is the correct number of rows, as well as the text boxes belonging to each row of the table images.

Fine-Tuning Horizontal Clustering. Given the *min_samples* parameter, the accuracy of horizontal clustering heavily depends on its ϵ parameter, which is the maximum distance between two neighboring centroids processed by the **Algorithm 1**. We assumed that the height (i.e., H) of table rows are equivalent. Thus, the ϵ parameter can be approximated to any value around the median height of the character bounding boxes, which is calculated as in the **Algorithm 2**.

However, there are cases where rows may include multiple lines. We therefore proposed a probing algorithm to find an appropriate ϵ parameter around the median height, which is calculated in **Algorithm 2**. We represented $f_r(\epsilon)$ as the function of the number of rows, r, found by the horizontal clustering where ϵ parameter ranging from $(0.1 * median\_height)$ to $(1 * median\_height)$. $f_r(\epsilon)$ is

Algorithm 1. *Horizontal Clustering Algorithm*

Data: $N \leftarrow$ the total number of character bounding boxes
$\{(x^i_{min}, y^i_{min}), (x^i_{max}, y^i_{max})\} \leftarrow$ the coordinates of the upper-left and the lower-right corners of the $i^{th}, \forall i \in [1; N]$ character bounding boxes
$min\_samples \leftarrow$ the number of pixels of the smallest character in a specific language
$\epsilon \leftarrow$ to be fine-tuned
Result: Number of rows
$i \leftarrow 1$
while $i \leq N$ **do**
  $x^i_c = (x^i_{min} + x^i_{max})/2$
  $y^i_c = (y^i_{min} + y^i_{max})/2$
  $x^i_n = 0$
  $y^i_n = y^i_c$
  $i \leftarrow i + 1$
end
$num\_clusters$ = DBSCAN((x^i_n, y^i_n), ϵ, $min\_samples$)

Algorithm 2. *Median Height Calculation Algorithm*

Data: $N \leftarrow$ the total number of character bounding boxes
$(y^i_{min}, y^i_{max}) \leftarrow$ the upper-left and lower-right y-coordinate of the $i^{th}, \forall i \in [1; N]$ bounding box
Results: ϵ as the median height
$H_i = |y^i_{max} - y^i_{min}|$;
$Sort(H_i, \forall i \in [1; N])$;

$$\epsilon = \begin{cases} H_{(\frac{N}{2})} & \text{if } N \text{ is odd} \\ \left(H_{(\frac{N}{2})} + H_{(\frac{N}{2}+1)}\right)/2 & \text{if } N \text{ is even} \end{cases} \quad (1)$$

then calculated as in the **Algorithm 3**. Next, the density distribution of $f_r(\epsilon)$ for each table image was plotted. Then, we applied a peak detection algorithm [21] on the density distribution to find the best ϵ parameter for the horizontal clustering.

2.4 Column Detection

In the column detection process, we based on the following vertical clustering algorithm to calculate the number of columns, as well as to group the detected text boxes into their corresponding columns.

Vertical Clustering. Algorithm 4 describes our vertical clustering algorithm.

Algorithm 3. *Probing Algorithm for Horizontal Clustering*

Data: median height from **Algorithm 2**
Result: $f_r(\epsilon)$
$\epsilon \leftarrow$ median height
$k \leftarrow 1$
while $k \geq 0.1$ **do**
  $\epsilon = k * \epsilon$
  $num\_clusters$ = HorizontalClustering($min\_samples, \epsilon$) $k \leftarrow k - 0.01$
end

Algorithm 4. *Vertical Clustering Algorithm*

Data: $N \leftarrow$ the total number of character bounding boxes
$\{(x^i_{min}, y^i_{min}), (x^i_{max}, y^i_{max})\} \leftarrow$ the coordinates of the upper-left and the lower-right corners of the $i^{th}, \forall i \in [1; N]$ character bounding boxes
$min\_samples \leftarrow$ the number of rows found in the previous step
$\epsilon \leftarrow$ to be fine-tuned
Result: Number of columns $i \leftarrow 1$
while $i \leq N$ **do**
  $x^i_c = (x^i_{min} + x^i_{max})/2$
  $y^i_c = (y^i_{min} + y^i_{max})/2$
  $x^i_n = x^i_c$
  $y^i_n = 0$
  $i \leftarrow i + 1$
end
$num\_clusters$ = DBSCAN((x^i_n, y^i_n), ϵ, $min\_samples$)

Fine-Tuning Vertical Clustering. Since the *min_samples* parameter is fixed to the number of rows found from the previous step, we tried to find the best ϵ parameter for our vertical clustering algorithm. We noticed that the width of columns in the table images are not equivalent as in the case of row's height. We therefore proposed another technique to probe for the converged value of ϵ parameter. We represented $f_c(\epsilon)$ as the function of the number of clusters, c, found by the above vertical clustering with regard to ϵ. $f_c(\epsilon)$ is then calculated as in the **Algorithm 5**. The curvature of a continuous $f_c(\epsilon)$ can be defined as follows [20]:

$$K_{f_c}(\epsilon) = \frac{f''_c(\epsilon)}{(1 + f'_c(\epsilon)^2)^{\frac{3}{2}}} \tag{2}$$

Furthermore, there exist a critical point in this curve called knee, where the curvature is a local maximum [20]. We observed that this point gives the best accuracy for column detection in scanned images. An example of knee point is shown in Fig. 4, where the point (40,10) is the knee point of the curve.

In this case, since $f_c(\epsilon)$ is a discrete function, the knee point can be detected using Kneedle algorithm [20]. Kneedle alculates the distance from all discrete points to the straight segment formed by the first and the last point of the curve. Local maxima (or knees) are considered as the points of the curve which

Algorithm 5. *Probing Algorithm for Vertical Clustering*

Data: $\epsilon_{min}, \epsilon_{max}$
Result: $f_c(\epsilon)$
$\epsilon \leftarrow \epsilon_{min}$
while $\epsilon \leq \epsilon_{max}$ **do**
| $num\_clusters$ = VerticalClustering($min\_samples, \epsilon$) $\epsilon \leftarrow \epsilon + 1$
end

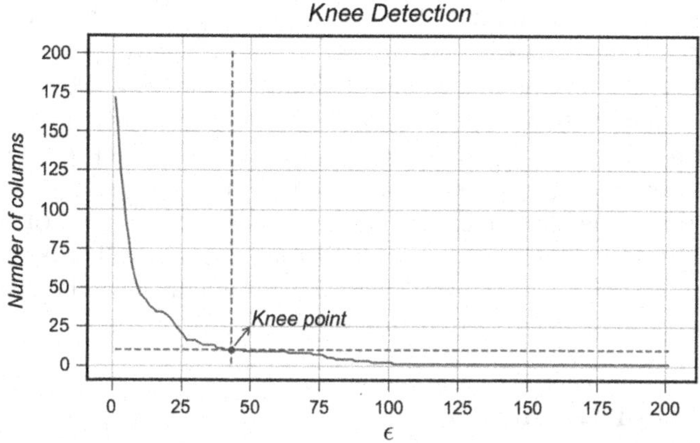

Fig. 4. The curve representing the dependence of the number of columns on ϵ parameter for DBSCAN and its corresponding knee point.

are the most distant to this straight segment. Knees can be detected faster or slower depending on a predefined sensitivity parameter S [20]. This is a measure of how the required number of knee points, for the best result, is was set to 10.

Column Detection in Low Resolution Table Image. Knee detection is applicable in most cases to determine the best ϵ parameter for the vertical clustering. However, in low resolution table images where the distance between columns in terms of pixel count is relatively small, this algorithm is not efficient. Supposing that there exist vertical lines separating table columns in such table image, we proposed another technique to recognize these lines, and then the columns can be detected.

Particularly, after row detection, we extracted text boxes for every rows, then after applying a binary filter, we determined the pixel count of each row as the summation of pixel counts of its text boxes. We then chose the row with the lowest pixel count since it is the least noisy. The morphological closing and Gaussian blur filtering [23] are then applied to this row so that the vertical lines are exposed. Next, the pixels in the resulting row are vertically clustered with ϵ set to one pixel. Finally, the vertical lines separate the columns correspond to the clusters with the smallest width, and with the same height to the table row.

2.5 Cell Reconstruction

Cells can be reconstructed by determining their actual width, height and coordinates. Specifically, after row detection, the height of a row and the y-axis coordinate of the row's center are approximated as the median height and the median y-axis coordinate of all the text boxes and their centroids, respectively. Similarly, after column detection, the width of the columns and their center's x-axis coordinate are also computed.

However, in the table images, there may be empty cells which can not be detected by the text recognition technique. Therefore, we rebuilt an anchor row which are fully filled. The anchor row is built by normalizing coordinates of all detected text boxes so that the y-axis coordinate of their centroids is the same (say *zero*), and then merging together to the normalized text boxes of the same column. In the merging process, the width of the cells is updated as the difference between the maximal and minimal x-axis coordinate of the text boxes in the same column. The empty cells of all rows are then filled by moving the anchor row along the y-axis to every rows in the table image. The final result of text boxes detection can be seen as in Fig. 5.

3 Evaluation Results

In this section, we evaluated the accuracy of table structure recognition using our proposed CluSTi method on three different datasets. We also compared the performance of CluSTi to DeepDeSRT [22], which is known as the best recent method for table structure recognition on the ICDAR 2013 and ICDAR 2019 competition's datasets.

3.1 Performance Evaluation of CluSTi

CluSTi is firstly evaluated on 397 table images, which were selected from a public dataset of 403 scanned images [1]. Six images were removed because of their lack of table. An example of table scanned image is presented in Fig. 5. The table is composed of 6 columns and 37 rows. However, the first two columns are typically sparse while the other columns are fully filled. Moreover, since the column's names consist of multiple lines in this table, the height of the first row is greater than the remaining rows. In this case, the table structure can not be recognized using DeepDeSRT due to the fact that there exist many empty cells. In contrast, CluSTi can detect the structure with an accuracy of 100%. Corresponding table cells are represented by color rectangle bounding boxes in Fig. 5.

CluSTi's performance on the 397 scanned table dataset is shown in Table 1. The overall $\mathbf{F_1}$**-score** [19], which is the harmonic mean of precision and recall of CluSTi on this dataset, is 87.5%. The accuracy of the row detection (i.e., 92.9%) is higher compared to the column detection (i.e., 82.0%) since the height of rows is mostly uniform while the width of columns is different.

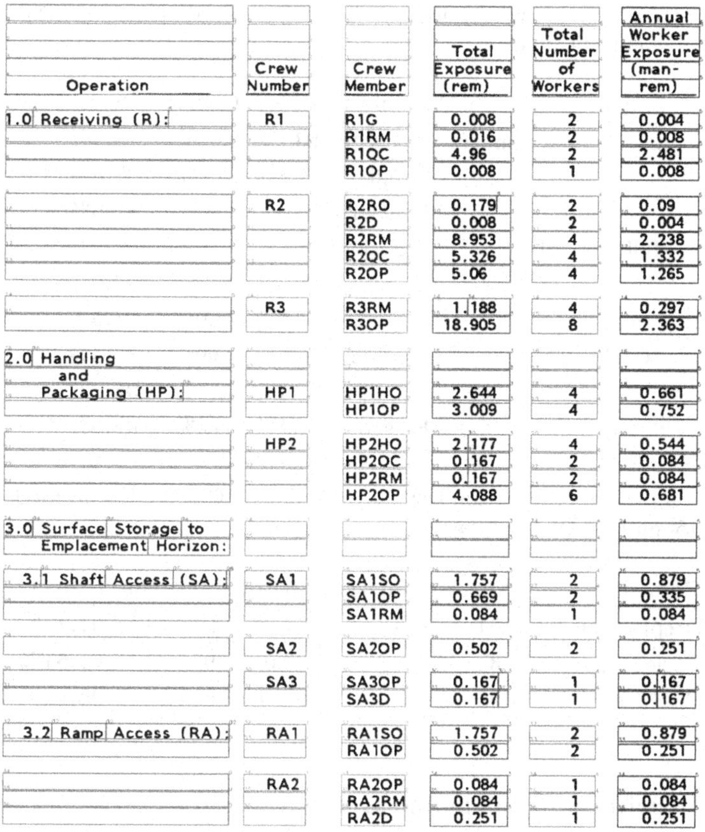

Operation	Crew Number	Crew Member	Total Exposure (rem)	Total Number of Workers	Annual Worker Exposure (man-rem)
1.0 Receiving (R):	R1	R1G	0.008	2	0.004
		R1RM	0.016	2	0.008
		R1QC	4.96	2	2.481
		R1OP	0.008	1	0.008
	R2	R2RO	0.179	2	0.09
		R2D	0.008	2	0.004
		R2RM	8.953	4	2.238
		R2QC	5.326	4	1.332
		R2OP	5.06	4	1.265
	R3	R3RM	1.188	4	0.297
		R3OP	18.905	8	2.363
2.0 Handling and Packaging (HP):	HP1	HP1HO	2.644	4	0.661
		HP1OP	3.009	4	0.752
	HP2	HP2HO	2.177	4	0.544
		HP2QC	0.167	2	0.084
		HP2RM	0.167	2	0.084
		HP2OP	4.088	6	0.681
3.0 Surface Storage to Emplacement Horizon:					
3.1 Shaft Access (SA):	SA1	SA1SO	1.757	2	0.879
		SA1OP	0.669	2	0.335
		SA1RM	0.084	1	0.084
	SA2	SA2OP	0.502	2	0.251
	SA3	SA3OP	0.167	1	0.167
		SA3D	0.167	1	0.167
3.2 Ramp Access (RA):	RA1	RA1SO	1.757	2	0.879
		RA1OP	0.502	2	0.251
	RA2	RA2OP	0.084	1	0.084
		RA2RM	0.084	1	0.084
		RA2D	0.251	1	0.251

Fig. 5. Cell reconstruction result. (Color figure online)

CluSTi is also evaluated on the ICDAR 2013 competition's dataset, which contains 193 document images. Note that these are PDF born-digital images, noiseless, and not scanned documents. Unsurprisingly, the overall $\mathbf{F_1}$**-score** of CluSTi achieved on ICDAR 2013 dataset is significantly higher compared to the scanned images dataset (i.e., 98.5% and 87.5%, respectively; Table 1). In fact, these document images have considerably high resolution, and the column's width is relatively equivalent. Thus, the CluSTi's $\mathbf{F_1}$**-score** corresponding to column detection in this case is 96.9%, which is much higher than 82.0% on scanned images. Similary, on the 281 ICDAR 2019 document images, CluSTi also achieved a very high $\mathbf{F_1}$**-score** accuracy of 94.5%.

Table 1. Detection accuracy (%) of CluSTi on 397 scanned images, ICDAR 2013 and ICDAR 2019 document images.

Dataset	Accuracy	Row	Column	Overall
397 Scanned Images	Precision	93.2%	83.2%	**88.3%**
	Recall	92.8%	82.4%	**87.6%**
	F_1-score	92.9%	82.0%	**87.5%**
ICDAR 2013	Precision	99.9%	97.0%	**98.5%**
	Recall	99.9%	97.2%	**98.6%**
	F_1-score	99.9%	96.9%	**98.5%**
ICDAR 2019	Precision	99.8%	92.9%	**96.4%**
	Recall	99.7%	87.6%	**93.7%**
	F_1-score	99.8%	89.3%	**94.5%**

Table 2. Comparison of detection accuracy (%) among CluSTi, DeepDeSRT [22], and TableNet [15] on ICDAR 2013 dataset

Method	Number of images	Recall	Precision	F1-score
DeepDeSRT [22]	34	87.36%	95.93%	**91.44%**
TableNet [15]	34	90.01%	93.07%	**91.51%**
CluSTi	193	98.60%	98.51%	**98.48%**

3.2 Comparison of CluSTi and Other Methods

CluSTi outperforms DeepDeSRT and TableNet with an overall **F_1-score** of 98.48% on 193 document images compared to 91.44% on 34 images (Table 2). In fact, CluSTi concentrates on the detection of characters since these are the most essential elements in table cells. Then, the table structure can be deduced and filled thanks to its horizontal and vertical clustering. In contrast, DeepDeSRT is based on Faster R-CNN, a semantic segmentation model which focuses on cell object detection [18]. That's why when cells are empty or not large enough, they are still recognized by CluSTi but not by DeepDeSRT. This approach also overcome the limitations of DeepDeSRT method, which segments table cells relying on their boundaries [22]. TableNet showed comparable results to DeepDeSRT method [22], and their model is end-to-end which means further improvements can be made with richer semantic knowledge, and additional branches for learning row-based segmentation.

4 Conclusion

This paper introduced CluSTi, an efficient approach for table structure recognition problem in scanned images, which have not been addressed in the literature. This is a *bottom-up* method, which emphasizes that the table structure is formed by relative positions of text cells, and not by inherent boundaries. Therefore, CluSTi firstly detects the character regions with an accurate scene text detector called CRAFT. Then, the detected text boxes are spatially clustered into their corresponding rows and columns using the Horizontal and Vertical clustering methods, respectively. Finally, every table cells are correctly aligned and extracted according to their detected rows and columns. CluSTi is evaluated on both scanned images and document images, and the achieved $\mathbf{F_1}$**-score** are 87.5%, 98.5%, and 94.5% on three datasets including 397 scanned images, ICDAR 2013 and ICDAR 2019, respectively. This is the highest accuracy for table structure recognition problem executed on scanned image datasets.

However, CluSTi bears certain inconveniences, especially for complicated table structures where exist spreading rows or columns. In such cases, the columns' (or rows) names may not be aligned to the texts of the other rows (or columns) in the same column (or row). These columns (or rows) need to be recognized and processed separately.

References

1. Table-detection-dataset. https://github.com/sgrpanchal31/table-detection-dataset
2. Baek, Y., Lee, B., Han, D., Yun, S., Lee, H.: Character region awareness for text detection. In: Proceedings of the IEEE Conference on Computer Vision and Pattern Recognition, pp. 9365–9374 (2019)
3. Clinchant, S., Déjean, H., Meunier, J.L., Lang, E.M., Kleber, F.: Comparing machine learning approaches for table recognition in historical register books. In: 2018 13th IAPR International Workshop on Document Analysis Systems (DAS), pp. 133–138. IEEE (2018)
4. Deng, D., Liu, H., Li, X., Cai, D.: Pixellink: detecting scene text via instance segmentation. In: 32nd AAAI Conference on Artificial Intelligence (2018)
5. Ester, M., Kriegel, H.P., Sander, J., Xu, X., et al.: A density-based algorithm for discovering clusters in large spatial databases with noise. In: KDD, pp. 226–231 (1996)
6. Farahmand, A., Sarrafzadeh, H., Shanbehzadeh, J.: Document image noises and removal methods. In: Proceedings of the International MultiConference of Engineers and Computer Scientists, pp. 436–440. Newswood Ltd. (2013)
7. Hartigan, J.A., Wong, M.A.: Algorithm as 136: a k-means clustering algorithm. J. Roy. Stat. Soc.: Ser. C (Appl. Stat.) **28**(1), 100–108 (1979)
8. He, T., Tian, Z., Huang, W., Shen, C., Qiao, Y., Sun, C.: An end-to-end textspotter with explicit alignment and attention. In: Proceedings of the IEEE Conference on Computer Vision and Pattern Recognition, pp. 5020–5029 (2018)
9. He, W., Zhang, X.Y., Yin, F., Liu, C.L.: Deep direct regression for multi-oriented scene text detection. In: Proceedings of the IEEE International Conference on Computer Vision, pp. 745–753 (2017)
10. Hu, J., Kashi, R.S., Lopresti, D.P., Wilfong, G.: Table structure recognition and its evaluation. In: Document Recognition and Retrieval VIII, vol. 4307, pp. 44–55. International Society for Optics and Photonics (2000)
11. Kboubi, F., Chabi, A.H., Ahmed, M.B.: Table recognition evaluation and combination methods. In: 8th International Conference on Document Analysis and Recognition (ICDAR 2005), pp. 1237–1241. IEEE (2005)

12. Liu, W., et al.: SSD: single shot multibox detector. In: Leibe, B., Matas, J., Sebe, N., Welling, M. (eds.) ECCV 2016. LNCS, vol. 9905, pp. 21–37. Springer, Cham (2016). https://doi.org/10.1007/978-3-319-46448-0_2
13. Liu, X., Liang, D., Yan, S., Chen, D., Qiao, Y., Yan, J.: FOTS: fast oriented text spotting with a unified network. In: Proceedings of the IEEE Conference on Computer Vision and Pattern Recognition, pp. 5676–5685 (2018)
14. Long, J., Shelhamer, E., Darrell, T.: Fully convolutional networks for semantic segmentation. In: Proceedings of the IEEE Conference on Computer Vision and Pattern Recognition, pp. 3431–3440 (2015)
15. Paliwal, S.S., Vishwanath, D., Rahul, R., Sharma, M., Vig, L.: Tablenet: deep learning model for end-to-end table detection and tabular data extraction from scanned document images. In: 2019 International Conference on Document Analysis and Recognition (ICDAR), pp. 128–133. IEEE (2019)
16. Qasim, S.R., Mahmood, H., Shafait, F.: Rethinking table recognition using graph neural networks. In: 2019 International Conference on Document Analysis and Recognition (ICDAR), pp. 142–147. IEEE (2019)
17. Rashid, S.F., Akmal, A., Adnan, M., Aslam, A.A., Dengel, A.: Table recognition in heterogeneous documents using machine learning. In: 2017 14th IAPR International Conference on Document Analysis and Recognition (ICDAR), vol. 1, pp. 777–782. IEEE (2017)
18. Ren, S., He, K., Girshick, R., Sun, J.: Faster R-CNN: towards real-time object detection with region proposal networks. In: Advances in Neural Information Processing Systems, pp. 91–99 (2015)
19. Sasaki, Y., et al.: The truth of the f-measure. Teach Tutor mater **1**(5), 1–5 (2007)
20. Satopaa, V., Albrecht, J., Irwin, D., Raghavan, B.: Finding a "kneedle" in a haystack: detecting knee points in system behavior. In: 2011 31st International Conference on Distributed Computing Systems Workshops, pp. 166–171. IEEE (2011)
21. Scholkmann, F., Boss, J., Wolf, M.: An efficient algorithm for automatic peak detection in noisy periodic and quasi-periodic signals. Algorithms **5**(4), 588–603 (2012)
22. Schreiber, S., Agne, S., Wolf, I., Dengel, A., Ahmed, S.: Deepdesrt: deep learning for detection and structure recognition of tables in document images. In: 2017 14th IAPR International Conference on Document Analysis and Recognition (ICDAR), vol. 1, pp. 1162–1167. IEEE (2017)
23. Soille, P.: Morphological Image Analysis: Principles and Applications. Springer Science & Business Media, Heidelberg (2013)
24. Sudana, O., Putra, D., Sudarma, M., Hartati, R.S., Wirdiani, A.: Image clustering of complex balinese character with dbscan algorithm. J. Eng. Technol. **6**(1), 548–558 (2018)
25. Xu, R., Wunsch, D.: Survey of clustering algorithms. IEEE Trans. Neural Netw. **16**(3), 645–678 (2005)

Distributed Watermarking for Cross-Domain of Semantic Large Image Database

Le Danh Tai, Nguyen Kim Thang, and Ta Minh Thanh(✉)

Le Quy Don Technical University, 239 Hoang Quoc Viet, Cau Giay, Ha Noi, Vietnam
ledanhtai@gmail.com, thangnk1990@gmail.com, thanhtm@mta.edu.vn

Abstract. This paper proposes a new method of distributed watermarking for large image database that is used for deep learning. We detect the semantic meaning of set of images from the database and embed the a part of watermark into such images set. A part of watermark is one shadow generated from the original watermark by using (n, n) secret sharing scheme. Each shadow is embedded into DCT-SVD domain of one image from the dataset. Since the image sets have multiple image and are distributed in the whole of multiple database, we expect that the proposed method is robust against several attacks.

Keywords: Distributed watermarking · Multiple image database · Image sets

1 Introduction

1.1 Overview

With the rapidly increasing use of Internet, various form of digital data is proposed to share digital multimedia for everyone over the wold. Normal users can access to digital contents like electronic advertising, video, audio, digital repositories, electronic libraries, web designing, and so on. Also, users can copy the digital contents and distribute it easily via network. Digital copyright violations happen frequently because of the increased importance of digital contents. That makes the content providers need to focus on the protection that of copyright. The techniques for copyright protection are developed and researched more and more nowadays.

Digital watermarking technique is the promising technique for protecting the copyright of the valuable digital contents. This technique embeds the copyright information (*e.g* digital logo, author's name, identifier number, ...) into the digital contents without quality degradation. The embedded contents, also called cover contents, can be used to distribute to the users who bought the contents. When copyright disputes happened, the embedded information is needed

N.-S. Vo and V.-P. Hoang (Eds.): INISCOM 2020, LNICST 334, pp. 163–180, 2020.
https://doi.org/10.1007/978-3-030-63083-6_13

to extract from the embedded contents to verify the copyright of the related parties. Therefore, the watermarking techniques are required to be robust against the common attacks on the embedded contents such as digital content processing, compression, RST (rotation, scaling, translation) attack, noise attacks, and so on. The copyright information should be successfully extracted even the embedded contents are adjusted by illegal users.

There are two big classifications of watermarking techniques such as spatial domain based techniques [1] and frequency domain based techniques [2] techniques. In general, the spatial domain techniques embed directly the copyright information into the pixels of the original digital contents. These techniques achieve high performance and do not degrade the quality of contents much. However, such techniques are not robust against even simple image processing attacks [3]. On the other hand, the frequency domain based watermarking techniques are employed to embed copyright information on the coefficients' values of the image after applying some frequency transforms (*i.e.* DCT, DFT, DWT, SVD). Frequency domain based watermarking techniques are mostly focused on real applications since it is robust and secure. The application of both techniques is data integrity, authentication, copyright protection, broadcast monitoring [4].

In the frequency domain, several digital watermarking methods are available in literature including Discrete Cosine Transform (DCT) [5], Discrete Wavelet Transform (DWT) [6], Singular Value Decomposition (SVD) [7], and so on. Such frequency domains are normally employed on the digital format that is employed in the real applications such as audio, image, text, and video. In this paper, we focus on the DCT-SVD based watermarking for a proposal of digital watermarking technique. DCT divides carrier signal into low, middle, and high frequency bands. DCT watermarking is classified into two types: Global DCT watermarking and Block-based DCT watermarking. That makes us possible to control the regions for watermark embedding and extraction. On the other hand, SVD transform decompose a information matrix into orthogonal matrices of singular values(eigen values). It is used to approximate the matrix decomposing the data into an optimal estimate of the signal and the noise components. That property is important for watermarking technique to be robust against noise filtering, compression, and forensic attacks. Based on this analytic, we choose DCT-SVD watermarking method to propose a new framework of distributed watermarking.

As mentioned above, general watermarking techniques are always applied on the normal digital format of multimedia contents. Recently, incorporating deep neural networks with image watermarking [8,9] has attracted increasing attention by many researchers. In this framework, researchers mostly focus on how to embed the watermarks into the trained or pre-trained model of deep neural networks (DNN) [10]. They almost do not focus on how to protect the copyright of image dataset provided for training and testing of DNN methods. That means there are not a watermarking method for deep learning image dataset such as

CIFAR-10[1], MNIST[2], MS-COCO[3], and so on. In our understanding, in order to make the dataset for deep learning method, the providers take much more time and effort to gather and to annotate the data. It also is updated to increase amount of dataset year by year. Therefore, the large datasets for deep learning are very valuable datasets to apply on real applications. In our knowledge, there is not a proposal of copyright protection applied on multimedia dataset. That means the copyright protection for large dataset is required as soon as possible.

1.2 Our Contributions

In this paper, we propose a new method of distributed watermarking for cross-domain of semantic large image database. We make a first version of watermarking method for dataset of deep learning algorithms. In order to keep the accuracy of deep learning model, we distribute the copyright information (watermark) on whole of dataset. The original watermark is separated into many scramble shadows to keep the secure of copyright information. Each shadow will be embedded into one image of dataset. The embeddable set of images are selected from the dataset by checking the semantic relationship of images. In summary, we briefly introduce our contributions as follows:

1. We propose a new distributed watermarking method for semantic image sets extracted from dataset published for deep learning applications. The semantic image set is defined as a set of images belonging one class from deep learning dataset. In case of video format, the semantic image can be defined as a set of frames from on shot video. This is the first consideration of copyright protection for deep learning publish dataset.
2. We have an idea to distribute the separated shadows over all image sets to keep the security of original watermark. Only the owners have the secret keys, can extract the original watermark to prove the ownership of dataset.
3. We try to apply the proposed method on the shots of video file in order to verify the efficiency of solution. Shot detection process can be simply performed by using some conventional method[4]. All frames from one shot are used to embed all scrambled shadows in order to improve the security.

We hope that our proposed method can be used for copyright protection of large deep learning dataset. Datasets providers can embed their information into whole of datasets and then publish it for everyone.

1.3 Roadmap

The rest of this paper is organized as follow: in Sect. 2, a brief overview of related techniques using in the paper. In additional, we define the concept of semantic

[1] http://www.cs.toronto.edu/~kriz/cifar.html.
[2] https://datahack.analyticsvidhya.com/contest/practice-problem-identify-the-digits/.
[3] http://cocodataset.org/.
[4] https://github.com/albanie/shot-detection-benchmarks.

large image dataset. In Sect. 3, the detailed steps of embedding and extraction watermark are explained. In Sect. 4, the simulation experimental results and discussion are shown. Section 5 gives conclusions of this paper.

2 Preliminary

In this study, we employ the combination of DCT and SVD transformation and generate the frequency domain, called DCT-SVD domain. After that, the scrambled shadows are embedded into DCT-SVD domain. Such kind of transformation is briefly explained as follows:

2.1 Discrete Cosine Transform

Discrete cosine transform (DCT) is a most popular linear transform domain that is used in processing of multimedia contents [11]. DCT is always applied on compression format of digital contents to remove statistical correlation. In addition, the frequency bands of DCT including high-frequency, middle-frequency, and low-frequency can be selected appropriately for data embedding. The DCT is defined as:

$$\hat{I}(u,v) = \frac{1}{\sqrt{M \times N}} C(u)C(v) \sum_{i=0}^{M-1} \sum_{j=0}^{N-1} I(i,j) \cos(\frac{(2i+1)u\pi}{2M}) \cos(\frac{(2j+1)v\pi}{2N}) \tag{1}$$

$$C(u) = \begin{cases} \frac{1}{\sqrt{2}}, & u = 0; \\ 1, & u > 0, \end{cases} \tag{2}$$

where $i, u = 0, 1, 2, \ldots, M-1$ and $j, v = 0, 1, 2, \ldots, N-1$. The inverse discrete cosine transform (iDCT) is defined as follows:

$$I(i,j) = \frac{1}{\sqrt{M \times N}} C(u)C(v) \sum_{i=0}^{M-1} \sum_{j=0}^{N-1} \hat{I}(u,v) \cos(\frac{(2i+1)u\pi}{2M}) \cos(\frac{(2j+1)v\pi}{2N}) \tag{3}$$

where $I(i,j)$ is the intensity of image and $\hat{I}(u,v)$ is the DCT coefficients. The DCT low-frequency information contains the main information of image. Therefore, the visual quality of an image will be degraded significantly if the watermark is embedded into DCT low-frequency. In the other hand, the DCT high-frequency information contains the details edges of image, then, it is easily removed in lossy data compression. Based on analysis above, the DCT middle-frequency is more suitable for watermark embedding to maintain the visual quality of image and to keep the robustness of watermark [12].

2.2 Singular Value Decomposition

The singular value decomposition (SVD) is an important factorizing technique to decompose one matrix into three matrices. The SVD is defined as:

$$A = USV^T, \tag{4}$$

where A is an $M \times N$ matrix, U and V are two orthonormal matrices, S is a diagonal matrix consisting of the singular values (SVs) of A. The singular values (SVs) satisfy $s_1 \geq s_2 \geq s_3 \ldots \geq s_N \geq 0$ and the superscript T denotes matrix transposition.

If matrix A is a pixel of image, the SVs can keep stable with a slight perturbation. SVs are almost keep stable even the image is adjusted the pixel value under some attacks. Therefore, in order to improve the robustness of watermark after embedding, we choose to SVs for guaranteeing the copyright information.

2.3 Semantic Large Image Dataset

The concept of semantic large image dataset is not defined beforehand. In this paper, we define the semantic large image dataset as a set of images belonging one class of one object. In case of video file, the semantic large image dataset can be defined as a set of frames from one shot.

For instance, as the explanation of Kaggle[5], "CIFAR-10 is an established computer-vision dataset used for object recognition. It is a subset of the 80 million tiny images dataset and consists of 60,000 32×32 color images containing one of 10 object classes, with 6000 images per class.". That means that CIFAR-10 contains 10 object classes, with 6000 images per class. Therefore, we can extract ten semantic large image datasets from CIFAR-10, then we can embed the copyright information into 6000 images for each class.

2.4 Distributed Watermark

In order to keep the secret of watermark information, we choose the (n, n) secret sharing scheme [13] to distribute the original watermark. That implies that, n-shadows S are generated from the original watermark W XOR-ing with n-secret key K. In order to reveal the original watermark, we need to collect all n-shadows S with XOR-ing it.

The (n, n) secret sharing scheme is described in a pseudo-code style below in terms of its input, output, the construction procedure, and revealing procedure. In the construction procedure, our algorithm shows the way to compute the shadow watermark. In the revealing procedure, the algorithm that reveals the original watermark, is explained how to reconstruct the original watermark from the shadows.

According to the Algorithm 1, we can distribute the original watermark W into n shadows S_i. That can keep the secret of the original watermark (copyright information). Only the owners, who has the secret key K_i, can reveal the original W after collecting all shadows S_i.

[5] https://www.kaggle.com/c/cifar-10.

Algorithm 1: Distributed watermark - (n, n) secret sharing scheme

INPUT: Number of shadows n, the original watermark W

OUTPUT: n distinct matrices $\{S_1, ..., S_n\}$, called shadows watermark.

CONSTRUCTION: Generate $n+1$ random secret keys $K = \{K_1, ..., K_n, K_{n+1}\}$. Compute n shadow watermark $\{S_1, ..., S_n\}$ with

  $last.shadow = W$
  for all random secret keys in K_i in K **do**
    $S_i = last.shadow \oplus K_i$
    $last.shadow = S_i$, for $i = 1, 2, ..., n$
  end for

REVEALING: Reveal original watermark W from shadows $S = \{S_1, ..., S_{n-1}\}$.

  $last.watermark = S_n$
  for all random secret keys in K_i in K **do**
    $last.watermark = last.watermark \oplus K_i$, for $i = 1, 2, ..., n-1$
  end for

3 Our Proposed Method

Our proposed watermarking method can be applied on the large image dataset for deep learning applications. However, the semantic large images meaning of one class from dataset can be considered as a shot in the video format. Therefore, in order to verify our idea simply, we apply our method on video format. Our method consists of three processes: shadows construction from watermark, video embedding, video extraction, and watermark revealing.

3.1 Shadows Construction Process

To keep the secret of watermark information, we scramble the original watermark by using the (n, n) secret sharing scheme described in Sect. 2.4. We use the process CONSTRUCTION in Algorithm 1 to generate n shadows S_i from the original watermark W with the secret keys K_i, where $i = 1, 2, ..., n$.

The shadows S_i can be seen as Fig. 1. All shadows are randomized based on the secret keys, therefore, only the owner, who keeps the key, can reveal the watermark information.

3.2 Video Embedding Method

Our proposed watermarking method is shown in Fig. 2. The process to embed the shadows into the frames of one shot can be described as follows:

Step 1: The shadows watermark image S_i of size $p \times p$ is converted to a $1 \times p^2$ binary watermark sequence S_i^o. In our paper, $p = 32$.

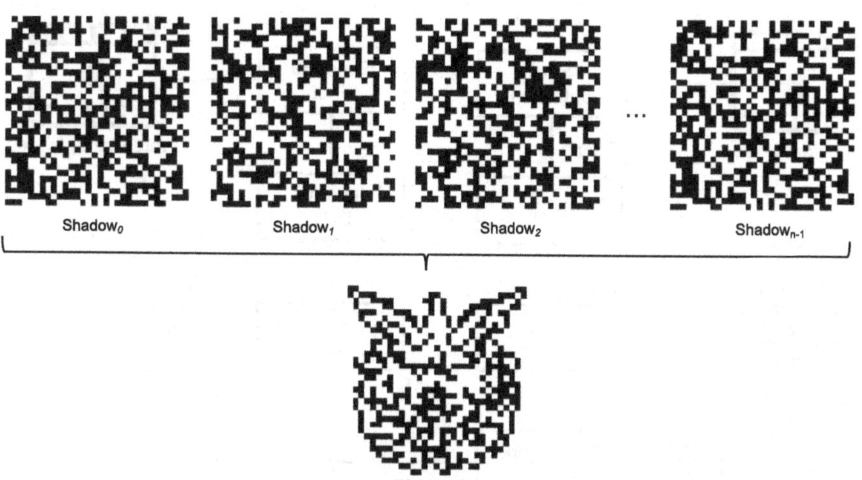

Fig. 1. (n, n) secret sharing scheme.

Step 2: Original video V is separated into multiple shots by using prepared algorithm [14] beforehand.
Step 3: DCT is performed on each frame from one shot to create the DCT frequency domain for all R, G, B plane. To improve the robustness and high quality of the proposed watermarking scheme, the low-frequency region is selected for watermark insertion.
Step 4: The low-frequency region is segmented into non-overlapping blocks C_i of size $4 \times 4, i = 1, 2, ..., N$.
Step 5: For each block:

(1) DCT is performed on each block C_i. The coefficient $C_i(0, 0)$ of each block is collected into the matrix A.
(2) The matrix A is segmented into non-overlapping blocks A_i of size $4 \times 4, i = 1, 2, ..., M$.
(3) SVD is applied on the matrix A_i and the largest singular value is extracted as follows:

$$[U_k, S_k, V_k] = SVD(A_k), k = 1, 2, ...M/4 \tag{5}$$

where U_k, V_k and S_k are the results of SVD operation, respectively. Let $x = S_k(0, 0)$ represents the largest SV of matric A_k.
(4) The shadows watermark sequence S_i^o is embedded by modifying the values of x. If $S_i^o = 1$, then $x = (\lfloor (x * Q)/2 \rfloor * 2)/Q$. If $S_i^o = 0$, then $x = (\lfloor (x * Q + 1)/2 \rfloor * 2 - 1)/Q$. In this term, Q is the embedding strength factor.
(5) The modified matrix S_k' is generated by altering their largest singular values with x.

$$S_k'(0, 0) = x \tag{6}$$

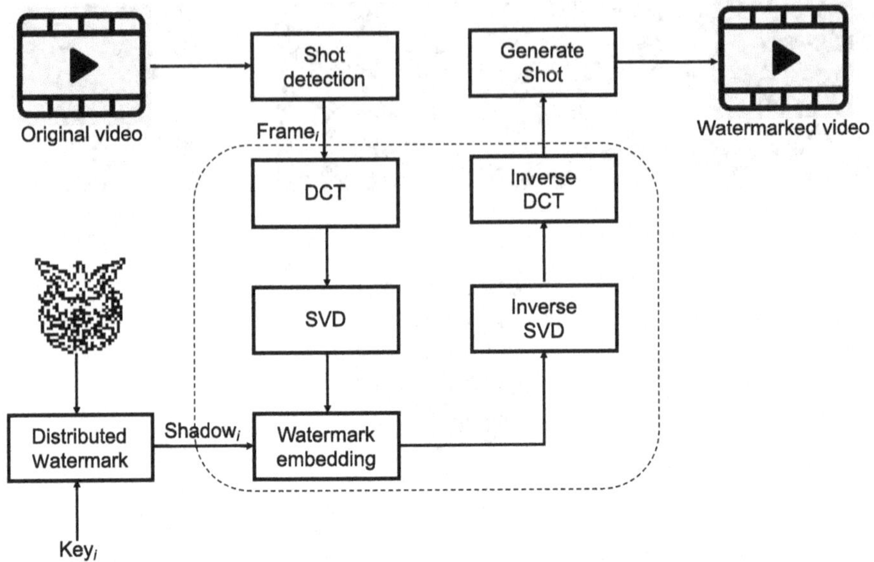

Fig. 2. Video embedding method

(6) Modified matrices A'_k are constructed by the inverse SVD operation.

$$A'_k = U_k S'_k V_k, k = 1, 2, ...M/4 \tag{7}$$

All elements of matrices A'_k are mapped back to their original positions in coefficient matrix A. Then a watermarked block C'_i is produced by the inverse DCT.

Step 6: After Step 5, the embedded frame are generated. Collect all frames and re-create all shots to generate a watermarked video V'.

3.3 Video Extraction Method

To extract the watermark information from a watermarked video V', we do not need the original video V. The secret key K_i, and the number of shadows n are required. The process of video extraction is shown in Fig. 3.

Step 1: The watermarked video V^* is separated into multiple shots by using prepared algorithm [14] beforehand.
Step 2: By applying the DCT operation on each frame from one shot, the DCT frequency domain for all R, G, B plane is obtained.
Step 3: The low-frequency region is segmented into non-overlapping blocks C_i^* of size $4 \times 4, i = 1, 2, ..., N$.
Step 4: For each block:
(1) By applying DCT operation on each block C_i^*, the matrix A^* is generated by collecting all the coefficient $C_i^*(0, 0)$ of each block.

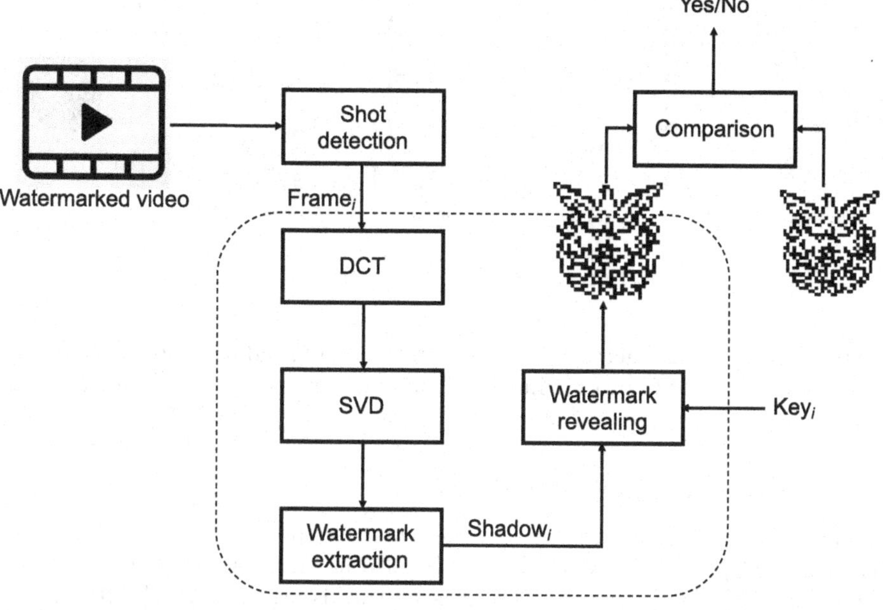

Fig. 3. Video extraction method

(2) The matrix A^* is segmented into non-overlapping blocks A_i^* of size $4 \times 4, i = 1, 2, ..., M$.

(3) SVD operation is applied on the matrix A_i^* and the largest singular value is extracted as follows:

$$[U_k^*, S_k^*, V_k^*] = SVD(A_k^*), k = 1, 2, ...M/4 \tag{8}$$

where U_k^*, V_k^* and S_k^* are the results of SVD operation, respectively. The values $x^* = S_k^*(0,0)$ represents the largest SV of matrix A_k^*.

(4) The embedded watermark bits can be extracted as follows:

$$votes = \begin{cases} votes + 1, & \text{if } \lfloor x^* * Q \rfloor \%2 = 0; \\ 0, & \text{other}, \end{cases} \tag{9}$$

where $votes$ is the number of even value of x^*. This method called "voting method". Therefore, the shadows watermark sequence $S_i^{o'}$ can be extracted from $votes$ as follows:

$$S_i^{o'} = \begin{cases} 1, & \text{if } votes > T; \\ 0, & \text{other}, \end{cases} \tag{10}$$

where T is threshold value that is predefined beforehand.

Step 5: The watermark bits from all watermarked blocks of all frames are extracted by repeating Step 4.

(a) Akiyo: 300 frames (b) Container: 300 frames (c) Foreman: 300 frames (d) Watermark

Fig. 4. Test video and watermark

Step 6: From all extracted $S_i^{o'}$, we can construct all shadows watermark S_i'. By using REVEALING process in the Algorithm 1, we can reveal the watermark W'.

3.4 Watermark Revealing Process

To obtain the watermark from the extracted shadows $S_i^{o'}$, we employ the REVEALING process in Algorithm 1. We also use the information n of (n, n) secret sharing scheme and the secret keys K_i that is used in CONSTRUCTION process. If we apply on all extracted shadows $S_i^{o'}$, we can obtain the watermark W' which is shown in Fig. 1.

4 Experimental Results and Analysis

4.1 Experimental Environment

Our proposed method is implemented in Python version 3.7.6. All experimental results are obtained in MacBook Pro, macOS version 10.13.

To evaluate the performance of the proposed watermarking method, some video "Akiyo", "Container", and "Foreman" [6] with 300 frames are chosen as the test videos. It is shown in Fig. 4 (a)∼(c). A binary watermark W of size 32×32 shown in Fig. 4 (d) is selected as the watermark image. In order to make all frames are similar to deep learning dataset, we scale the size of all frames with the same size as 1024×1024.

In order to generate the shadows watermark S_i^o, we employ the (4, 4) secret sharing scheme. That means $n = 4$. To generate the secret key K_i, we use the function $randint(0, 255)$[7] with secret $seed = 101$. The embedding strength factor Q is set at 15. The threshold T is set at 2.

4.2 Imperceptibility Measure

To evaluate the imperceptibility of watermaked video, the peak signal-to-noise ratio (PSNR) is adopted. The PSNR can measure the quality of all watermarked frames [15], then average PSNR value for the embedded video can be calculated.

[6] http://trace.eas.asu.edu/yuv/.
[7] https://docs.python.org/3/library/random.html.

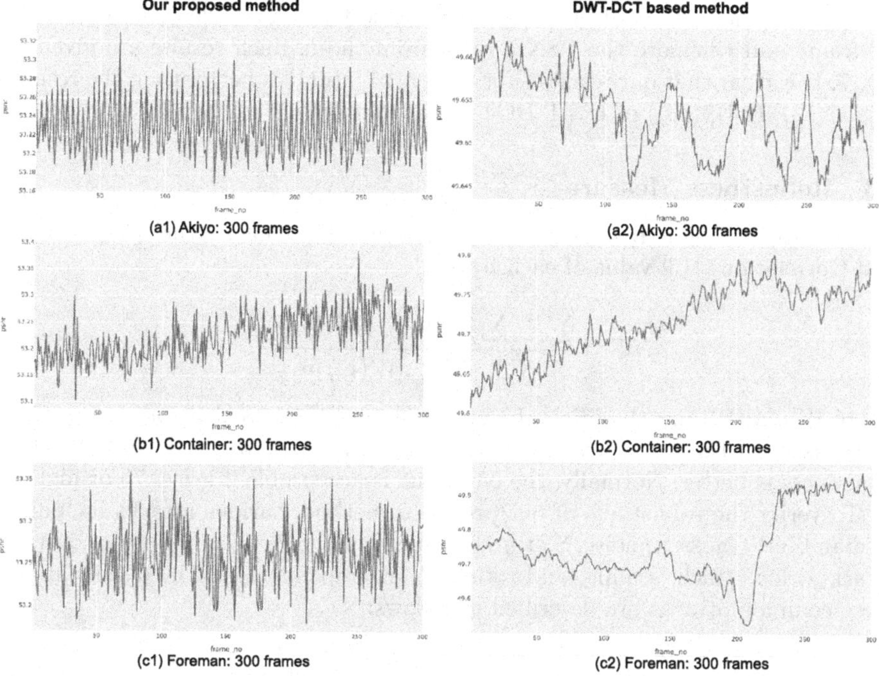

Fig. 5. All PSNR values of all frames

$$PSNR = 10\log_{10}\frac{MAX^2}{MSE}, \tag{11}$$

where MAX is the maximized value of pixel, *e.g.* $MAX = 255$. And MSE is mean squared error.

$$MSE = \frac{1}{H * W}\sum_{i=0}^{H-1}\sum_{j=0}^{W-1}[I(i,j) - \hat{I}(i,j)]^2, \tag{12}$$

where H and W are the height and width of frame. I and $\hat{I}$ are the original frame and the watermarked frame, respectively.

In order to compare the efficiency of our proposed method, we also implement the similar embedding process and extraction process on the DWT-DCT domain, which is usually employed for image or video watermarking field [18–20], called **DWT-DCT based method**.

We calculated all PSNR values of all frames from test videos and obtained the average PSNR values. All PSNR values of all frames from three test videos are shown in Fig. 5. The average PSNR values from those of all frames are shown in Fig. 6. Based on our experimental results, our proposed method achieved better results than that of DWT-DCT based method.

To show the visualized quality of one frame from test videos, we extracted 1st frame and calculate the PSNR value from those. Such results are given in Fig. 7. It is clear that our results (Fig. 7(a1) (b1), (c1)) is better than the results (Fig. 7(a3), (b3), (c3)) of DWT-DCT based method.

4.3 Robustness Measure

To evaluate the robustness of extraction method, we calculate the NC (Normalized Correlation) [16] value of each frame.

$$NC = \frac{\sum_{i=0}^{31}\sum_{j=0}^{31}[W(i,j)W'(i,j)]}{\sum_{i=0}^{31}\sum_{j=0}^{31}[W(i,j)]^2}, \tag{13}$$

where W and W' are the original watermark and the revealed watermark that is reconstructed in Sect. 3.4. If the value of NC is close to 1, it means that the robustness is better. Normally, the NC value is acceptable if it is 0.75 or higher.

To verify the robustness of our proposed method, various attacks including median blur, Gaussian noise, Salt and Pepper noise, JPEG compression, overlay attack, color attack, scaling, and rotation, are implemented. The details about these common attacks are described as follows:

Filtering Attack. In this attack, the watermarked frames are corrupted with Gaussian filtering and median filtering, respectively. Gaussian filtering removes the high frequency information of watermarked frames by using blurring based on Gaussian function. Median filtering replaces the original pixel value of watermarked frames with the median value in the 3×3 window. The results of extracted watermark are given in Fig. 8. As can be seen, the performance of the proposed algorithm under filtering attacks is better than DWT-DCT based method.

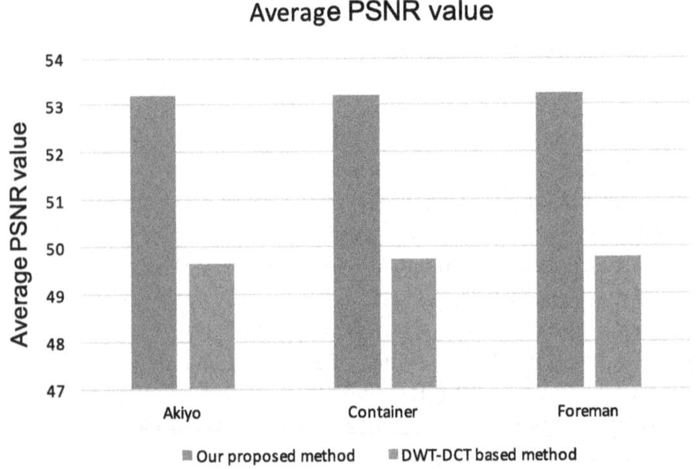

Fig. 6. Average PSNR value of three test videos

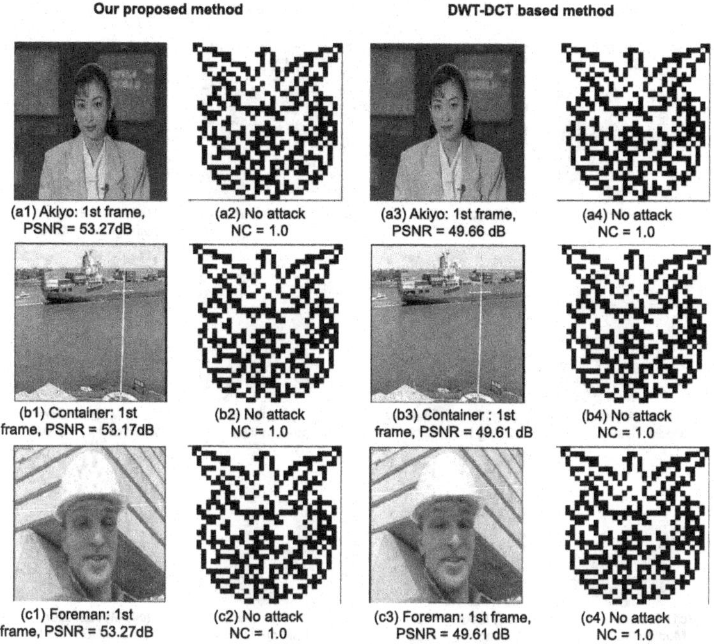

Fig. 7. The PSNR value and NC value of 1st frame from three test videos

Noise Attack. In the noise attack experiment, the watermarked frames are degraded by two kinds of noises attack such as Gaussian noise with variance 0.005, Salt and Pepper with variance 0.01. The results of extracted watermarks are shown in Fig. 9. As can be seen, both methods are not robust against Salt and Pepper attack. However, those are robust against Gaussian noise attack.

Geometric Attacks. Two geometric attacks are employed in this paper. For scaling operation, the watermarked frames are scaled down to 50%, then are scaled up to 100%. In the rotation experiment, the watermarked frames are rotated by 5 in the counterclockwise direction, then are re-rotated by 5 in the opposite direction. The results of our experiment are shown in Fig. 10. As can be seen from Fig. 10, our proposed method can extract exactly the watermark information. On the other hand, DWT-DCT based method is not robust against strong scaling attacks.

JPEG Attack. In general, the JPEG (Joint Photographic Experts Group) compression[8] is a most popular image compression technique in digital watermarking. In this experiment, the watermarked frames are compressed with quality factor set 75 that is set for normal JPEG.

[8] https://jpeg.org/.

Our proposed method
DWT-DCT based method
(a1) Akiyo: Median filter
(a2) Median filter
Votting NC = 0.99
(a3) Akiyo: Median filter
(a4) Median filter
Average NC = 0.79
(b1) Container:
Median filter
(b2) Median filter
Votting NC = 0.89
(b3) Container:
Median filter
(b4) Median filter
Average NC = 0.70
(c1) Foreman:
Median filter
(c2) Median filter
Votting NC = 0.98
(c3) Foreman:
Median filter
(c4) Median filter
Average NC = 0.81
(d1) Akiyo:
Gaussian filter
(d2) Gaussian filter
Votting NC = 0.92
(d3) Akiyo:
Gaussian filter
(d4) Gaussian filter
Average NC = 0.08
(e1) Container:
Gaussian filter
(e2) Gaussian filter
Votting NC = 0.53
(e3) Container:
Gaussian filter
(e4) Gaussian filter
Average NC = 0.06
(f1) Foreman:
Gaussian filter
(f2) Gaussian filter
Votting NC = 0.71
(f3) Foreman:
Gaussian filter
(f4) Gaussian filter
Average NC = 0.01

Fig. 8. Filtering attacks: Median and Gaussian filtering

Fig. 9. Noise attacks: Salt & Pepper and Gaussian noise

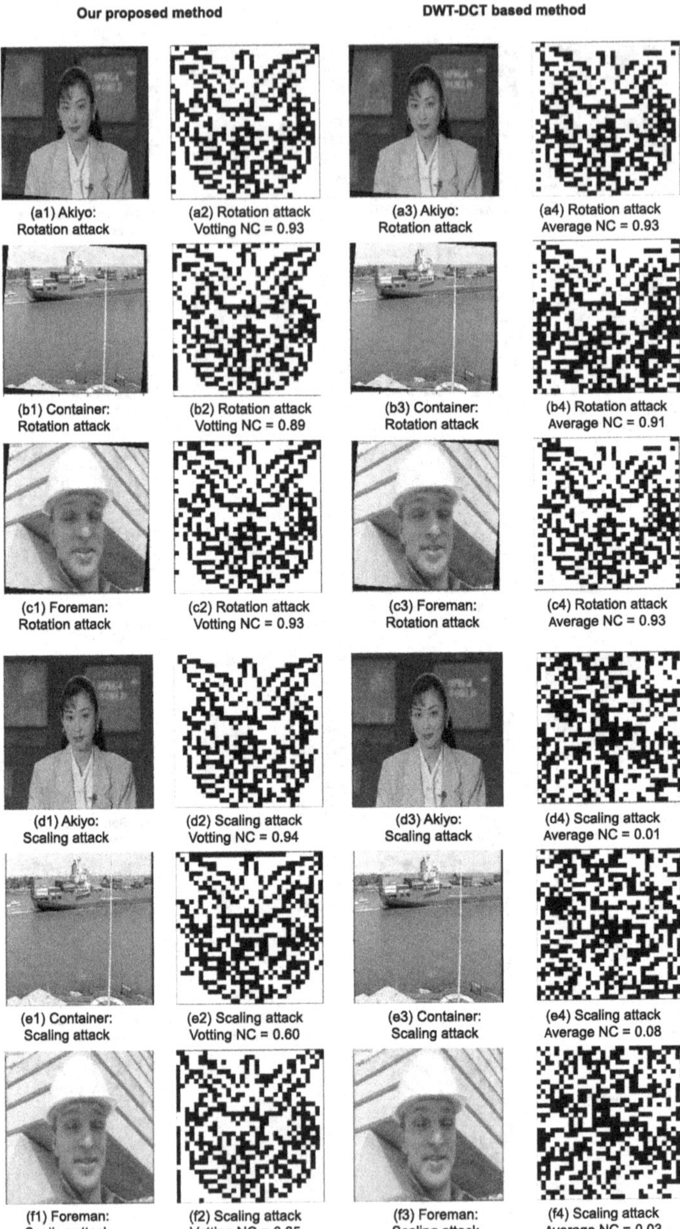

Fig. 10. Geometric attacks: Scaling and Rotation attacks

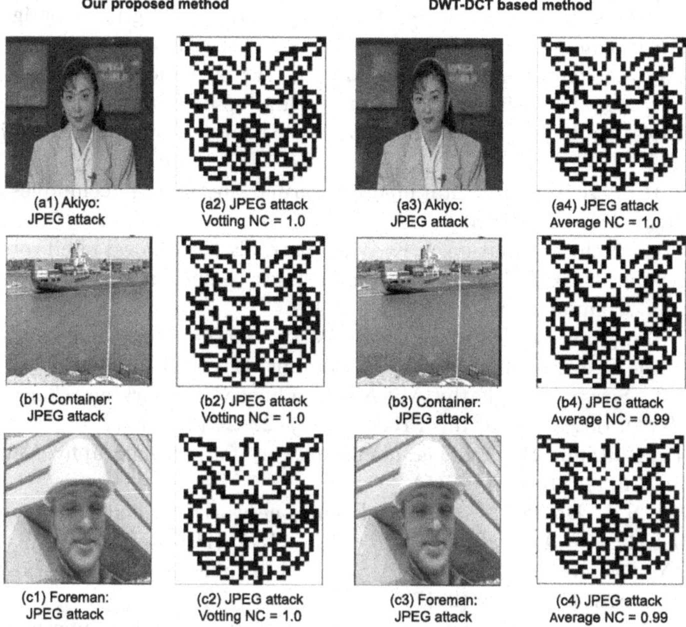

Fig. 11. JPEG attacks

Figure 11 shows that both methods robust against the JPEG compression. Based on these results, we can conclude that our proposed method and DWT-DCT based method are acceptable for JPEG compression.

5 Conclusions

In this paper, we have proposed a method to distribute original watermark to generate scrambled shadows, then embed such shadows into the set of frames from shots. We embed the watermark information into DCT-SVD domain, therefore, it is robust against some attacks such as median blur, Gaussian noise, Salt and Pepper noise attacks, JPEG compression, and geometric attacks. Compared with other related works, our proposed watermarking method performs better in terms of invisibility and robustness. Our proposed watermarking algorithm can be extended for audio signal processing, distributed database.

Acknowledgement. This research is funded by Vietnam National Foundation for Science and Technology Development (NAFOSTED) under grant number 102.01-2019.12.

References

1. Mukherjee, D.P., Maitra, S., Acton, S.T.: Spatial domain digital watermarking of multimedia objects for buyer authentication. IEEE Trans. Multimedia **6**(1), 1–15 (2004)

2. Lin, S.D., Chen, C.: F, "A robust DCT-based watermarking for copyright protection,". IEEE Trans. Consum. Electron. **46**, 415–421 (2000)
3. Ghadi, M., Laouamer, L., Nana, L., Pascu, A.: A blind spatial domain-based image watermarking using texture analysis and association rules mining. Multimedia Tools Appl. **78**(12), 15705–15750 (2018). https://doi.org/10.1007/s11042-018-6851-2
4. Vasudev, R.: A review on digital image watermarking and its techniques. J. Image Graph. **4**(2), 150–153 (2016)
5. Das, C., Panigrahi, S., Sharma, V.K., Mahapatra, K.K.: A novel blind robust image watermarking in DCT domain using inter-block coefficient correlation. AEU-Int. J. Electron. Commun. **68**(3), 244–253 (2014)
6. Srivastava, A., Saxena, P.: DWT? DCT? SVD based semiblind image watermarking using middle frequency band. IOSR J. Comput. Eng. **12**(2), 63–66 (2013)
7. Lagzian, S., Soryani, M., Fathy, M.: A new robust watermarking scheme based on RDWT-SVD. Int. J. Intell. Inf. Process. **2**(1), 22–29 (2011)
8. Sabah, H., Haitham, B.: Artificial neural network for steganography. Neural Comput. Appl. **26**(1), 111–116 (2015)
9. Alexandre, S.B., David, C.J.: Artificial neural networks applied to image steganography. IEEE Latin Amer. Trans. **14**(3), 1361–1366 (2016)
10. Uchida, Y., Nagai, Y., Sakazawa, S., Satoh, S.: Embedding watermarks into deep neural networks. In: ICMR (2017)
11. Leng, L., Zhang, J., Xu, J.: Dynamic weighted discrimination power analysis in DCT domain for face and palmprint recognition. In: 2010 International Conference on Information and Communication Technology Convergence (ICTC), pp. 467–471. IEEE (2010)
12. Kaur, B., Kaur, A., Singh, J.: Steganographic approach for hiding image in DCT domain. Int. J. Adv. Eng. Technol. **1**(3), 72–78 (2011)
13. Wang, D., Zhang, L., Ma, N., Li, X.: Two secret sharing schemes based on Boolean operations. Pattern Recogn. **40**(10), 2776–2785 (2007)
14. https://github.com/albanie/shot-detection-benchmarks
15. Thanh, T.M., Heip, P.T., Tam, T.M., Tanaka, K.: Robust semi-blind video watermarking based on frame-patch matching. AEU Int. J. Electr. Commun. **68**(10), 1007–1015 (2014)
16. Thanh, T.M., Tanaka, K.: An image zero-watermarking algorithm based on the encryption of visual map feature with watermark. J. Multimedia Tools Appl. **76**, 13455–13471 (2017)
17. Thanh, T.M., Iwakiri, M.: A proposal of digital rights management based on incomplete cryptography using invariant Huffman code length feature. Multimedia Syst. **20**(2), 127–142 (2013). https://doi.org/10.1007/s00530-013-0327-z
18. Abdulrahman, A.K., Ozturk, S.: A novel hybrid DCT and DWT based robust watermarking algorithm for color images. Multimedia Tools Appl. **78**(12), 17027–17049 (2019). https://doi.org/10.1007/s11042-018-7085-z
19. Chow, Y.-W., Susilo, W., Tonien, J., Zong, W.: A QR code watermarking approach based on the DWT-DCT technique. Faculty of Engineering and Information Sciences - Papers: Part B. 389 (2017)
20. Hu, H.-T., Hsu, L.-Y.: Collective blind image watermarking in DWT-DCT domain with adaptive embedding strength governed by quality metrics. Multimedia Tools Appl. **76**(5), 6575–6594 (2016). https://doi.org/10.1007/s11042-016-3332-3

Depth Image Reconstruction Using Low Rank and Total Variation Representations

Van Ha Tang(✉) and Mau Uyen Nguyen

Faculty of Information Technology, Le Quy Don Technical University, Hanoi, Vietnam
hatv@lqdtu.edu.vn, nguyenitt2005@gmail.com

Abstract. Rapid advancement and active research in computer vision applications and 3D imaging have made a high demand for efficient depth image estimation techniques. The depth image acquisition, however, is typically challenged due to poor hardware performance and high computation cost. To tackle such limitations, this paper proposes an efficient approach for depth image reconstruction using low rank (LR) and total variation (TV) regularizations. The key idea is LR incorporates non-local depth information and TV takes into account the local spatial consistency. The proposed model reformulates the task of depth image estimate as a joint LR-TV regularized minimization problem, in which LR is used to approximate the low-dimensional structure of the depth image, and TV is employed to promote the depth sparsity in the gradient domain. Furthermore, this paper introduces an algorithm based on alternating direction method of multipliers (ADMM) for solving the minimization problem, whose solution provides an estimate of the depth map from incomplete pixels. Experimental results are conducted and the results show that the proposed approach is very effective at estimating high-quality depth images and is robust to different types of data missing models.

Keywords: Depth imaging · Depth image reconstruction · Low-rank matrix factorization · Sparse representation

1 Introduction

Recent development of numerous computer vision (CV) applications and 3D modeling have increased the research and development of depth sensing technologies. The capabilities of generating depth information together with the conventional 2D image of the desired scene, and thereby yielding the 3D model of the desired scene is very useful in a wide range in CV applications, from autonomous driving, robot navigation, to augmented reality and action recognition [6,7,18,19]. However, depth imaging acquisition faces with several technical difficulties, including poor hardware performance, prolonged data collection, and high computational cost, though much effort has been made to the development

N.-S. Vo and V.-P. Hoang (Eds.): INISCOM 2020, LNICST 334, pp. 181–194, 2020.
https://doi.org/10.1007/978-3-030-63083-6_14

of 3D cameras, such as Google Tango, Intel RealSense, and Microsoft Kinect [20,21]. Thus, new efficient techniques for fast data acquisition and efficient depth reconstruction are very vital for numerous depth technology applications.

Much interest in depth image estimation has been reported in several studies. Techniques exploiting the shapes of objects in the depth image were reported in [19] and [18]. In [19], the depth image quality has been improved by incorporating shading shapes. In [18] the defocusing shape of objects was considered to restore the missing values on the depth map. The other approaches that combine hand-tuned models and surface orientations were proposed in [14,17]. Image inpainting-based methods have also applied to depth estimate. In [5], missing depth values were restored by region growing and bilateral filtering. In [3], a Kalman filter was employed for enhancing smoothness of the depth image, whereas morphological operators were used in [8]. However, it is worth noting that the majority of such methods rely on color information of the scene. In other words, they require collecting the color image, which prolongs the time for data acquisition and makes a burden to data storage.

Recent approaches based on modern sensing and representations were proposed for depth estimate [11,12]. In [11], a sparse representation (SR) was used to estimate a depth map from the incomplete observed pixels. This approach relies on the theory of the powerful compressive sensing (CS) framework, which states that with high probability, a sparse signal/image can be reconstructed precisely from incomplete measurements/pixels provided that it is compressible or has a SR in proper bases [4,9]. In addition, CS enables the reconstruction and compression to be performed simultaneously, resulting in simple and cost-effective hardware sensing systems. Note that in the sparsity-based technique, the task of depth reconstruction is posed as a sparsity-regularized least squares (LS) minimization problem. A further extension of the sparsity-regularized model was presented in [12], where the color information was incorporated for enhancing the quality of image restoration.

Inspired by the SR-based models, this paper introduces a joint LR and TV regularizations for depth image estimation from incomplete observed pixels. Together with SR, the LR representation is incorporated into the imaging model to capture the low-dimensional structure of the depth images. Moreover, the TV regularizer is used to promote the local spatial consistency for the depth map. To this end, the task of depth image estimate is reformulated as a joint LR-TV regularized minimization problem, whose solution provides an estimate of the depth map. We present an algorithm based on ADMM technique for solving the minimization problem. The proposed model is validated by several experimental evaluations. Analysis and comparisons with other imaging models are also provided in this study.

The remainder of the paper is organized as follows. Section 2 introduces the depth image acquisition and the SR-based depth estimation. Section 3 describes the proposed LR-TV model for depth image reconstruction from incomplete measurements. Section 4 presents the experimental results. Section 5 gives concluding remarks.

2 Depth Sparse Imaging Model

This section first presents a brief mathematical model for depth image acquisition. It then describes a sparse representation approach for depth image reconstruction.

2.1 Depth Image Acquisition

Let $\mathbf{Z} \in \mathbb{R}^{h \times w}$ denote a depth map of a scene imaged by an active sensor. In the case of full sensing operations or ideal imaging, the sensor receives $m = h \times w$ depth pixels that represent the distances from the sensor to the surrounding obstacles. In fact, due to fast sensing or poor hardware performance, only n ($n \ll m$) measurements are acquired. Let $\mathbf{z} \in \mathbb{R}^m$ be a vector obtained by applying a vectorization operator to the unknown image $\mathbf{Z}$, $\mathbf{z} = \text{vec}(\mathbf{Z})$, where $\text{vec}(\cdot)$ is the vectorization operator forming a composite column vector by stacking the columns of a matrix in lexicographic order. Note also that the image $\mathbf{Z}$ can be obtained from its corresponding vector $\mathbf{z}$ through $\mathbf{Z} = \text{mat}(\mathbf{z})$, where $\text{mat}(\cdot)$ is the operator converting a column vector having $h \times w$ entries into a $h \times w$ matrix. Since the conversion between the image vector and matrix is simple, hereafter we use $\mathbf{z}$ to represent for both the vector and matrix depth image.

Let $\mathbf{y} \in \mathbb{R}^n$ be the vector containing the observed measurements. The incomplete image $\mathbf{y}$ can be related to the full unknown image $\mathbf{z}$ as

$$\mathbf{y} = \boldsymbol{\Phi}\, \mathbf{z}, \tag{1}$$

where $\boldsymbol{\Phi} \in \mathbb{R}^{n \times m}$ is the sampling matrix, mathematically representing the depth acquisition protocol. In this representation, $\boldsymbol{\Phi}$ is a diagonal binary matrix used to indicate which pixels are selected. Now, given the incomplete and thus corrupted image $\mathbf{y}$ and matrix $\boldsymbol{\Phi}$, our goal is to reconstruct a full and high-quality depth map $\mathbf{z}$.

2.2 Depth Estimation with Sparse Representation

The full depth image $\mathbf{z}$ can be reconstructed by exploring its structures. The most common structure that has been exploited is its sparsity in a transform domain. This is because the depth image $\mathbf{z}$ typically contains large homogenous objects with only a few discontinuity at their transitions. This characteristic leads to a SR for the depth image in proper domains, such as wavelets [12,13]. The sparsity and the wavelet representation can be justified in the aspect that large homogeneous regions can be compressively represented by only a small number of significant coefficients. Mathematically, the sparse representation of image $\mathbf{z}$ can be expressed as

$$\mathbf{z} = \boldsymbol{\Psi}\, \mathbf{x}, \tag{2}$$

where $\boldsymbol{\Psi} \in \mathbb{R}^{m \times q}$ is the dictionary matrix with its columns q ($q \geq m$) being the wavelet bases, and $\mathbf{x} \in \mathbb{R}^q$ is a vector of coefficients having only s nonzero components, i.e., $s = \|\mathbf{x}\|_0$. Due to the sparseness of $\mathbf{x}$, s is much smaller than q ($s \ll q$). Here the ℓ_0-norm is used to measure the number of nonzero entries in vector $\mathbf{x}$. In practice, its counterpart, the ℓ_1-norm, is used to regularize the sparsity as it is a convex relaxation for the ℓ_0-norm [16].

It follows from (1) and (2) that $\mathbf{y} = \boldsymbol{\Phi}\,\boldsymbol{\Psi}\,\mathbf{x}$. Given the observation vector $\mathbf{y}$, an estimate of the coefficient vector $\mathbf{x}$ can be obtained by solving the following ℓ_1-regularized LS minimization problem:

$$\underset{\mathbf{x}}{\text{minimize}} \quad \frac{1}{2}\|\mathbf{y} - \boldsymbol{\Phi}\,\boldsymbol{\Psi}\,\mathbf{x}\|_2^2 + \lambda\,\|\mathbf{x}\|_1, \tag{3}$$

where λ is a regularization parameter used to balance between the LS and the ℓ_1 penalty terms. Since the dictionary matrix $\boldsymbol{\Psi}$ is typically orthogonal, i.e., $\boldsymbol{\Psi}\,\boldsymbol{\Psi}^T = \mathbf{I}$. Thus, $\mathbf{z} = \boldsymbol{\Psi}\,\mathbf{x}$ and $\mathbf{x} = \boldsymbol{\Psi}^T\,\mathbf{z}$. This means that Problem (3) can be written equivalently as

$$\underset{\mathbf{z}}{\text{minimize}} \quad \frac{1}{2}\|\mathbf{y} - \boldsymbol{\Phi}\,\mathbf{z}\|_2^2 + \lambda\,\|\boldsymbol{\Psi}^T\,\mathbf{z}\|_1. \tag{4}$$

The solution to Problem (4) gives an estimate for a full depth image. This sparse regularization approach, however, is effective only if the observed image $\mathbf{y}$ is corrupted with random missing values. In other words, the obtained estimate is good if the image does not contain large continuous missing regions. For example, if pixels are missing along entire rows or columns, then the sparse-regularized model cannot cope well with this issue. This limitation can be overcome by introducing more effective regularizations into the imaging model. In this study, we investigate two more regularizations of LR and TV. The new LR-TV model is described in the next section.

3 LR-TV Depth Reconstruction

This section first presents the proposed LR-TV model for depth estimate problem in Subsect. 3.1, followed by an algorithm based on ADMM to solve the LR-TV regularized minimization problem in Subsect. 3.2.

3.1 LR-TV Problem Formulation

LR and TV regularizers are introduced to further strengthen the depth imaging model. The motivation of using LR is due to the fact that natural images can be well approximated by their LR components. Furthermore, principal image structures typically reside in a low-dimensional subspace, thereby having a LR representation. Therefore, incorporating LR depth structure can improve the reconstruction quality. To this end, we extend the minimization problem (4) as

$$\underset{\mathbf{z}}{\text{minimize}} \quad \frac{1}{2}\|\mathbf{y} - \boldsymbol{\Phi}\,\mathbf{z}\|_2^2 + \lambda\,\|\boldsymbol{\Psi}^T\,\mathbf{z}\|_1 + \beta\,\|\mathbf{z}\|_*. \tag{5}$$

Here, $\|\mathbf{z}\|_*$ is the nuclear-norm of the matrix $\mathbf{z}$, which the sum of its singular values, $\|\mathbf{z}\|_* = \sum_{i=1}^{p} \lambda_i(\mathbf{z})$ with $\lambda_i(\mathbf{z})$ being the ith largest singular value of matrix $\mathbf{z}$ of rank at most p. It is worth noting that the nuclear-norm is the convex relaxation for the LR constraint imposed on a matrix [2].

The LR is regarded as a non-local (global) regularizer since it considers the information throughout the image. As a result, LR may neglect the useful local information that often relates to the detail map. To fill this gap, a TV regularizer is added to guarantee the local spatial consistency. The proposed LR-TV model becomes

$$\underset{\mathbf{z}}{\text{minimize}} \quad \frac{1}{2}\|\mathbf{y} - \boldsymbol{\Phi}\,\mathbf{z}\|_2^2 + \lambda\,\|\boldsymbol{\Psi}^T\,\mathbf{z}\|_1 + \beta\,\|\mathbf{z}\|_* + \gamma\,\|\mathbf{z}\|_{\mathrm{TV}}. \tag{6}$$

Here, $\|\mathbf{z}\|_{\mathrm{TV}}$ is the anisotropic TV defined as

$$\|\mathbf{z}\|_{\mathrm{TV}} = \|\mathbf{D}\,\mathbf{z}\|_1 = \|\mathbf{D}_x\,\mathbf{z}\|_1 + \|\mathbf{D}_y\,\mathbf{z}\|_1, \tag{7}$$

where $\mathbf{D} = [\mathbf{D}_x; \mathbf{D}_y]$ is the first-order forward finite difference operator in the x-horizontal and y-vertical directions. In (6), the penalty parameters β and γ are used to control the importance of LR and TV terms, respectively. The remaining task is to solve Problem (6), which yields an estimate of the depth image.

3.2 ADMM-Based Algorithm

This subsection presents an algorithm to solve Problem (6) using ADMM technique [1,10]. The ADMM is a powerful framework for solving regularized optimization problems as it allows the entire problem to be decomposed into subproblems that can be handled more effectively. For efficiently handling the subproblems, it is vital to choose suitable auxiliary variables. Here, we introduce four auxiliary variables $\mathbf{s} = \mathbf{z}$, $\mathbf{r} = \boldsymbol{\Psi}^T\,\mathbf{z}$, $\mathbf{t} = \mathbf{z}$, and $\mathbf{u} = \mathbf{D}\,\mathbf{z}$. This way, the minimization problem in (6) is rewritten as

$$\begin{aligned} &\underset{\mathbf{z},\mathbf{s},\mathbf{r},\mathbf{t},\mathbf{u}}{\text{minimize}} && \frac{1}{2}\|\mathbf{y} - \boldsymbol{\Phi}\,\mathbf{s}\|_2^2 + \lambda\,\|\mathbf{r}\|_1 + \beta\,\|\mathbf{t}\|_* + \gamma\,\|\mathbf{u}\|_1, \\ &\text{subject to} && \mathbf{s} = \mathbf{z}, \quad \mathbf{r} = \boldsymbol{\Psi}^T\,\mathbf{z}, \quad \mathbf{t} = \mathbf{z}, \quad \mathbf{u} = \mathbf{D}\,\mathbf{z}. \end{aligned} \tag{8}$$

Problem (8) has its augmented Lagrangian function given by

$$\begin{aligned} &\mathcal{L}(\mathbf{z},\mathbf{s},\mathbf{r},\mathbf{t},\mathbf{u},\mathbf{w},\mathbf{b},\mathbf{h},\mathbf{v}) = \\ &\frac{1}{2}\|\mathbf{y} - \boldsymbol{\Phi}\,\mathbf{s}\|_2^2 + \lambda\,\|\mathbf{r}\|_1 + \beta\,\|\mathbf{t}\|_* + \gamma\,\|\mathbf{u}\|_1 \\ &- \mathbf{w}^T(\mathbf{s} - \mathbf{z}) - \mathbf{b}^T(\mathbf{r} - \boldsymbol{\Psi}^T\,\mathbf{z}) - \mathbf{h}^T(\mathbf{t} - \mathbf{z}) - \mathbf{v}^T(\mathbf{u} - \mathbf{D}\,\mathbf{z}) \\ &+ \frac{\mu}{2}\|\mathbf{s} - \mathbf{z}\|_2^2 + \frac{\rho}{2}\|\mathbf{r} - \boldsymbol{\Psi}^T\,\mathbf{z}\|_2^2 + \frac{\xi}{2}\|\mathbf{t} - \mathbf{z}\|_2^2 + \frac{\kappa}{2}\|\mathbf{u} - \mathbf{D}\,\mathbf{z}\|_2^2. \end{aligned} \tag{9}$$

In (9), $\mathbf{w}$, $\mathbf{b}$, $\mathbf{h}$, and $\mathbf{v}$ are the Lagrange multipliers associated with their constraints, and μ, ρ, ξ, and κ are regularization parameters associated with the corresponding quadratic penalty terms.

The idea of ADMM is to find a saddle point of the Lagrangian function $\mathcal{L}(\cdot)$, that is also the solution to Problem (6). The stationary point can be found by solving the following sequence of subproblems, in which the next estimate of each variable at the $(k+1)$th iteration is obtained by fixing the current estimates of the other counterparts,

$$\begin{aligned}
\mathbf{z}_{k+1} &= \arg\min \mathcal{L}(\mathbf{z}, \mathbf{s}_k, \mathbf{r}_k, \mathbf{t}_k, \mathbf{u}_k, \mathbf{w}_k, \mathbf{b}_k, \mathbf{h}_k, \mathbf{v}_k),\\
\mathbf{s}_{k+1} &= \arg\min \mathcal{L}(\mathbf{z}_k, \mathbf{s}, \mathbf{r}_k, \mathbf{t}_k, \mathbf{u}_k, \mathbf{w}_k, \mathbf{b}_k, \mathbf{h}_k, \mathbf{v}_k),\\
\mathbf{r}_{k+1} &= \arg\min \mathcal{L}(\mathbf{z}_k, \mathbf{s}_k, \mathbf{r}, \mathbf{t}_k, \mathbf{u}_k, \mathbf{w}_k, \mathbf{b}_k, \mathbf{h}_k, \mathbf{v}_k),\\
\mathbf{t}_{k+1} &= \arg\min \mathcal{L}(\mathbf{z}_k, \mathbf{s}_k, \mathbf{r}_k, \mathbf{t}, \mathbf{u}_k, \mathbf{w}_k, \mathbf{b}_k, \mathbf{h}_k, \mathbf{v}_k),\\
\mathbf{u}_{k+1} &= \arg\min \mathcal{L}(\mathbf{z}_k, \mathbf{s}_k, \mathbf{r}_k, \mathbf{t}_k, \mathbf{u}, \mathbf{w}_k, \mathbf{b}_k, \mathbf{h}_k, \mathbf{v}_k).
\end{aligned} \tag{10}$$

The Lagrange multipliers are updated as

$$\mathbf{w}_{k+1} = \mathbf{w}_k - \mu(\mathbf{s}_{k+1} - \mathbf{z}_{k+1}), \tag{11}$$

$$\mathbf{b}_{k+1} = \mathbf{b}_{k+1} - \rho(\mathbf{r}_{k+1} - \boldsymbol{\Psi}^T \mathbf{z}_{k+1}), \tag{12}$$

$$\mathbf{h}_{k+1} = \mathbf{h}_k - \xi(\mathbf{t}_{k+1} - \mathbf{z}_{k+1}), \tag{13}$$

$$\mathbf{v}_{k+1} = \mathbf{v}_k - \kappa(\mathbf{u}_{k+1} - \mathbf{D}\,\mathbf{z}_{k+1}). \tag{14}$$

Now, our task is to solve Subproblems in (10). For notation simplicity, we drop the loop index k in the following description.

z-Subproblem: The $\mathbf{z}$-subproblem is obtained by keeping only terms involving $\mathbf{z}$ in (9) yielding

$$\begin{aligned}
\mathbf{z}_{k+1} = &\arg\min\{-\mathbf{w}^T(\mathbf{s}-\mathbf{z}) - \mathbf{b}^T(\mathbf{r}-\boldsymbol{\Psi}^T\mathbf{z}) - \mathbf{h}^T(\mathbf{t}-\mathbf{z}) - \mathbf{v}^T(\mathbf{u}-\mathbf{D}\,\mathbf{z})\\
&+\frac{\mu}{2}\|\mathbf{s}-\mathbf{z}\|_2^2 + \frac{\rho}{2}\|\mathbf{r}-\boldsymbol{\Psi}^T\mathbf{z}\|_2^2 + +\frac{\xi}{2}\|\mathbf{t}-\mathbf{z}\|_2^2 + \frac{\kappa}{2}\|\mathbf{u}-\mathbf{D}\,\mathbf{z}\|_2^2\}.
\end{aligned} \tag{15}$$

Applying the first-order optimal condition to Problem (15), we have

$$\begin{aligned}
[(\mu+\rho+\xi)\mathbf{I} + \kappa\mathbf{D}^T\mathbf{D}]\mathbf{z}_{k+1} = &\boldsymbol{\Psi}^T(\rho\mathbf{r}-\mathbf{b}) + (\mu\mathbf{s}-\mathbf{w})\\
&+(\xi\mathbf{t}-\mathbf{h}) + \mathbf{D}^T(\kappa\mathbf{u}-\mathbf{v}).
\end{aligned} \tag{16}$$

Note that $\mathbf{D}^T\mathbf{D}$ is a circulant matrix, and thus the matrix $(\mu+\rho+\xi)\mathbf{I}+\kappa\mathbf{D}^T\mathbf{D}$ is diagonalizable. Therefore, the solution to Problem (16) can be found efficiently using fast Fourier transforms. In particular, a closed form solution is obtained by

$$\mathbf{z}_{k+1} = \mathcal{F}^{-1}\left[\frac{\mathcal{F}(\mathbf{g})}{(\mu+\rho+\xi)\mathbf{I} + \kappa|\mathcal{F}(\mathbf{D})|^2}\right]. \tag{17}$$

In (17), $\mathbf{g}$ is the right hand side of (16), i.e., $\mathbf{g} = \boldsymbol{\Psi}^T(\rho\mathbf{r}-\mathbf{b}) + (\mu\mathbf{s}-\mathbf{w}) + (\xi\mathbf{t}-\mathbf{h}) + \mathbf{D}^T(\kappa\mathbf{u}-\mathbf{v})$. The notation $\mathcal{F}(\cdot)$ is the 2D Fourier transform, whereas $\mathcal{F}^{-1}(\cdot)$ is the 2D inverse Fourier transform. The term $|\mathcal{F}(\mathbf{D})|^2$ is the square magnitude of the eigenvalues of the differential operator matrix $\mathbf{D}$.

s-Subproblem: the $\mathbf{s}$-subproblem is obtained by holding only $\mathbf{s}$-related terms in (9):

$$\frac{1}{2}\|\mathbf{y} - \boldsymbol{\Phi}\,\mathbf{s}\|_2^2 - \mathbf{w}^T(\mathbf{s} - \mathbf{z}) + \frac{\mu}{2}\|\mathbf{s} - \mathbf{z}\|_2^2. \tag{18}$$

Performing the first-order optimal condition produces

$$(\boldsymbol{\Phi}^T\,\boldsymbol{\Phi} + \mu\mathbf{I})\,\mathbf{s} = \boldsymbol{\Phi}^T\,\mathbf{y} + \mathbf{w} + \mu\mathbf{z}. \tag{19}$$

As $\boldsymbol{\Phi}$ is a diagonal binary matrix, the solution for $\mathbf{s}_{k+1}$ is computed efficiently through an element-wise manner.

r-Subproblem: The $\mathbf{r}$-subproblem is obtained by keeping only $\mathbf{r}$-related terms in (9) yielding

$$\lambda\,\|\mathbf{r}\|_1 - \mathbf{b}^T(\mathbf{r} - \boldsymbol{\Psi}^T\,\mathbf{z}) + \frac{\rho}{2}\|\mathbf{r} - \boldsymbol{\Psi}^T\,\mathbf{z}\|_2^2. \tag{20}$$

Define an element-wise soft-thresholding operator,

$$\mathcal{T}(x,\tau) = \text{sign}(x)\max(|x| - \tau, 0) = \frac{x}{|x|}\,\max(|x| - \tau, 0). \tag{21}$$

This way, Problem (20) has a closed-form solution,

$$\mathbf{r}_{k+1} = \mathcal{T}(\boldsymbol{\Psi}^T\,\mathbf{z} + \frac{\mathbf{b}}{\rho}, \frac{\lambda}{\rho}) = \text{sign}(\boldsymbol{\Psi}^T\,\mathbf{z} + \frac{\mathbf{b}}{\rho})\,\max(\left|\boldsymbol{\Psi}^T\,\mathbf{z} + \frac{\mathbf{b}}{\rho}\right| - \frac{\lambda}{\rho}, 0). \tag{22}$$

Hereafter, the soft-thresholding operator $\mathcal{T}(\cdot)$ is used in element-wise manner.

t-Subproblem: likewise, the $\mathbf{t}$-subproblem is obtained by holding only $\mathbf{t}$-terms in (9) having

$$\beta\,\|\mathbf{t}\|_* - \mathbf{h}^T(\mathbf{t} - \mathbf{z}) + \frac{\xi}{2}\|\mathbf{t} - \mathbf{z}\|_2^2. \tag{23}$$

This problem can be solved efficiently through a singular value soft-thresholding (SVT) operator $\mathcal{S}(\cdot)$,

$$\mathbf{t}_{k+1} = \mathcal{S}(\mathbf{z} + \frac{\mathbf{h}}{\xi}, \frac{\beta}{\xi}). \tag{24}$$

Here, the SVT operator in the form $\mathcal{S}(\mathbf{x},\tau)$ is computed by applying the soft-thresholding operator with τ to the singular values of the input matrix $\mathbf{x}$,

$$\mathcal{S}(\mathbf{x},\tau) = \mathbf{u}\,\mathcal{T}(\boldsymbol{\lambda},\tau)\,\mathbf{v}^T, \tag{25}$$

where $\mathbf{x} = \mathbf{u}\,\boldsymbol{\lambda}\,\mathbf{v}^T$ is the singular value decomposition of the input matrix $\mathbf{x}$.

u-Subproblem: the $\mathbf{u}$-subproblem is obtained by keeping only $\mathbf{u}$-related terms in (9) yielding

$$\gamma\,\|\mathbf{u}\|_1 - \mathbf{v}^T(\mathbf{u} - \mathbf{D}\,\mathbf{z}) + \frac{\kappa}{2}\|\mathbf{u} - \mathbf{D}\,\mathbf{z}\|_2^2. \tag{26}$$

The solution to Problem (26) is obtained by,

$$\mathbf{u}_{k+1} = \mathcal{T}(\mathbf{D}\,\mathbf{z} + \frac{\mathbf{v}}{\kappa}, \frac{\gamma}{\kappa}) = \text{sign}(\mathbf{D}\,\mathbf{z} + \frac{\mathbf{v}}{\kappa})\,\max(\left|\mathbf{D}\,\mathbf{z} + \frac{\mathbf{v}}{\kappa}\right| - \frac{\gamma}{\kappa}, 0). \tag{27}$$

In summary, the ADMM-based algorithm is presented in Algorithm 1. This algorithm starts by the initialization of the variables and iteratively updates them until convergence. Here, the stopping condition is satisfied if the relative change of the image $\mathbf{z}$ is negligible, i.e., $\|\mathbf{z}_{k+1} - \mathbf{z}_k\|_2 / \|\mathbf{z}_k\|_2 < \text{tol}$. The algorithm is computational efficient because the variable updates have closed-form solutions with element-wise computation.

Algorithm 1. LR-TV regularized depth estimation.

1: Input: $\mathbf{y}$, $\boldsymbol{\Phi}$
2: Initialize $\mathbf{z}_0 = \boldsymbol{\Phi}^T \mathbf{y}$, $\mathbf{s}_0 = \mathbf{z}_0$, $\mathbf{r}_0 = \boldsymbol{\Psi}^T \mathbf{z}_0$, $\mathbf{t}_0 = \mathbf{z}_0$, $\mathbf{u}_0 = \mathbf{D}\,\mathbf{z}_0$.
3: **repeat**
4: Update $\mathbf{z}$ using (17).
5: Update $\mathbf{s}$ by solving (19).
6: Update $\mathbf{r}$, $\mathbf{t}$, and $\mathbf{u}$ using (22), (24), and (27), respectively.
7: Update the multipliers $\mathbf{w}$, $\mathbf{b}$, $\mathbf{h}$, and $\mathbf{v}$ using (11)–(14), respectively.
8: **until** $\|\mathbf{z}_{k+1} - \mathbf{z}_k\|_2 / \|\mathbf{z}_k\|_2 < \text{tol}$.

4 Experimental Results

This section gives experimental evaluations on depth image benchmark datasets. Subsection 4.1 presents the setup for experiments, followed by imaging results, performance analysis and comparisons in Subsects. 4.2 and 4.3.

4.1 Experimental Setup

The proposed approach is evaluated using the Middlebury Stereo Dataset[1] [15], where the ground truth depth maps are available. To measure the performance accuracy, the peak signal-to-noise ratio (PSNR) is used (in dB):

$$\text{PSNR} = 10 \log_{10} \left(\frac{I_{\text{peak}}^2}{\text{MSE}} \right), \tag{28}$$

where I_{peak} is the peak intensity of reconstructed image I, and MSE is the mean-square-error between the reconstructed image I and the ground-truth image I_g defined as

$$\text{MSE} = \frac{1}{h \times w} \sum_{i=1}^{h} \sum_{j=1}^{w} |I(i,j) - I_g(i,j)|^2 . \tag{29}$$

The algorithm requires a set of parameters that need to be set appropriately. The basis matrix $\boldsymbol{\Psi}$ is constructed from wavelets with Daubechies 2 and 3 decomposition levels. Although the algorithm gives satisfactory results for a range of parameters, the typical settings for the regularization parameters are given in Table 1. The algorithm converges if the relative change of the reconstructed image is smaller than $\text{tol} = 10^{-4}$ (see Step 8 in Algorithm 1).

[1] http://vision.middlebury.edu/stereo/data/.

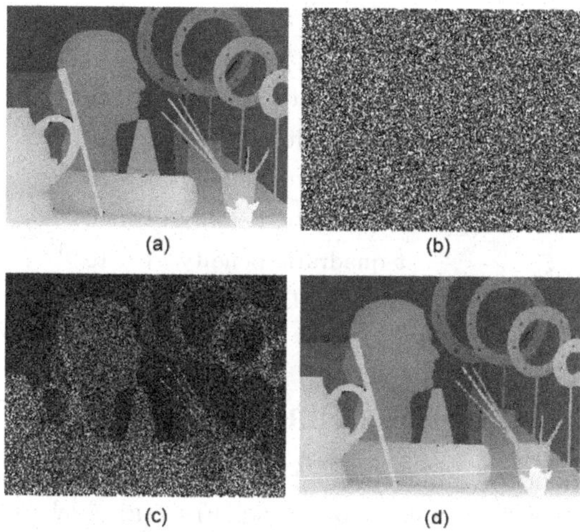

Fig. 1. Depth image reconstructed by the proposed LR-TV algorithm: (a) ground-truth depth image of art, (b) the random missing mask keeping only 50% pixels, (c) corrupted depth image with only 50% data measurements (PSNR = 7.08 dB), and (d) reconstructed image by the proposed LR-TV model (PSNR = 31.28 dB).

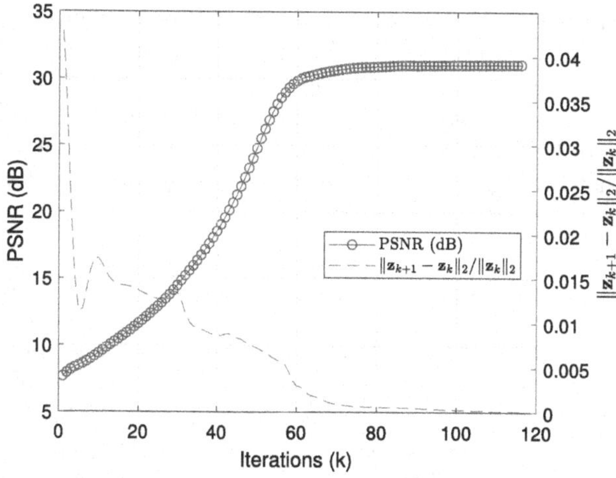

Fig. 2. PSNRs of the depth image reconstructed by the proposed LR-TV algorithm (solid line) and the relative change of the image $\mathbf{z}$, $\|\mathbf{z}_{k+1} - \mathbf{z}_k\|_2/\|\mathbf{z}_k\|_2$ (dashed line) recorded during the minimization using 50% of total measurements.

Table 1. Parameter settings.

Parameters	Related terms	Values
λ	Sparsity penalty	4×10^{-5}
β	LR penalty	1×10^{-2}
γ	TV penalty	1×10^{-3}
μ	**s**-quadratic penalty	1×10^{-2}
ρ	**r**-quadratic penalty	1×10^{-3}
ξ	**t**-quadratic penalty	1×10^{-3}
κ	**u**-quadratic penalty	1×10^{-1}

4.2 Evaluation and Analysis of LR-TV Model

In the first experiment, we aim to evaluate the performance of the LR-TV model for depth image reconstruction. In doing so, Fig. 1(a) shows the ground-truth depth image of art with a size of $h \times w = 277 \times 347$ used to evaluate the performance. Now, only 50% measurements of the total pixels are randomly selected using the mask shown in Fig. 1(b). Because only 50% measurements are kept, the depth image is corrupted as shown in Fig. 1(c) with a PSNR = 7.08 dB. Using the 50% data measurements, the depth image reconstructed by the LR-TV model is presented in Fig. 1(d). It can be observed that the LR-TV model estimates the depth image well and yields a high quality image with a PSNR = 31.28 dB.

For further insights into the proposed algorithm, we report here the PSNR values of the estimated image and the relative change of the depth image, $\|\mathbf{z}_{k+1} - \mathbf{z}_k\|_2/\|\mathbf{z}_k\|_2$, during the minimization. Figure 2 shows the PSNRs of the images estimated during minimization as a function of iterations. It can be observed from the figure that the image quality is enhanced during the update, and the quality in terms of PSNR is not changed much after 60 iterations. Furthermore, it can be seen that the algorithm is deemed to converge after 116 iterations when the relative change reaches 9.5×10^{-5}.

4.3 Comparison with Other Imaging Approaches

In the second experiment, we aim to compare the performance of the proposed LR-TV model with those by other imaging models. Two other imaging models are considered here. The first approach is the sparsity-based method for depth estimate that exploits only the sparseness of depth representation [11]. The second method considers both the sparsity and TV regularization for depth reconstruction [12]. We evaluate these models using two missing scenarios: random missing and entire (large) column missing. The random missing is generally related to compressive sensing data acquisition, whereas the large column missing is typically due to poor hardware performance. The ground-truth depth image and the corrupted images due to the masks for the two missing cases are shown

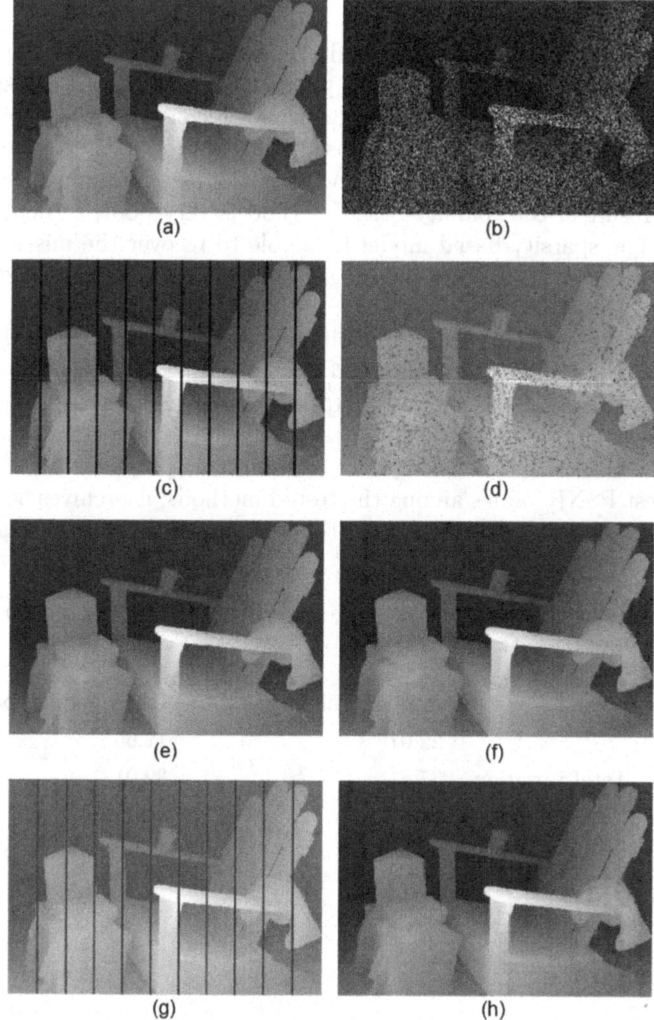

Fig. 3. Depth image reconstructed by the different imaging models with two missing data patterns: (a) the ground-truth depth image, (b) the corrupted image by the missing mask representing missing values (black pixels), (c) the corrupted image by the columns-missing mask representing missing values (black pixels); for the random missing case, the depth images reconstructed by (d) the sparsity-based model, (e) the sparsity-TV model, and (f) the LR-TV model; for the large column missing case, the depth images reconstructed by (g) the sparsity-based model, and (g) the proposed LR-TV model.

in Figs. 3(a)–(c), respectively. Here, the random missing mask is generated by keeping 50% pixels, and the missing column width is 5 pixels. The same input corrupted data and the same input parameters are used for all the evaluated models.

For the random missing case, Figs. 3(d)–(f) show the images recovered by the sparsity-based, the sparsity-TV, and the proposed LR-TV method, respectively. It can be observed that all the three models can reconstruct the depth images well, but the sparsity-TV and the proposed LR-TV produce better reconstructions compared with the sparsity method. For the large column missing case, the imaging result of the sparsity-based method is very poor, as demonstrated in Fig. 3(g). The sparsity-based model is unable to recover the missing column pixels. The proposed LR-TV, on the other hand, reconstructs the image well, as shown in Fig. 3(h).

To quantify the performances of the different imaging approaches, the PSNRs of the reconstructed images are computed and listed in Table 2. The most noticeable feature from the table is the considerable improvement of the sparsity + TV and the LR+TV over the sparsity model, especially for the large missing data case. Furthermore, for both missing data patterns, the proposed LR+TV model has the highest PSNR values among the tested methods; it archives 38.22 dB for the random missing data case, and 40.39 dB for large missing data case.

Table 2. PSNRs in dB of the depth images reconstructed by the different imaging models for two missing data patterns.

Imaging methods	Random missing pattern	Large missing pattern
Sparsity	22.07	17.65
Sparsity + Total variation	37.81	39.01
Low rank + Total variation	**38.22**	**40.39**

5 Conclusion

This paper presented a new LR-TV imaging model for depth image reconstruction from incomplete data measurements. The proposed approach formulates the task of depth image estimate as a joint LR and TV regularized minimization problem and proposes an algorithm based on ADMM to solve it, yielding a reconstructed depth image. By incorporating the LR and TV regularizers, the proposed model yields high quality image reconstruction for different missing data cases. Experimental evaluations are provided and the results show the proposed model is very promising that it enhances the quality of image estimation and outperforms the other evaluated state-of-the-art imaging models in terms of PSNR metric.

References

1. Boyd, S., Parikh, N., Chu, E., Peleato, B., Eckstein, J.: Distributed optimization and statistical learning via the alternating direction method of multipliers. Found. Trends Mach. Learn. **3**(1), 1–122 (2011)
2. Cai, J.F., Candes, E.J., Shen, Z.: A singular value thresholding algorithm for matrix completion. SIAM J. Optim. **20**(4), 1956–1982 (2010)
3. Camplani, M., Salgado, L.: Adaptive spatio-temporal filter for low-cost camera depth maps. In: IEEE International Conference on Emerging Signal Processing Applications, pp. 33–36 (2012)
4. Candes, E.J., Wakin, M.B.: An introduction to compressive sampling. IEEE Signal Process. Mag. **25**(2), 21–30 (2008)
5. Chen, L., Lin, H., Li, S.: Depth image enhancement for Kinect using region growing and bilateral filter. In: IEEE International Conference on Pattern Recognition, pp. 3070–3073 (2012)
6. Chen, W., Fu, Z., Yang, D., Deng, J.: Single-image depth perception in the wild. In: Advances in Neural Information Processing Systems, pp. 730–738 (2016)
7. Chodosh, N., Wang, C., Lucey, S.: Deep convolutional compressed sensing for lidar depth completion. In: Jawahar, C.V., Li, H., Mori, G., Schindler, K. (eds.) ACCV 2018. LNCS, vol. 11361, pp. 499–513. Springer, Cham (2019). https://doi.org/10.1007/978-3-030-20887-5_31
8. Dimitrievski, M., Veelaert, P., Philips, W.: Learning morphological operators for depth completion. In: Blanc-Talon, J., Helbert, D., Philips, W., Popescu, D., Scheunders, P. (eds.) ACIVS 2018. LNCS, vol. 11182, pp. 450–461. Springer, Cham (2018). https://doi.org/10.1007/978-3-030-01449-0_38
9. Donoho, D.L.: Compressed sensing. IEEE Trans. Inf. Theory **52**(4), 1289–1306 (2006)
10. Han, D., Yuan, X.: A note on the alternating direction method of multipliers. J. Optim. Theory Appl. **155**(1), 227–238 (2012)
11. Hawe, S., Kleinsteuber, M., Diepold, K.: Dense disparity maps from sparse disparity measurements. In: IEEE International Conference on Computer Vision, pp. 2126–2133 (2011)
12. Liu, L.K., Chan, S.H., Nguyen, T.Q.: Depth reconstruction from sparse samples: representation, algorithm, and sampling. IEEE Trans. Image Process. **24**(6), 1983–1996 (2015)
13. Mallat, S.: A Wavelet Tour of Signal Processing, Third Edition: The Sparse Way. Academic Press, Inc., Orlando (2008)
14. Saxena, A., Chung, S.H., Ng, A.Y.: Learning depth from single monocular images. In: Advances in Neural Information Processing Systems, pp. 1161–1168 (2006)
15. Scharstein, D., et al.: High-resolution stereo datasets with subpixel-accurate ground truth. In: Jiang, X., Hornegger, J., Koch, R. (eds.) GCPR 2014. LNCS, vol. 8753, pp. 31–42. Springer, Cham (2014). https://doi.org/10.1007/978-3-319-11752-2_3
16. Scherzer, O.: Handbook of Mathematical Methods in Imaging. Springer, New York (2011). https://doi.org/10.1007/978-0-387-92920-0
17. Sun, M., Ng, A.Y., Saxena, A.: Make3D: learning 3D scene structure from a single still image. IEEE Trans. Pattern Anal. Mach. Intell. **31**(5), 824–840 (2009)
18. Suwajanakorn, S., Hernandez, C., Seitz, S.M.: Depth from focus with your mobile phone. In: IEEE Conference on Computer Vision and Pattern Recognition, pp. 3497–3506 (2015)

19. Zhang, R., Tsai, P.S., Cryer, J.E., Shah, M.: Shape from shading: a survey. IEEE Trans. Pattern Anal. Mach. Intell. **21**(8), 690–706 (1999)
20. Zhu, Z.Y., Chan, S., Shum, H.Y.: Real-time depth image acquisition and restoration for image based rendering and processing systems. J. Sig. Process. Syst. **79**, 1–18 (2013)
21. Zou, N., Xiang, Z., Chen, Y.: RSDCN: a road semantic guided sparse depth completion network. Neural Process. Lett. **51**(3), 2737–2749 (2020). https://doi.org/10.1007/s11063-020-10226-7

Deep Learning Based Hyperspectral Images Analysis for Shrimp Contaminated Detection

Minh-Hieu Nguyen[1], Xuan-Huyen Nguyen-Thi[1], Cong-Nguyen Pham[1], Ngoc C. Lê[2], and Huy-Dung Han[1(✉)]

[1] Electronics and Computer Engineering Department, School of Electronics and Telecommunications, Hanoi University of Science and Technology, Hanoi, Vietnam
dung.hanhuy@hust.edu.vn

[2] School of Applied Mathematics and Informatics, Hanoi University of Science and Technology, Hanoi, Vietnam

Abstract. In this paper, a deep learning based hyperspectral image analysis for detecting contaminated shrimp is proposed. The ability of distinguishing shrimps into two classes: clean and contaminated shrimps is visualized by t-distributed Stochastic Neighbor Embedding (t-SNE) using spectral feature data. Using only some small data set of hyperspectral images of shrimps, a simple processing technique is applied to generate enough data for training a deep neural network (DNN) with high reliability. Our results attain the accuracy of 98% and F1-score over 94%. This works confirms that with only few data samples, Hyperspectral Imaging processing technique together with DNN can be used to classify abnormality in agricultural productions like shrimp.

Keywords: Hyperspectral Imaging · Abnormality classification · t-SNE · Deep neural network

1 Introduction

Recently, Hyperspectral Imaging/Image (HSI) shows its potential in many sectors such as medical applications, remote sensing imagery or food processing [2–6]. Employing capability of taking images of multiple wavelength bands, including the visible and near-infrared region (VNIR), the HSI system can provide much higher details and extra information compare to conventional RGB (Red Green Blue) images [2]. In HSI, each pixel of the image contains spectral information representing the electromagnetic strength of a narrow wavelength range reflected from the object beamed under a full spectrum light. The wavelength is added as a third dimension to the two-dimensional spatial image, producing a three-dimensional data cube [3]. Thanks to the expanded wavelength range in HSI data cube, HSI captures richer information invisible to the human

N.-S. Vo and V.-P. Hoang (Eds.): INISCOM 2020, LNICST 334, pp. 195–205, 2020.
https://doi.org/10.1007/978-3-030-63083-6_15

eyes, enabling possibilities of data computation and analysis [7]. The rich information in HSI data cube also opens plethora of new applications.

In a report of the U.S. Food and Drug Administration (FDA)[1] in year 2019, three out of the 85 (3.5%) total seafood entry line refusals were of shrimp for reasons related to banned antibiotics. Contaminated shrimp is a big trouble to Vietnamese export seafood sector at the moment. There are several impurities such as agar (jelly-like) or chemical materials like veterinary drug residues. Therefore, the detection of impurities in shrimps is a necessary task to protect health of consumers, to prevent contaminated shrimps from the sources, to ensure the quality and the reputation of Vietnamese seafood production. This work investigates the shrimps suffering from being injected contaminated substance for the purpose of increasing shrimps' weight. To our best knowledge, the problem of classifying clean and contaminated shrimps have not been addressed before in Vietnam.

In this research, HSI of is utilized for shrimps contaminated by unexpected substance detection. The used HSI photography provides detailed spectral data of the shrimps in visible and near-infrared region (VNIR) [7]. The SPECIM FX camera provides 224 spectra spreading from 400 nm to 1000 nm for each pixel of a shrimp image. With pushbroom line scanning mode, the target shrimp is scanned one spatial line at a time and all spectral data from that line is acquired simultaneously. By collecting the unique and detailed spectral information of shrimps, HSI could reveal different materials and physical characteristics. Therefore, the clean shrimps and contaminated shrimps can be classified. Data augmentation by t-Distributed Stochastic Neighbor Embedding (t-SNE) technique has shown that it is possible to classify the two types of shrimps. Training based DNN technique is applied to the collected HSI data and shows the capability of distinguishing clean shrimps and contaminated shrimps with high accuracy.

In the next section, we will summary some works related to the issue. The system model will be presented in Sect. 3. Section 4 is a description about the data preprocessing techniques used in this research. Section 5 presents the experiments set up and the results while Sect. 6 is devoted for some conclusion about the research.

2 Related Works

In agriculture and food industry, HSI can be used for detecting the unique spectral information of meat, fruits [4–6]. It can be used to identify, measure, and locate different materials as well as their chemical and physical properties. Machine learning, especially deep learning algorithms, which are popular for image processing, can be applied to these applications. Recently, deep learning algorithms have gained many attentions as they can solve many difficult problems [8–10]. For HSI, DNN are applied to classify food, the model achieves the best performance with a 94.4% overall classification accuracy independent of the

[1] https://www.shrimpalliance.com/fda-refuses-antibiotic-contaminated-shrimp-from-china-and-vietnam-in-july/.

state of the products, with high accuracy [5]. A research [5] presents an inclusive analysis of the performance of Hyperspectral Imaging for detecting adulteration in red-meat products. In [6], the developments and applications of HSI in the nondestructive assessment of fruits and vegetables are presented. Partial least squares regression is used as a multivariate method in calibration of spectroscopy data. In [4], deep learning algorithm is applied to visible and near-infrared HSI of shrimps for distinguishing the cleanness of shrimp during cold storage.

In food industry, shrimp is one of the rich in protein production, high quality food and the tasks of guaranteeing the quality and safety of s are crucial. In [4], the cleanness of shrimp during cold storage is investigated using visible light and near-infrared (VIS/NIR) hyperspectral images. Deep learning algorithm is applied to analyze the HSI for calculating the cleanness of shrimps. The stacked Autoencoders - Linear Regression algorithm achieves satisfactory total classification accuracy of 96.55% and 93.97% for freshness grade of shrimp in calibration (116 samples) and prediction (116 samples) sets, respectively. In [11], the authors aimed to develop an image-based method for detection of improperly deveined shrimps. A sequence of image processing techniques in this work is subjected before extracting significant parameters from gray scale images. Some parameters including shape measurements and pixel value measurements are drawn from the image histogram. Then, deveined shrimps were identified by two classification techniques: linear discriminant analysis and support vector machine (SVM). Despite the excellent capability of the HSI in food industries, there are still not many works applying HSI to shrimp productions.

In this paper, the classification of shrimps under limited number of HSI data cubes is presented. Our system includes the following steps: collecting, preprocessing and extracting data, then building a DNN model to classify these data into two classes (contaminated shrimps and clean shrimps). Due to limited a number of samples, the data augmentation approach in [4] is applied to enrich our training data. By extracting spectral information of HSI, the difference between clean shrimps and contaminated ones can be explored. These data will be used for detecting contaminated shrimps using a DNN model. The results shows that DNN can classify shrimps using small number of HSI data cube with high accuracy.

3 System Model

The system for detecting contaminated shrimps from their HSIs using deep neural network was shown in Fig. 1. Let the HSI data cube be $s = [s_{ijk}]$ which is three dimensional data matrix of size $H \times W \times F$, where H, W are the height and the width of the shrimp image in pixel, and F is the number of wave length bins representing the recordable bandwidth of the HSI machine. Here, i,j are the width and height indices of the shrimp image, and k is the wave length index starting 1 to 224 to represent the wavelength of the HSI from 400 nm to 1000 nm. Going together with each HSI is a RGB image of the same size. The meat parts of the shrimps are extracted. Segmentation is applied to the input

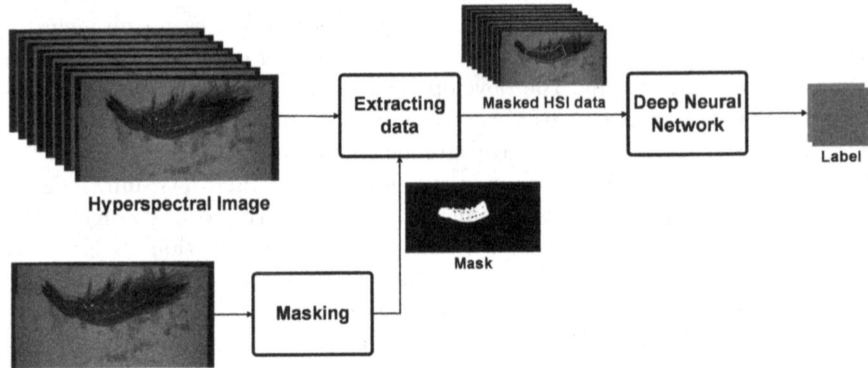

Fig. 1. System model for classifying using hyperspectral image, RGB image and deep neural network.

HSI data cubes to separate the shrimp image part and the background. Noise filtering techniques are applied to remove unwanted noise caused by the shooting environment. The denoised data is extracted and fed into the neural network. The training HSI data cubes labeled as "clean" and "contaminated" are used to finding the optimum weights of the DNN. After training, our system can be used to classify clean and contaminated shrimps.

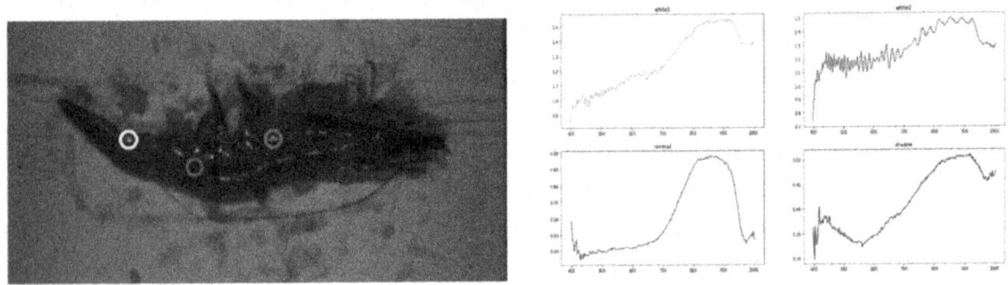

Fig. 2. Visualize noise data. (a) white noise (yellow and red), shadow of shrimp (blue), normal point (pink). (b) spectral data of fours points in (a) (Color figure online)

First, the useful part of data cube, i.e., the meat parts of the shrimps, is extracted in the data pre-processing phase. Figure 2(a) shows a shrimp image and Fig. 2(b) shows the hyper spectrum at four locations: good data area (marked as pink), white area (marked as yellow and red), shadow of the shrimp (marked as blue). It can be seen that the spectrum of the good data area has different properties compared to that of white area and shadows, especially in infrared regime. Due to the strong light condition in HSI machine, some parts of the shrimp image appear as white marks as the reflection from the light. As the marks are pure white, a simple threshold-based method can be applied to the data cube to exclude the white marks from the good data [12]. The background of the images is also white, therefore, this method also can separate the shrimp and the background. Figure 3(a) shows the results of white area removal. The white marked and the background is removed from the image.

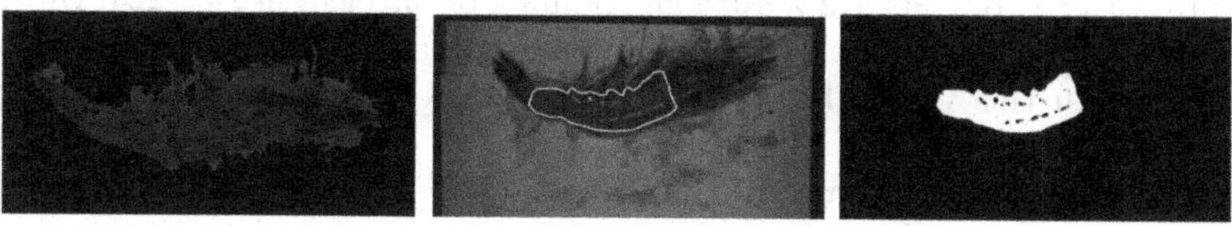

Fig. 3. Preprocessing and extracting data of shrimp from hyperspectral image. (a) our first mask which removes white noise and white background. (b) boundary of shrimp body after using Labelme. (c) the last mask achieved by combining our first mask (a) and boundary (b)

However, the data cube still mix the meat part of shrimp, the shrimp legs, and the shadow. The shrimp legs and the shadow part can be removed by the segmentation process. For focusing to the problem of shrimp classification, in this work, the data segmentation is performed manually with the associated RGB images to have good segmentation results. The shrimp images are segmented to extract only the body parts. As can be seen in Fig. 3(a), the boundary of shrimps should be created to avoid shrimp shadow generated by the light from the spectroscopy machine. Because impurities or chemical changes usually happens on shrimp bodies, only the bodies of shrimps where shrimps' flesh is concentrated is taken into account. Using the Labelme tool [13], a new boundary for shrimp is achieved as in Fig. 3(b). Combining the first mask and the boundary, a better mask as in Fig. 3(c) is produced.

Figure 4(a) shows the resulting hyperspectrum of several data cubes after extraction for clean shrimps (marked as blue) and contaminated shrimps (marked as red). The hyper spectrum curves of each HSI data cube are averaged from the hyperspectrum curves of every pixels belonging to the data cube. The average hyperspectrum can be calculated as

$$s_k = \frac{1}{HW} \sum_i \sum_j s_{ijk} \tag{1}$$

The HSI data cube now becomes the vector $x = [s_1, s_2, ..., s_F]$ of size F. This new data can be used to train the DNN to classify the clean shrimps and contaminated shrimps.

4 Data Preprocessing

4.1 Hyperspectral Data Visualization

In general, DNN can be applied to classify the two type of shrimps using processed HSI data. Usually, the required number of inputs for training a DNN can be up to thousands. If the number of training data is not large enough, DNN can not learn correctly most of object features and this may lead to incorrect prediction.

Our data set includes 20 HSIs of clean shrimps and 15 HSIs of contaminated shrimps with size of HSIs 512 × 1024 and number of wavelength $F = 224$. Due to the limited number of shrimp HSI data cubes, data augmentation for HSI must be applied. In this work, the shrimp bodies' images are divided into smaller square areas. Let the size of the square be S, we can divide a shrimp HSI of size $H \times W \times F$ into many shrimp HSI images of size $S \times S \times F$ depending on the result of the segmentation process. Using (1) to the square images, we obtained a quite large data set of hyperspectrum vectors of size F used for classification process using neural network.

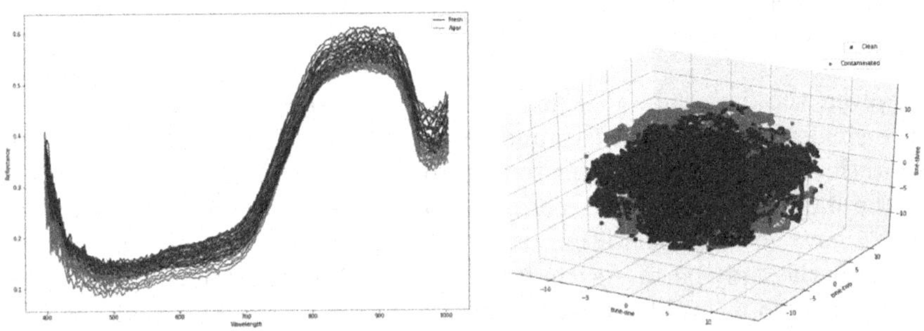

Fig. 4. t-SNE anlysis on shrimp's hyperspectrum. (a) mean spectra of each of two classes: contaminated shrimp (red) and clean shrimp (blue). (b) scatter plot 3D for contaminated shrimp class and clean shrimp class. (Color figure online)

As can be seen in Fig. 4(a), the spectrum curves of clean shrimps and contaminated shrimps are quite similar, making it difficult for classification. The spectrum of each pixel of data cubes for two types of shrimps are even more intermingled. Therefore, we apply a non-linear visualization technique to determine the possibility of classification. Recently, t-SNE has been popular for data visualization as it can reduces the high dimensional data set into 3-dimensional data set where the dissimilarity properties of two data set is preserved statistically [1]. Let the hyperspectral data set of some shrimp be $X = \{x_1, x_2, ..., x_N\}$, where N is the total number of hyperspectral data vectors calculated from (1) for all augmented data cubes. The t-SNE output is the data set $Y = \{y_1, y_2, ..., y_N\}$ of 3-dimensional data vectors. As such, the hyperspectral data of F dimensions can be visualized.

The perplexity parameter is set at 40 according to the recommendation of skLearn framework. To ensure the convergence of the Kullback-Leibler divergence loss function, the number of iterations is set at 400. Other parameters are set as default in sklearn.manifold.TSNE function of skLearn framework.

Figure 4(b) shows the best output of t-SNE in 3D coordinators. It can be seen that the data points corresponding to the contaminated shrimp (marked as red) is in the outer sphere while the clean shrimp data points gathered in the inner sphere. Hence, it is possible to differentiate the contaminated shrimps from the clean ones. Due to the t-SNE property that the visualization results are non-deterministics, it can not be used for classification. In the following section, contaminated/clean shrimps classification using Deep Neural Network (DNN) is presented.

4.2 DNN for Classification

In this sub-section, we describe the deep learning model for classifying contaminated shrimps and clean shrimps. To the best of our knowledge, this is the first time that DNN has been applied to HSI imaging for contaminated shrimp detection.

Different dataset created by different sizes of window would affect the results of the model.

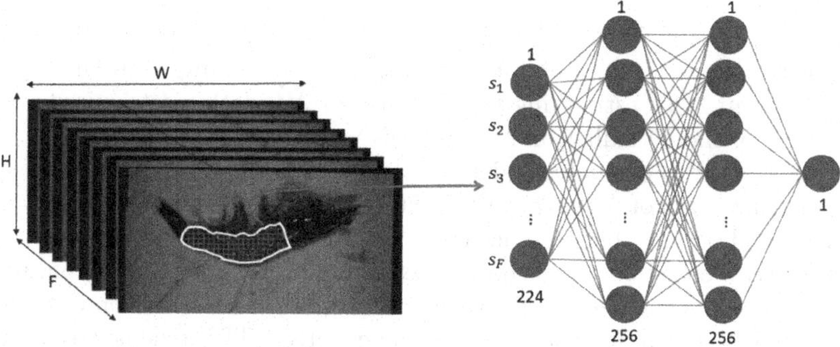

Fig. 5. Applying deep neural network in hyperspectral image with many size of window data.

A fully connected layers DNN with two hidden layers is used. Table 1 summarizes the specifications of the DNN model and Fig. 5 describes the DNN with shrimps' hyperspectral vectors as inputs. The input layer has 224 units corresponding to 224 spectral bin of the HSI and takes the HSI vector x as input. The output layer has 1 unit as the results is binary classification. The output is 1 if contaminated shrimp is detected and 0 if clean shrimp is detected. At each hidden layer, Rectified Linear Unit function (ReLU) [14] to non-linearity transformation are applied for accelerating the training processes. In the classification layer, i.e., output layer, the Sigmoid activation [15] is used.

For training the DNN, we use Adam algorithm [16] with the back-propagation algorithm for minimizing the loss function and updating the weight matrix and the bias vector of each layer. The binary cross entropy algorithm [17] is chosen as loss function:

$$J = -\frac{1}{M}\sum_{i=1}^{N}(z_i \times log(\hat{z}_i) + (1 - z_i) \times log(1 - \hat{z}_i)), \tag{2}$$

Table 1. Specifications of DNN model

Layer	No. unit	Activation function	Parameters
Input layer	224	-	-
Hidden layer 1	256	ReLU	57600
Hidden layer 2	256	ReLU	65792
Output layer	1	Sigmoid	257

where M is the batch size for each DNN's weight update, z_i is label of ith data in the batch and $\hat{z}_i$ is the model prediction of ith data.

5 Experiments and Results

5.1 DNN Setup and Metrics

These input data set X with their labels is divided into three sets: Training set (60%), Validating set (20%) and Testing set (20%). Training set is the input of DNN for learning, for estimating the parameters. Validating set is used to give an estimate of model skill while tuning DNN model's hyper-parameters. The testing set is used to evaluate model.

The learning rate of Adam algorithm is set at 0.001. The batch size $M = 128$ to have a good balance between accuracy and running time.

The metrics based on confusion matrix including accuracy, precision, recall, F1-score is used to evaluate the experimental results. The definition of the confusion matrix is shown in Table 2, where true positive (TP), true negative (TN), false positive (FP), and false negative (FN) are the number of correctly recognized contaminated shrimp, the number of correctly recognized clean shrimp, the number of incorrectly recognized clean shrimp, and the number of incorrectly recognized contaminated shrimp, respectively.

Accuracy, precision, recall, F1-score for evaluating the effectiveness of the DNN model can be calculated as follows

$$Accuracy = \frac{TP + TN}{TP + TN + FP + FN} \tag{3}$$

$$Precision = \frac{TP}{TP + FP} \tag{4}$$

Table 2. Confusion matrix

		Predicted	
		Contaminated shrimp	Clean shrimp
Real	Contaminated shrimp	TP	FN
	Clean shrimp	FP	TN

$$Recall = \frac{TP}{TP + FN} \tag{5}$$

$$F1 = \frac{2 \times Precision \times Recall}{Precision + Recall} \tag{6}$$

5.2 Results and Discussion

In the first experiments, we examine the effect of window size S on the training process of DNN. The window size under test are 1, 3, 5 and 8. In these training processes, early stopping is set as 20. Early stopping results are used for evaluate results on validating set. If the accuracy of validating set does not improve after 20 epochs, the training processes are interrupted. The accuracy and the loss of the training processes are shown in Fig. 6(a) and Fig. 6(b), respectively. All of the training processes converge and achieve reasonable training accuracy and loss after about 50 epochs. The training accuracy and loss of window size of 5 gains the highest results while that of window size 1 exhibits the worst values.

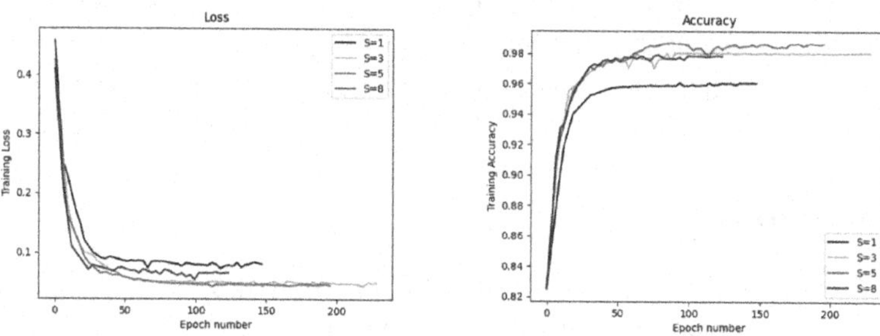

Fig. 6. The training process. (a) loss. (b) accuracy

Table 3 show the loss, accuracy, recall, precision and F1-score of experiments with different window size and number of data points. Note that, the number of data points does not reduce with square rate as the window size S increases. For having enough data points to train the DNN, for window size larger than 1, a sliding window technique is applied to sample the data. The influence of the window size on the accuracy and F1-score is also shown in Fig. 7. The losses are small enough to decide that the DNN model has converged. The other metrics shows that the DNN works well for all cases and the classification of contaminated shrimps and clean shrimps can be accomplished using DNN. The results also show that the data point of size 5×5 provides the best result. The accuracy is 98.11% and the F1-score is 94.29%. It is reasonable because the averaging of spectra in each window helps to reject several types of noise, such as salt-and-pepper noise. Furthermore, averaging over large size of windows may leads to feature loss.

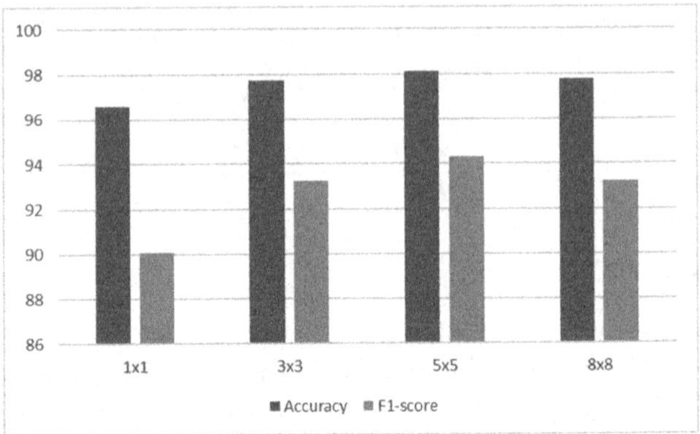

Fig. 7. The effect of window size on the accuracy and F1-score of DNN.

Table 3. Result for the window size of data task

Window size	No. data	Loss	Accuracy	Recall	Precision	F1-score
1 × 1	514639	0.0874	96.57	87.50	92.77	90.06
3 × 3	470472	0.0572	97.73	89.99	96.75	93.25
5 × 5	429665	0.0496	98.11	90.27	98.70	94.29
8 × 8	371163	0.0578	97.78	90.83	95.72	93.21

6 Conclusion

In this paper, we examine the possibility of contaminated/clean shrimps classification using Deep Learning with HSI data. Data visualization using t-SNE has shown that the classification is feasible. With only few HSI data samples, DNN successfully differentiate the contaminated shrimps and the clean ones with good F1 score. In the future, the better result might be achieved by the detection only contaminated part of shrimps.

Acknowledgement. We thank Minh Phu seafood corporation for providing hyperspectral imaging data and inspiring us to realize this work.

References

1. van der Maaten, L., Hinton, G.: Visualizing data using t-SNE. J. Mach. Learn. Res. **9**, 2579–2605 (2008)
2. Schelkanova, I., Pandya, A., Muhaseen, A., Saiko, G., Douplik, A.: 13 - early optical diagnosis of pressure ulcers. In: Igor, M. (ed.) Biophotonics for Medical Applications, pp. 347–375. Woodhead Publishing (2015)
3. Vasefi, F., MacKinnon, N., Farkas, D.L.: Chapter 16 - hyperspectral and multispectral imaging in dermatology. In: Hamblin, M.R., Avci, P., Gupta, G.K. (eds.) Imaging in Dermatology, pp. 187–201. Academic Press, Boston (2016)

4. Yu, X., Tang, L., Wu, X., Lu, H.: Nondestructive freshness discriminating of shrimp using visible/near-infrared hyperspectral imaging technique and deep learning algorithm. J. Food Anal. Methods **11**, 768–780 (2018)
5. Al-Sarayreh, M., Reis, M., Yan, W., Klette, R.: Detection of red-meat adulteration by deep spectral-spatial features in hyperspectral images. J. Imaging **4**, 63 (2018)
6. Li, X., Li, R., Wang, M., Liu, Y., Zhang, B., Zhou, J.C.: Hyperspectral imaging and their applications in the nondestructive quality assessment of fruits and vegetables (2017). https://doi.org/10.5772/intechopen.72250
7. Specim: Specim FX10 - user guide 1.0. Specim imaging Oy Ltd
8. Li, Y., Zhang, H., Xue, X., Jiang, Y., Shen, Q.: Deep learning for remote sensing image classification: a survey. In: Wiley Interdisciplinary Reviews: Data Mining and Knowledge Discovery, p. e1264 (May 2018). https://doi.org/10.1002/widm.1264
9. Wang, W., et al.: Medical image classification using deep learning. In: Chen, Y.-W., Jain, L.C. (eds.) Deep Learning in Healthcare. ISRL, vol. 171, pp. 33–51. Springer, Cham (2020). https://doi.org/10.1007/978-3-030-32606-7_3
10. Lu, Y.: Food image recognition by using convolutional neural networks (CNNs) (December 2016)
11. Thanasarn, N., Chaiprapat, S., Waiyakan, K., Thongkaew, K.: Automated discrimination of deveined shrimps based on grayscale image parameters. J. Food Process Eng. **42**, e13041 (2019). https://doi.org/10.1111/jfpe.13041
12. Sural, S., Qian, G., Pramanik, S.: Segmentation and histogram generation using the HSV color space for image retrieval. In: Proceedings of International Conference on Image Processing, vol. 2, pp. II-589 (February 2002). https://doi.org/10.1109/ICIP.2002.1040019
13. Russell, B., Torralba, A., Murphy, K., Freeman, W.: LabelMe: a database and web-based tool for image annotation. Int. J. Comput. Vis. **77**, 157–173 (2008). https://doi.org/10.1007/s11263-007-0090-8
14. Hahnloser, R., Sarpeshkar, R., Mahowald, M.A., Douglas, R.J., Seung, H.S.: Digital selection and analogue amplification coexist in a cortex-inspired silicon circuit. Nature **405**(6789), 947–951 (2000)
15. Han, J., Moraga, C.: The influence of the sigmoid function parameters on the speed of backpropagation learning. In: Mira, J., Sandoval, F. (eds.) IWANN 1995. LNCS, vol. 930, pp. 195–201. Springer, Heidelberg (1995). https://doi.org/10.1007/3-540-59497-3_175
16. Kingma, P., Lei Ba, J.: Adam: a method for stochastic optimization. arXiv:1412.6980v9 (2014)
17. Goodfellow, I., Bengio, Y., Courville, A.: Deep Learning. MIT Press, Cambridge (2016)

A Predictive System for IoTs Reconfiguration Based on TensorFlow Framework

Tuan Nguyen-Anh[1,2](✉) and Quan Le-Trung[1]

[1] Faculty of Computer Networks and Communications, University of Information Technology, Vietnam National University, Ho Chi Minh City, Vietnam
natuan@vku.udn.vn, quanlt@uit.edu.vn

[2] Vietnam - Korea University of Information Technology and Communications, The University of Danang, Danang City, Vietnam

Abstract. IoTs are rapidly growing with the addition of new sensors and devices to existing IoTs. The demand of IoT nodes keeps increasing to adapt to changing environment conditions and application requirements, the need for reconfiguring these already existing IoTs is rapidly increasing. It is also important to manage the intelligent context to execute when it will trigger the appropriate behavior. Yet, many algorithms based on different models for time-series sensor data prediction can be used for this purpose. However, each algorithm has its own advantages and disadvantages, resulting in different reconfiguration behavior predictions for each specific IoTs application. Developing an IoTs reconfiguration application has difficulty implementing many different data prediction algorithms for different sensor measurements to find the most suitable algorithm. In this paper, we propose IoTs Reconfiguration Prediction System (IRPS), a tool that helps IoT developers to choose the most suitable time-series sensor data prediction algorithms for trigger IoTs reconfiguration actions.

Keywords: IoTs · Reconfiguration · Intelligent context management · IoTs prediction system

1 Introduction

In the broad overview of the Internet of Things, reconfiguring and reprogramming applications deployed on IoTs devices is one of the most important and urgent content. Because IoT devices can be deployed in places where people cannot access to work such as rugged terrain (such as deep forests, volcanoes …) or operating in contexts where information is unpredictable. A reconfiguration of an IoTs network includes alterations to the devices connected, changing the behavioral patterns of the devices and modifying the software modules that control the IoTs network and devices. Reconfiguring an already existing IoTs network is a challenge due to the amount of data loss when carrying out a reconfiguration procedure in a limited power supply environment and reconfiguration time is often slow to respond to real-time IoTs applications. Many technical solutions using artificial intelligence have been proposed to solve the problem

N.-S. Vo and V.-P. Hoang (Eds.): INISCOM 2020, LNICST 334, pp. 206–218, 2020.
https://doi.org/10.1007/978-3-030-63083-6_16

of unreliable sensor data collection and how to shorten the time to reconfigure IoTs applications.

Intelligent context management is influential in the process of triggering the reconfiguration of IoTs applications. This will help the system know when to reconfigure the IoTs application and how that behavior will be. For the problem of reconfiguring IoTs applications when sensor data is unreliable and improving reconfiguration time, many artificial intelligence techniques have been proposed to build this component, bring the problem back to solve the classification and prediction problems.

To address the challenges posed, we have proposed R-IoT, RFL-IoT and RoB-IoT frameworks [1, 5, 6], for the reconfiguration of IoTs applications based on the changes of the context and the intelligent context management. The results show that our proposal framework is suitable for remote reconfiguration of IoTs applications with intelligent context management components for triggering reconfiguration. However, the preceding works did neither apply for complex contexts and unreliable data. For example, during the working time, if someone is in the room and the room temperature is higher than 30 °C, the air conditioner will be turned on and the cool mode will be changed. Actually the time-series sensor data are collected in IoTs environment often do not accurately reflect the context information, which is unstable and unreliable. Therefore, when we meet a situation like this, not sure whether context data is true or not, we cannot rely on it to make a decision. This is one of the reasons why the decision to reconfigure is incorrect. To address this issue, many suggestions on the use of machine learning algorithms approach. Mehdi Mohammadi [8] reviewed the characteristics of IoTs data and its challenges for machine learning (ML) and deep learning (DL) methods. We recognize that the next work needs to apply more machine learning algorithms to the intelligent context management component in the reconfiguration framework. It can be seen that the implementation of many IoTs projects with different data collection and characteristics requires a different technique and algorithm to help the smart context management component make the best predictions. Many research groups have developed algorithms for time-series sensor data prediction, and evaluated them on different dataset. However, there is no golden standard for choosing the best algorithm, since it is an application specific task. Finding techniques and algorithms that fit the criteria {accuracy of decision making, reconfiguration time} also takes a lot of time for developers of IoTs applications to test and build algorithm models, individual elements of that project and experiment. It is necessary to propose a tool to select the appropriate AI technique for any problem to reduce development costs of reconfiguration projects.

This research aims to help developers of IoTs solution to choose the best algorithm for time-series sensor data prediction regarding their application data. For this proposal, we developed IoTs Reconfiguration Prediction System (IRPS), which is also a component of Framework R-IoT, a web-based online tool that implements four different algorithms for time-series sensor data prediction. Users can upload historic sensor measurement from their application, and IRPS can analyze which time-series sensor data prediction algorithm fits best regarding two evaluation metrics: prediction accuracy and time reconfiguration. This system uses Google open-source TensorFlow framework to build the model of gesture recognition, introduces the platform characteristics of TensorFlow, and puts one deep learning algorithms (FFNN), two algorithms are unsupervised learning (RF, NB) algorithms based on TensorFlow framework. Teachable Machine[1] is a browser application that you can train with your webcam to recognize objects or expressions. All training is done in the browser using the deeplearn.js library. TensorFlow Serving[2] is another tool was proposed by many developers. This is an open source toolkit, used to deploy models TensorFlow which trained on production environment. With TensorFlow Serving, the process of deploying the model to the system will become much faster and easier with normal model deployment and loading. TensorFlow Serving deploys the model independently, separates from the backend code, supports two protocols: gRPC and RESTful API, easily add and update new models according to each version without causing images to affect other parts of the system or other services. When the problem is to help the developers of the IoTs reconfiguration solution make the choice of algorithms to train the model, those tools is not appropriate for IoTs reconfiguration. Teachable machine tool mainly focuses on training and testing support for image processing problems and it does not support for IoTs reconfiguration. The TensorFlow Serving tool is great, but it focuses on deploying primarily after the model is available, which is a disadvantage. To the best of our knowledge, IRPS is the first tool designed for the IoTs reconfiguration developers to help them create efficient IoTs reconfiguration applications. This is also a module in the intelligent context management component of the reconfigured R-IoT framework. This paper is organized as follows. Time-series sensor data algorithms are presented in Sect. 2. Section 3 describes the development of the IRPS. Two case studies are presented in Sect. 4. Finally, the conclusion is given in Sect. 5.

2 Related Work

2.1 IoTs Reconfiguration Framework

In reconfiguration mechanisms, the reprogramming is handled at the physical device directly used as much as the researches results [9, 10]. IoTs applications will be able to change their behavior by sending parameters or uploading new binary firmware to change application behavior. However, this approach is thought to be a manual method, and this incurs costs for deployment and maintenance. According to [3], Nikolov, N. proposed a reconfiguration approach by Firmware Update Over The Air (FOTA) for ESP8266 device. Its procedure will be executed after new firmware bin files are uploaded to cloud. When a reconfiguration is triggered by the user, the ESP8266 device requires the FOTA control system on Cloud to return the corresponding firmware. However, this solution has some shortcomings in terms of security protocols, reconfiguration time, especially intelligent context management that does not meet the needs of current reconfiguration. The reconfiguration middleware is a new approach to reconfiguration with recent researches [9, 11–15]. However, the proposals are limited in terms of solving fully automatic reconfiguration, big data stores sensor information, intelligent context management and these do not support real-time applications. In the previous study [1, 7], we proposed the RFL-IoT framework to change the behavior of IoTs applications based on Fuzzy Logic (FL). Smart

[1] https://teachablemachine.withgoogle.com/.

[2] https://www.tensorflow.org/tfx/guide/serving.

Context Management with FL approach analyzes the contextual information collected by IoTs applications. Based on user-defined FL rules, it determines the time that will trigger behavior change and transmits the decision to reconfigure and control the corresponding end-devices. One of our other proposed approaches to intelligent context management is based on Ontology for context modeling and the Bayesian network for context reasoning [6]. A Bayesian network that uses statistical methods can receive uncertain information to make predictions. Our empirical studies have limitations when assessing the environment with more complex contextual scenarios. To overcome this limitation, we continue to improve the framework by building prediction system using a machine learning approach. It uses some well-known machine learning algorithms to handle the vast amount of contextual information collected by IoTs applications.

2.2 Time-Series Sensor Data Prediction Algorithms

In this section we are going to explain the basics of time-series sensor data prediction. Then, we briefly introduce the algorithms used in IRPS. Time-series sensor data prediction is important in IoTs reconfiguration as it helps to make better decisions for trigger reconfiguration action. Different models are used in the literature for time-series data prediction. Today, there is increased interest in using Artificial Intelligence (AI) techniques to analyze complex data sets to extract useful information that can be used in the prediction of time-series sensor data trends and the effect of intervention actions on data trend predictions. Such information then can be used to change behavior. ML [2] and DL [4] techniques are the specific sub-sets of AI that are of particular current interest to academia and industry. ML and DL are applied to different situations where large and complex data sets are available and metadata, as well as more detailed information, could be extracted from the provided data if suitable means to extract and link the data were available.

For deep learning, the Feedforward Neural Network (FFNN) algorithm, also known as the multi-layer neural network, is a fundamental algorithm so that we can advance to more advanced algorithms such as convolutional neural network or recurrent neural network when the algorithms improve. This height is also a combination of many different layers in the entire neuron network. The two unsupervised learning algorithms are RF and NB that are also two famous algorithms in automatic control of IoTs applications. Random Forest is another machine-learning method that is successfully implemented in a wide variety of fields, including animal science. This regression or classification method makes use of decision trees: a sequence of rules that split the data in a way that most optimally reduces variation. Each tree receives a random subset of training samples, and then the algorithm randomly selects a subset of variables at each split in the tree. These trees, which are relatively poor classifiers individually combined into an ensemble of trees called a random forest, which is used for prediction. The prediction results of a random forest are a summation of the prediction outcomes of many individual trees. Naive Bayes is a family of classifiers that implements Bayesian techniques to form a simple network based on previous probabilities. Its method relies on independence between the input variables, but it performs surprisingly well even under conditions that might be considered suboptimal for the algorithm. Despite the relative simplicity of its algorithm, naive Bayes is still widely used. To implement re-configuring IoTs applications using Fuzzy Logic, system reads data from sensor as crisp input for fuzzyfication then fuzzy input from fuzzyfication processed with inference system (fuzzy rule bases), then fuzzy output processed with defuzzyfication and produces crisp output to change the behaviors. By discovering the advantages and disadvantages of each technique in reconfiguration action case study, it is hoped to gain a better understanding of the wide variety of available tools for predicting complex contexts to trigger changing behavior actions.

3 Development of IRPS

The platform we use to build algorithms on models is to use the TensorFlow library. TensorFlow is an open source software developed by Google to perform machine learning, which is widely used by many researchers and large companies. TensorFlow provides many libraries for programmers to demonstrate machine learning algorithms and an application to implement these algorithms. A calculation expressed by TensorFlow can be performed with little or no modification in a range of heterogeneous systems mobile devices such as phones and tablets or distributed systems large scale spread of hundreds of machines or various computing devices like GPU cards. One of the important reasons we chose the TensorFlow platform is the ability to integrate the framework. The algorithms implemented in Tensorflow can be easily transplanted on many heterogeneous systems. In this section, we will explain the process of developing IRPS, the implementation of the algorithms.

3.1 Methodology

Figure 1 represents the methodology of research process can be learned for training data which are analyzed by a classification algorithm. The test data are used to calculate the accuracy of classification algorithms. There are many algorithms that can be used for classification for IoTs application such as RF, BN, rule based, neural network algorithms. In our system, we implement four algorithms: one algorithm is deep learning (FFNN), two algorithms are unsupervised learning (RF, NB) and the logical algorithm (FL). The classification of data has two step processes are learning and prediction.

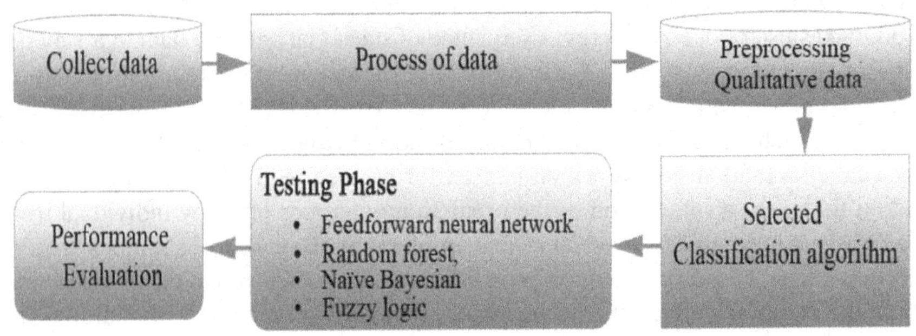

Fig. 1. Methodology of classification algorithms used in IRPS

3.2 Design and Implementation of IRPS

IRPS Training Model

In order to predict the importance of IoTs data prediction as well as show the results of its algorithms, we create a web-based system that shows the calculations in a more comprehensible way for the IoTs application developer. On that basis, IoTs application developers can refine input features, output features, algorithms to get the most suitable results for their specific context. Since it is a system on the web, the client-server architecture is the most common and appropriate. It is designed in a way that the user sends all the necessary data to the server. The server analyses and processes the request, and later builds to available model and predicts the results. Different parts of the system are implemented with different technologies. Express.JS, is used for making this API Restful API system and library TensorFlowJS (Tf.js). Tf.js is faster for small models, but when model becomes large training becomes 10–15x slower[3]. In the problem of reconfiguring IoTs applications which often involves a lot of sensor time-series data. The value of these features is usually not complicated, so we use the platform mainly as TensorFlow.JS instead of the python platform. We implemented IRPS on the server with the following hardware: Intel Xeon E3-1225 v5 @3.30 GHz (4 CPUs) ~ 3.3 GHZ, 32 GB RAM.

In this section, the system has three steps: *Step 1:* upload the Dataset; *Step 2:* select the input features and output features that recognized from the dataset; *Step 3:* select the algorithm among the 04 mentioned algorithms in Sect. 3.1 to train. Once the model has been trained, the user can download the model to use for prediction and see results like accuracy and loss function. The system is based on TensorFlow framework with associated libraries to access online data. The operation of the system is shown in Fig. 2.

IRPS Predicting Data

For predictive systems on models created from training, we use ReactJS[4] and TensorFlow to build web applications that run on the Node.JS server. We experiment on the Edge layer instead of the Cloud layer with the Raspberry Pi 4. The device is configured as a quad-core ARM Cortex-A72 CPU clocked at 1.5 GHz, 2 GB RAM. Data is transferred between Express.JS Restful API and ReactJS (Fig. 3).

The result of the training process according to the data, characteristics, algorithms of the user selected are two files {*model.json; weights.bin*}. The predictive system is deployed at the edge and the gateway which do not need powerful hardware. The prediction only needs to select the model corresponding to the previously selected algorithm, the data series to test according to the given syntax. Figure 4 shows the prediction of controlling room fan opening based on the parameter {room temperature, outside temperature, whether or not people in the room exist}, input data 24.11, 26.66 1. IRPS will immediately predict the output: Fan on/off mode respectively ~1%/~99%. Based on the predicted results, the system will trigger application behavior changes to the R-IoT Framework to perform the next steps of the reconfiguration process.

[3] https://www.tensorflow.org/js/tutorials.

[4] https://reactjs.org/.

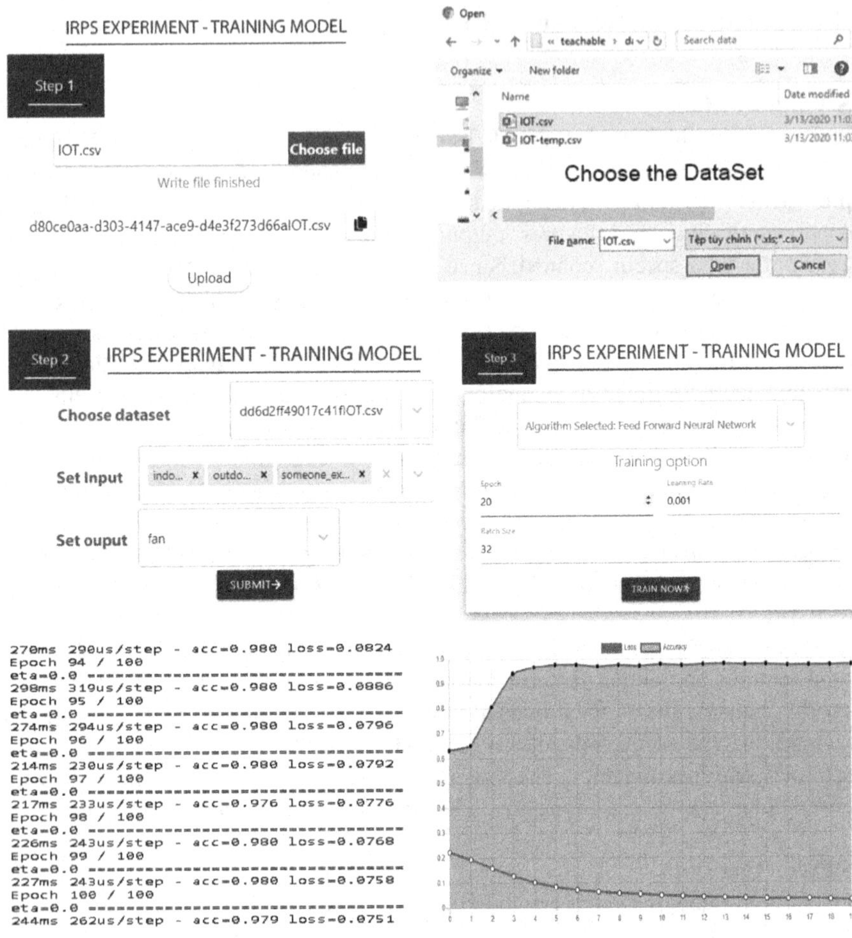

Fig. 2. Steps to training model on user dataset

IRPS EXPERIMENT - TEST YOUR MODEL

Test your model

model.json
weights.bin

model.json-weights.bin Choose file

UPLOAD

Ready for predict new model

Type your input with space

24.11 26.66 1

PREDICT

[[0.019986357539892197,0.9800136089324951]]

Fig. 3. Steps to perform prediction behavior based on model

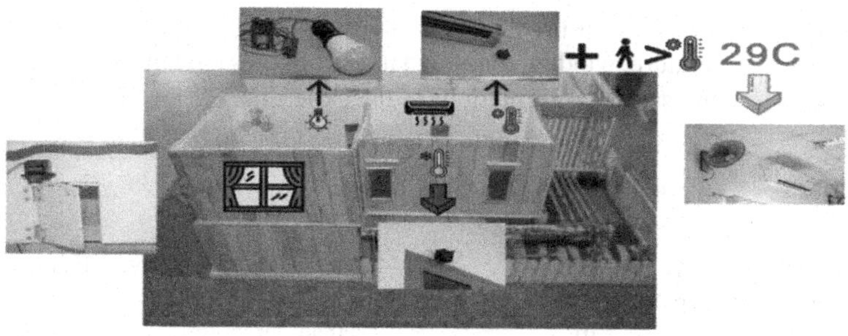

Fig. 4. Overall each smart room: temperature, fan and light control service

4 Case Study

In this section, we will show the actual output of the system and see how it performs on two case studies. We consider two different datasets in order to compare the results. The first one is a dataset for the smart room and the second one has data that shows the smart hydroponic cultivation. The details of how to setup the data, how many input, output features and how to choose the parameters for each experiment are described in the contextual information of each case study.

4.1 Smart Room

In the first case, IRPS is used to predict to change the state of fan and light in each room of the smart room. The framework is also designed in order to collect data from

Table 1. Services and smart scenarios in the smart room

Service	Scenario information
Temperature control service	The temperature control service performs switching on/off of fans and opening/closing windows based on information such as working time, indoor temperature/outdoor temperature, noise level, and the presence of people in the room. For example, if it is during working hours, the outdoor temperature is higher than 30 °C and someone is in the room, temperature control service will be performed. By comparing the indoor and outdoor temperatures, if the indoor temperature is higher, the windows open, other wires, turn on the fan. If the indoor temperature is lower than the outdoor temperature and there are no people in the room, turn off the fan
Light control service	Service control on/off lights in the room is affected by information such as working time, occupants and lighting levels. By comparing the light levels inside the room, we can make a decision to turn on/off the light. For example if no one is in the room during working time, the light will turn off, if there is a person in the room and the lighting level is low, the room lights will be turned on

sensors to monitor and store in the database {noise level, lighting level, exist person, outdoor/indoor temperature, window state} and it reconfigures the fan/light state. Table 1 shows the scenario and service information for the experiment (Fig. 6).

Table 2. Example of some smart room scenario

Noise level	Lighting level	Exist person	Indoor temp over 30	Temp control	Higher temp in than out	Window	Fan	Light
high	high	true	true	on	true	open	on	off
low	low	false	true	off	false	close	off	off
high	low	true	false	off	false	close	off	on
low	high	false	false	off	false	close	off	off
low	low	false	true	off	false	close	off	off
low	high	false	false	off	false	close	off	off

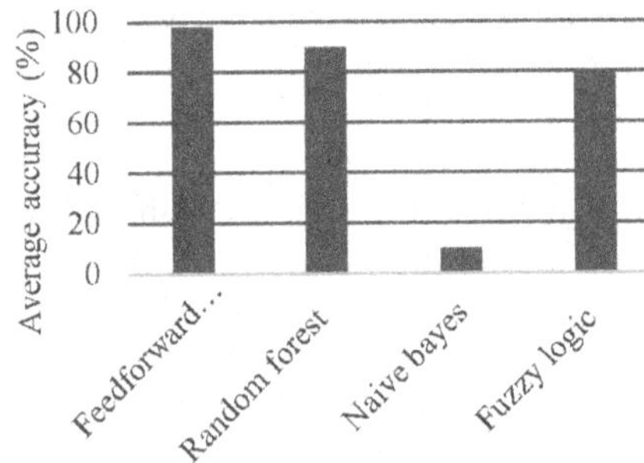

Fig. 5. Average accuracy (percent)

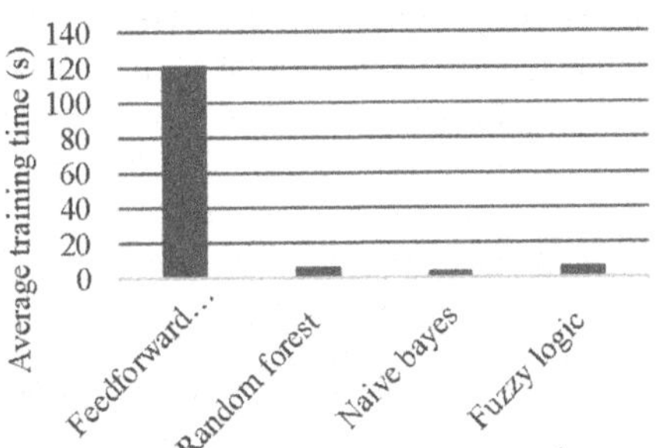

Fig. 6. Average training time (s)

We prepare the dataset for experiments from Kaggle dataset[5]. This is a time series data set measured at a smart room from 2016-11-27 17:22:45 until 2017-07-30 12:05:30, with a total of 14573 records. The above data set has a total of 10 features (date-time, Volume [mV], Light_Level [Ohms], Temperature-DHT [Celsius], Pressure [Hectopascal], Temperature-BMP [Celsius], Relative_Humidity [%], Air_Quality [Ohms], Carbon_Monoxide [Ohms], Nitrogen_Dioxide [Ohms]). Data pre-processing was applied to prepare the raw data. Pre-processing data is an important step to convert the raw data as shown in Fig. 5 into a format that can replace missing values to improve data quality. Next, we conducted divided into two sets of data with 80% for training and 20% for testing (Fig. 7).

[5] https://www.kaggle.com/saikatchoudhury/starter-smart-home-dataset-with-a8895cb2-c.

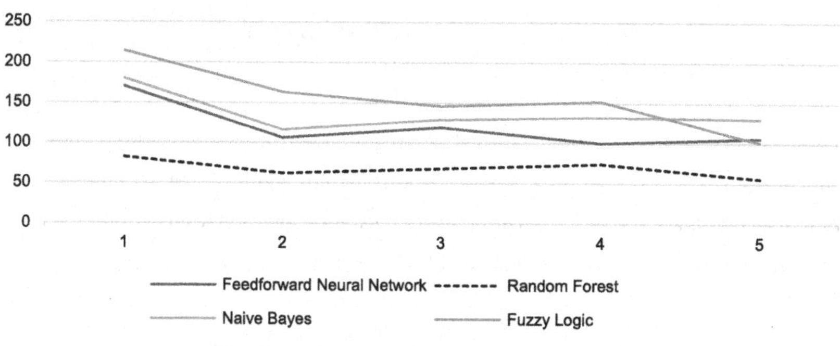

Fig. 7. Average reconfiguration time result

4.2 Smart Hydroponic Cultivation

In the second case, IRPS is used to predict which algorithm is suitable for the problem of hydroponic cultivation of 04 plants: tomatoes, lettuce, amaranth, vegetables. We prepare the dataset for experiments from process of ornamental dragon in Da Nang (vegetable garden). This is a time series data set measured from 2019-06-27 until 2019-12-30, with a total of 4574 records. The above data set has a total of 5 features (date time, type of plant, staging temperature, ambient temperature, humidity, PPM). The scenario of the hydroponic vegetable system is described as Table 2.

Table 3. Some scenarios in the context of a smart hydroponic cultivation

Service	Scenario content
Pump control service	The system collects real time indicators such as crop type, ppm, temperature, humidity of the planting system. Artificial intelligence techniques based on that will make a decision whether to change the reconfiguration behavior such as performing regression pump and regression pump time, or adding nutrients to the tank
Fan, light control service	Artificial intelligence techniques will use temperature and humidity data for 30 min continuously to predict the likelihood of heavy rain or sunshine. For example, with a possibility of rain above 80%, the system will make a decision that the roof will be pulled out. If the temperature is too high, the fan will be turned on, otherwise the lighting will turn on

The system will change behavior of pump, Fan, Light, outhouse control when meeting relevant sensor context information. The behaviors are shown as follows (Tables 3 and 4):

The devices and sensors used in the system are: Arduino WEMOS D1, Ds18b20, PH, relay - SLI, Module DHT22, Water pump motor, Engine sensor 775, mist kit, fan 5v. The output of the system is shown in Fig. 8.

Most of the analyzed datasets, as the two previous cases, show the following Table 5. For FFNN, with extremely high accuracy, the main reason is that our problem data is

Table 4. Some actions of smart hydroponic cultivation

Service	Scenario content
Circulating pump	Normal pump; pump with double time, no pumps
Nutrient pumps	Pumps A and B; pumps adding water, no pumps
Temperature-based behavior	Turn on/off {fans; misting; light}; no action
Rain-based behavior	Pull the blinds; retract the curtain; no action

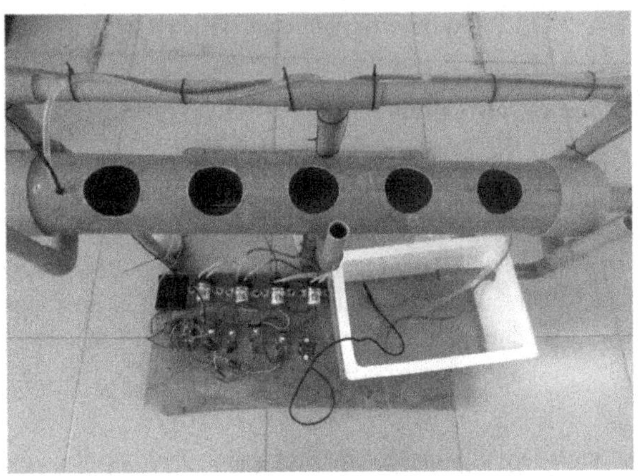

Fig. 8. Prototype of hydroponic cultivation system

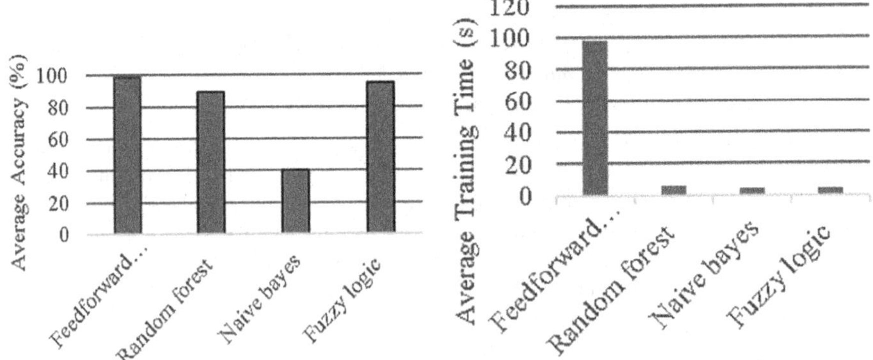

Fig. 9. Average accuracy (percent)

Fig. 10. Average training time (s)

not complicated, so the classification algorithm of neural network can work optimally in the distribution problem. However, training time and computing resources are limited to this algorithm. In order to overcome the above limitations, it is possible to use the server to optimize performance and use the save and reload feature that tensorflow.js provides to limit the training many times. However, for our problem, the need to train many times is unavoidable because as stated in the idea, we cannot impose a single rule. For random forest, the algorithm returned amazing predictions when we had never touched normalize data but the results were really acceptable. With over 90% of the correct answers, it also partly reflects the applicability of this algorithm. Building algorithmic models by voting in different decision trees has helped RF become a highly appreciated algorithm in complex classification problems from linear-data to non-linear data. For our reconfiguration prediction, RF has worked perfectly in both training time and in terms of accuracy, in addition to using it when programming is simple and easy to use with different test-bed. NB, as mentioned in the theory, is an algorithm with very fast execution time because the algorithm only calculates the probability of being input from the input data. For example, logistic regression uses weight and bias to optimize the calculation process and uses decent gradients to find the optimal solution. While Naïve Bayes does not have the optimal solution to occur that is the reason make the reconfiguration time is slower for the time series data set that we give. The prediction with not high accuracy was anticipated, but we can completely improve this algorithm by adding one several other factors to improve its accuracy include normalizing data, reducing noise, missing data or errors. Finally, we can use fuzzy logic algorithm when the rules can only be set once while reprogramming, but the results are predictive and fast. The processing of this algorithm is extremely good. However, it is really difficult to implement a specific test-case IoTs because the setting rules in R-IoT framework takes a lot of time. In fact, there are exceptions in this rule, confirmed in fewer processed datasets, which means that finally the results depend on the dataset itself (Figs. 9 and 10).

Table 5. Comparison of the performance experiments

Algorithms	Accuracy (%)		Training time (s)		Reconfiguration time (ms)	
	First case	Second case	First case	Second case	First case	Second case
Feedforward neural network	98	99	120.2	98.4	143.75	133.5
Random forest	90	89	5.5	6.1	68.4	108.4
Naive Bayes	10	40	3.4	3.9	135.4	125.7
Fuzzy logic	80	95	6.5	4.3	235.2	115.4

5 Conclusion

In this paper, we present IoTs Reconfiguration Prediction System (IRPS) that helps future developers IoTs application solutions to choose the most suitable data prediction algorithms for trigger their reconfiguration actions. IRPS performs data prediction for sensor readings, using four different algorithms, and compares their performances regarding two different evaluation metrics (accuracy to decision making, reconfiguration time). Additionally, the monitoring screen from IRPS visualizes the results obtained from the data prediction process. This proposed system helps IoTs developers to choose the most suitable time-series sensor data prediction algorithms for trigger IoTs reconfiguration actions. In future, we continue to our work on improving IRPS such as: implementation many AI algorithms into the system to support the intelligent context management of IoTs reconfiguration framework.

References

1. Nguyen-Anh, T., Le-Trung, Q.: RFL-IoT: an IoT reconfiguration framework applied fuzzy logic for context management. In: IEEE International Conference on Research, Innovation and Vision for the Future (RIVF). IEEE (2019)
2. Sharma, K., Nandal, R.: A literature study on machine learning fusion with IOT. In: 2019 3rd International Conference on Trends in Electronics and Informatics (2019)
3. Nikolov, N.: Research firmware update over the air from the cloud. In: International Scientific Conference Electronics (ET2018). IEEE, Bulgaria (2018)
4. Tang, J., Sun, D., Liu, S., Gaudiot, J.-L.: Enabling deep learning on IoT devices. Computer **50**, 92–96 (2017)
5. Anh, T.N., Le Trung, Q., Hai, B.T., Van, D.H: R-IoT: a framework for IoTs reconfiguration in cloud. In: The 6th Conference on Information Technology and Its (CITA) (2017)
6. Nguyen-Anh, T., Le-Trung, Q.: An IoT reconfiguration framework applied ontology-based modeling and bayesian-based reasoning for context management. In: 2019 6th NAFOSTED Conference on Information and Computer Science (NICS), IEEE NICS 2019 (2019)
7. Nguyen-Anh, T., Le-Trung, Q.: An IoTs reconfiguration framework with intelligent context management. In: IEEE Seventh International Conference on Communications and Electronics (ICCE). IEEE (2018)
8. Mohammadi, M., Al-Fuqaha, A., Sorour, S., Guizani, M.: Deep learning for IoT big data and streaming analytics: a survey. IEEE Commun. Surv. Tutor. **20**, 2923–2960 (2018)
9. Perera, C., Zaslavsky, A., Christen, P.: Context aware computing for the Internet of Things: a survey. IEEE Commun. Surv. Tutor. **16**, 414–454 (2013)
10. Craig, G., Adnan, Al., Quan, B.: OTAP arbitration effects in randomly deployed WSN's. In: International Telecommunication Networks and Applications. IEEE, Australia (2015)
11. Sivaharan, T., Blair, G., Coulson, G.: GREEN: a configurable and re-configurable publish-subscribe middleware for pervasive computing. In: Meersman, R., Tari, Z. (eds.) OTM 2005. LNCS, vol. 3760, pp. 732–749. Springer, Heidelberg (2005). https://doi.org/10.1007/11575771_46
12. Ruckebusch, P., Van Damme, J., De Poorter, E., Moerman, I.: Dynamic reconfiguration of network protocols for constrained Internet-of-Things devices. In: Mandler, B., et al. (eds.) IoT360 2015. LNICST, vol. 170, pp. 269–281. Springer, Cham (2016). https://doi.org/10.1007/978-3-319-47075-7_31
13. Aberer, K., Hauswirth, M., Salehi, A.: A middleware for fast and flexible sensor network deployment. In: Proceedings of 32nd International Conference on Very Large DataBase. ACM (2006)
14. Gámez, N., Fuentes, L.: FamiWare: a family of event-based middleware for ambient intelligence. Pers. Ubiquit. Comput. **15**(4), 329–339 (2011)
15. Henry, J., Marti, H.: Quick and efficient link quality estimation in wireless sensors networks. In: Wireless On-Demand Network Systems and Services. IEEE, France (2018)

Industrial Networks and Intelligent Systems

An Optimal Eigenvalue-Based Decomposition Approach for Estimating Forest Parameters Over Forest Mountain Areas

Nguyen Ngoc Tan and Minh Nghia Pham(✉)

Faculty of Radio-Electronics, Le Quy Don Technical University, Hanoi, Vietnam
nghiapm2018@mta.edu.vn

Abstract. This paper aims to provide a new method for retrieving forest parameters in forest mountain areas by using L-band polarimetric interferometry synthetic aperture radar (PolInSAR) data. Applying the model-based (ground, double-bounce, and volume scattering) decomposition techniques to PolInSAR data has opened a new way for vegetation parameters estimation. However, the modeling of the vegetation backscattering mechanisms is complicated due to the influences of the topographic slope variation and assumptions about the volume scattering component. In order to overcome these limitations, an eigenvalue-based decomposition technique is proposed. In which, a simple volume scattering model introduced by Neumann is used. The proposed method has improved 1.2664 m height of forest trees compared to the three-state inverse approach. In addition, evaluation results with simulated data generated from PolSARProSim software and PolInSAR data over the Kalimantan areas, Indonesia from ALOS/PALSAR L-band spaceborne radar system show that the proposed method produces reasonable and outstanding physical results in comparison with traditional decomposition methods.

Keywords: Polarimetric interferometry synthetic aperture radar · Forest height estimation · Coherence matrix · Three-stage

1 Introduction

Applying the model-based decomposition to polarimetric interferometry synthetic aperture radar (PolInSAR) data is an effective approach to separate scattering mechanisms in the natural environment. In recent years, this method has been widely applied to target detection and estimation of vegetation parameters. In general, model-based decomposition techniques are to accommodate a number of simple physical scattering models. The commonly applied decomposition techniques are mainly divided into two categories: (1) coherence decomposition based on the measured Sinclair matrix; (2) incoherent decomposition based on the measured coherency and covariance matrix [1–4]. In which, the three component decomposition of Freeman – Durden [5] is one of the most popular ones. Accordingly, the cross-correlation PolInSAR matrix obtained from observations can be described as the sum of three scattering sub-matrices, each representing the contributions of volume, ground, and double bounce scattering component. However, some major

N.-S. Vo and V.-P. Hoang (Eds.): INISCOM 2020, LNICST 334, pp. 221–232, 2020.
https://doi.org/10.1007/978-3-030-63083-6_17

shortcoming have been pointed out when applying this method, such as negative powers of ground scattering or double bounce scattering components and assumption of scattering reflection symmetry for volume scattering components. To effectively employed the Freeman–Durden decomposition algorithm, several assumptions are required for solving a nonlinear equation system. Recently, many modified methods have been proposed to overcome the mentioned shortcoming but mostly for modeling on relatively flat terrain and little research has been carried out on slope areas. On slope forest topography, local direction (OA) and local incidence angle are significantly changed by effect of slope terrain. This leads to a strong influence on the scattered signals. Therefore, scattering models in the Freeman – Durden decomposition are difficult to appropriately describe physical scattering mechanisms in sloping forest areas.

For these reasons, we proposed an eigenvalue-based decomposition approach for estimating forest parameters over forest mountain areas using single-baseline L-band PolInSAR data. In this approach, we shall suggest a simplified Neumann volume scattering model [6], which is characterized by a randomness parameter. This model is useful to overcome the limits of the scattering reflection symmetry assumption. This volume model includes both random and non-random volume situations, so it is suitable for volume scattering mechanisms on slope terrain. After that, the unknown parameters of the volume scattering matrix are determined as well as the parameters of the remain scattering mechanisms are estimated by using the optimal eigenvalue-based decomposition technique.

This paper is organized according to the following structure. In Sect. 2, basic scattering mechanisms and the simplified Neumann volume scattering model are introduced. Section 3 describes the optimal eigenvalue-based decomposition approach to estimate forest parameters. The estimated results by the proposed approach with simulated data and spaceborne data are given and discussed in Sect. 4. Finally, the conclusion and future work are presented in Sect. 5.

2 Basic Scattering Mechanisms in Forest Mountain Areas for PolInSAR and Simplified Neumann Volume Scattering Model

2.1 The PolInSAR Covariance Matrix

For PolInSAR, the obtained full polarimetric interferometry information data can be expressed in the form of two scattering matrices $[S_1]$ and $[S_2]$. The target vectors $\vec{k}_{L_1}$ and $\vec{k}_{L_2}$ in the lexicographic basis are measured at two ends of the baseline, as [7]

$$\vec{k}_{L_1} = [S_{1HH}, S_{1HV}, S_{1VV}]^T,\ \vec{k}_{L_2} = [S_{2HH}, S_{2HV}, S_{2VV}]^T \tag{1}$$

Accordingly, we can form the PolInSAR covariance matrix by multiplying the target vectors by their conjugate transpose

$$[T] = \left\langle \vec{k}\vec{k}^T \right\rangle = \begin{bmatrix} T_1 & \Omega \\ \Omega^T & T_2 \end{bmatrix} \text{ with } \vec{k} = \begin{bmatrix} \vec{k}_{L_1} \\ \vec{k}_{L_2} \end{bmatrix} \tag{2}$$

where $\langle . \rangle$ indicates the ensemble average in the data processing, the superscript $(.)^T$ denotes the complex conjugation. $[T_1]$ and $[T_2]$ are the normal Hermitian polarimetric covariance matrices for the two apertures, while $[\Omega]$ is a non- Hermitian cross-covariance matrix between the apertures and it contains both polarimetric information and interferometric information. As proposed in [8] and [9], the matrix $[\Omega]$ can be decomposed into three sub-matrices $[T_v]$, $[T_d]$ and $[T_g]$ which correspond to the volume, double bounce, and ground scattering mechanisms, respectively.

$$[\Omega] = e^{j\phi_g} f_v [T_v] + e^{j\phi_d} f_d [T_d] + e^{j\phi_v} f_g [T_g] \tag{3}$$

where f_v, f_d, f_g correspond to the contribution of the volume, double-bounce, ground scattering and ϕ_v, ϕ_d, ϕ_g are the phases of the three scattering components.

2.2 Volume Scattering Mechanism and Simplified Neumann Scattering Model

As suggested in [9, 10], the scattering from a tree canopy can be logically described by a multitude of randomly oriented particles. For each single particle, the normalized coherency matrix can be expressed as

$$T_\delta = \begin{bmatrix} 1 & \delta & 0 \\ \delta^* & |\delta|^2 & 0 \\ 0 & 0 & 0 \end{bmatrix} \tag{4}$$

with δ is the scattering anisotropy. When $|\delta|$ changes from 0 to 1, that mean the shape of particle varies from an isotropic sphere to a dipole, respectively. Neumann *et al.* [11] proposed the von Misses distribution for the orientation of particles. Accordingly, the coherency matrix is transformed as follows

$$T_v(\delta, \tau) = \begin{cases} \dfrac{1}{1+|\delta|^2} \begin{bmatrix} 1 & (1-\tau)\delta & 0 \\ (1-\tau)\delta^* & (1-\tau)|\delta|^2 & 0 \\ 0 & 0 & \tau|\delta|^2 \end{bmatrix} & \tau \le \dfrac{1}{2} \\ \dfrac{1}{1+|\delta|^2} \begin{bmatrix} 1 & (1-\tau)\delta & 0 \\ (1-\tau)\delta^* & |\delta|^2/2 & 0 \\ 0 & 0 & |\delta|^2/2 \end{bmatrix} & \tau > \dfrac{1}{2} \end{cases} \tag{5}$$

with $\tau \in [0, 1]$ denotes the normalized degree of orientation randomness.

Assume that forest areas consist mainly of dipoles ($\delta = \pm 1$) [12], the two coherency matrix are obtained as follows

$$T_v^H = \begin{cases} \dfrac{1}{2} \begin{bmatrix} 1 & 1-\tau & 0 \\ 1-\tau & 1-\tau & 0 \\ 0 & 0 & \tau \end{bmatrix} & \tau \le \dfrac{1}{2} \\ \dfrac{1}{2} \begin{bmatrix} 1 & 1-\tau & 0 \\ 1-\tau & 1/2 & 0 \\ 0 & 0 & 1/2 \end{bmatrix} & \tau > \dfrac{1}{2} \end{cases} ; \; T_v^V = \begin{cases} \dfrac{1}{2} \begin{bmatrix} 1 & \tau-1 & 0 \\ \tau-1 & 1-\tau & 0 \\ 0 & 0 & \tau \end{bmatrix} & \tau \le \dfrac{1}{2} \\ \dfrac{1}{2} \begin{bmatrix} 1 & \tau-1 & 0 \\ \tau-1 & 1/2 & 0 \\ 0 & 0 & 1/2 \end{bmatrix} & \tau > \dfrac{1}{2} \end{cases} \tag{6}$$

where T_v^H and T_v^V are the horizontal and vertical simplified Neumann volume scattering models. In case $\tau = 1$ then $T_v^H = T_v^V$.

According to the analysis presented in [6], the copolarization power ratio $\gamma(\Omega) = \langle |S_{hh}|^2 \rangle / \langle |S_{vv}|^2 \rangle$ is capable of distinguishing the two types of the volume scattering model that shown in Eq. (6) and this ratio is highly correlated with the orientation randomness degree τ. We can derive $\gamma(\Omega)$ from cross-covariance matrix Ω and for each value of τ in the range [0, 1], the ratio $\gamma(T_v(\tau))$ can be obtained from the volume scattering matrix $T_v(\tau)$. Therefore, the optimum τ can be determined when the difference between these two ratios is minimal, i.e.,

$$\min|\gamma(\Omega) - \gamma(T_v(\tau))| \tag{7}$$

When τ is computed for each pixel, the corresponding volume scatterring matrix T_V is determined. Figure 1 shows an algorithm flowchart for determining the volume scattering matrix.

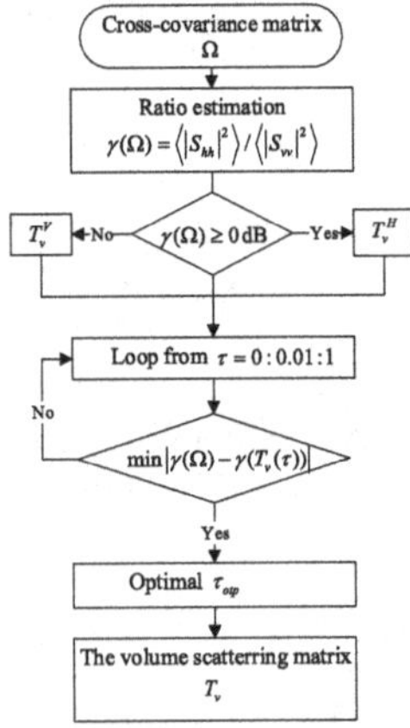

Fig. 1. Flowchart for determining volume scatter matrix

2.3 Ground and Double-Bounce Scattering Model for Sloped Terrain

For sloped terrain such as forest mountain areas, variations of topography and the local orientation angle due to the local terrain will lead to a change in the scattered signal. A local coordinate system is formed to accommodate the inclined ground surface. In order to represent the scattering components in the forest mountain areas, the flat double-bounce and ground scattering components are replaced by the scattering from the inclined ground plane [13]. In this case, the coherence scattering matrix for the ground and double-bounce contribution is created from the rotation of a local orientation angle χ represented as [14]

$$[T_d(\chi)] = \begin{bmatrix} |\alpha|^2 & \alpha\eta & 0 \\ \alpha^*\eta & \kappa & 0 \\ 0 & 0 & (1-\kappa) \end{bmatrix} \quad [T_g(\chi)] = \begin{bmatrix} 1 & \beta\eta & 0 \\ \beta^*\eta & |\beta|^2\kappa & 0 \\ 0 & 0 & |\beta|^2(1-\kappa) \end{bmatrix} \tag{8}$$

With $\kappa = \cos^2 2\chi p(\chi) d\chi$ and $\eta = \int \cos 2\chi p(\chi) d\chi$ denote the influence of terrain slope on the double-bounce and ground scattering mechanisms, respectively. The parameter κ is between 0.5 and 1 and η varies from 0 to 1 [11]. Two parameters α and β are the parameter of the double-bounce and ground scatterings as proposed by Freeman-Durden [5].

3 The Optimal Eigenvalue-Based Decomposition of PolInSAR Data

In this section, we propose an optimal eigenvalue-based decomposition approach for PolInSAR data to estimate vegetation parameters in mountain terrains. The approach is conducted in three steps. First, the terms of the volume scattering mechanism are identified. Next, an algorithm for analyzing the eigenvalue of the residual matrix is applied to determine the parameters of the ground and scattering mechanism. Finally, plant height is estimated by the canopy and ground phase difference.

From Eq. (3) and (8), we can transform as follows

$$\begin{aligned} [\Omega] - e^{j\phi_v} f_v [T_v] &= e^{j\phi_g} f_g [T_g] + e^{j\phi_d} f_d [T_d] \\ &= e^{j\phi_g} f_g \begin{bmatrix} 1 & \beta\eta & 0 \\ \beta^*\eta & |\beta|^2\kappa & 0 \\ 0 & 0 & |\beta|^2(1-\kappa) \end{bmatrix} + e^{j\phi_d} f_d \begin{bmatrix} |\alpha|^2 & \alpha\eta & 0 \\ \alpha^*\eta & \kappa & 0 \\ 0 & 0 & (1-\kappa) \end{bmatrix} \\ &= [\Omega_r] \end{aligned} \tag{9}$$

It is not difficult to recognize the unknown number identified as Eq. (10). To find the best fit forest parameters, $\{f_v, \phi_v, f_g, \phi_g, f_d, \phi_d, \alpha, \beta, \kappa, \eta\}$, the eigenvalues of two matrices $[\Omega_r]$ are determined by Eigen-decomposition techniques. Then, the eigenvalues are confined to zero, we can obtain canopy phase $\phi_v^i\{i = 1, 2, 3\}$ and the volume coefficients $f_v^i\{i = 1, 2, 3\}$

We assume that $\phi_v, f_v, \kappa, \eta$ are input loop parameters with $\kappa \in [0.5, 1]$ and $\eta \in [0, 1]$. For each value set $\{\phi_v, f_v, \kappa, \eta\}$ the residual matrix that contains the remaining two unknowns α, β can be obtained. We perform eigenvalue analysis for this residual matrix

$$[\Omega_r] = \lambda_1 \vec{k}_1 \vec{k}_1^* + \lambda_2 \vec{k}_2 \vec{k}_2^* \tag{10}$$

where λ_i and $\vec{k}_i = [u_{i1}, u_{i2}, u_{i3}]^T, (i = 1, 2)$ are eigenvalues and eigenvectors of residual matrix $[\Omega_r]$, respectively

We see that the minimum eigenvalue of the residual matrix is zero when the volume scattering component is completely removed from the PolInSAR data. As suggested by Cloude and Pottier [15], the parameter α is one of the main parameters for determining the dominant scattering component of independent targets. In this paper, we use an equivalent parameter μ_i to identify the dominant scattering mechanism. This parameter is determined as follows

$$\mu_i = \frac{|u_{i1}|}{\sqrt{|u_{i2}|^2 + |u_{i3}|^2}} \quad (i = 1, 2) \tag{11}$$

In case $\mu_i \geq 1$ for all i, the ground scattering component will be dominant and the parameters of the two scattering components are determined as follows

$$P_s = \lambda_1 + \lambda_2;\ P_d = 0;\ |\alpha| = \sqrt{\frac{P_s - \Omega_{r,11}}{\Omega_{r,11}}} \tag{12}$$

In case $\mu_i \leq 1$ for all i, the double-bounce scattering is the most dominant scattering and the parameters are determined in (13)

$$P_d = \lambda_1 + \lambda_2;\ P_s = 0;\ |\beta| = \sqrt{\frac{P_d - \Omega_{r,11}}{\Omega_{r,11}}} \tag{13}$$

On the contrary, if $\mu_i \geq 1$ and $\mu_j \leq 1$ $(i \neq j)$, both scattering components exist in the residual matrix $[\Omega_r]$, the parameters of the two scattering components are determined as follows

$$\alpha = \frac{u_{i1} + \sqrt{\lambda_r}\psi^{-1}u_{i3}e^{-j\zeta}}{u_{j1} + \sqrt{\lambda_r}\psi^{-1}u_{j3}e^{-j\zeta}};\ \beta = \frac{u_{i1} - \sqrt{\lambda_r}\psi u_{i3}e^{-j\zeta}}{u_{j1} - \sqrt{\lambda_r}\psi^{-1}u_{j3}e^{-j\zeta}};\ P_s = \lambda_i;\ P_d = \lambda_j;\ \lambda_r = \lambda_i/\lambda_j \tag{14}$$

The coefficients ψ and ζ are components of the identity matrix introduced by Yamaguchi [16]. Finally, the residual matrices are compared and the minimum residual matrix corresponds to the optimal parameter set. Based on the obtained parameters, the forest height can be estimated by the difference between canopy phase and surface phase, as in Eq. (15)

$$h_v = \frac{\phi_v - \phi_0}{k_z} \tag{15}$$

with ϕ_v, k_z are tree canopy phase, vertical wavenumber and ϕ_0 is surface phase obtained by using coherence set method [17].

The implementation steps of the proposed approach are summarized in Fig. 2.

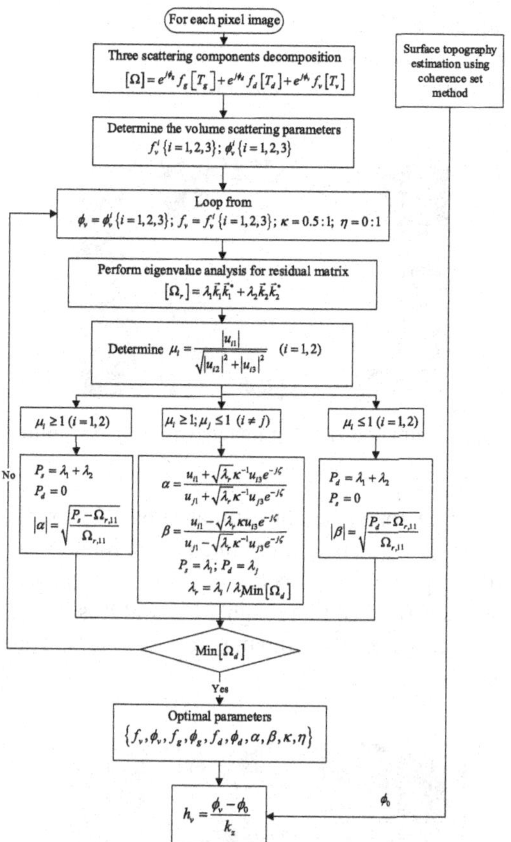

Fig. 2. Flowchart of the proposed approach

4 Applied Data Set and Experimental Results

To assess the effectiveness of the proposed method, a data set that is acquired from PolSARProSim software [18] is applied. The software performs data processing and it can generate simulated forest scenarios with similar parameters in actual scenarios. Accordingly, we created a forest scenario called HEDGE with an average tree height of 18 m. Forest area is 0.82745 Ha and tree density is 1000 stem/Ha. This is a sloped forest area with a slope in the azimuth of 16.7° (30%) and the slope in the direction of the range is 11.3° (20%). The interferometry polarimetric radar system is operated at frequency of 1.3 GHz and incidence angle of 45° when considering different soil conditions. The horizontal baseline is 15 m and the vertical baseline is 1.5 m. Azimuth resolution is 1.5 m whereas slant range resolution is 1 m.

The image of Pauli coded forest scenario is shown in Fig. 3 with an image size of 133 × 169 pixels in the range and azimuth direction, respectively. In this paper, we extract the results for analysis along the transect marked by the red line.

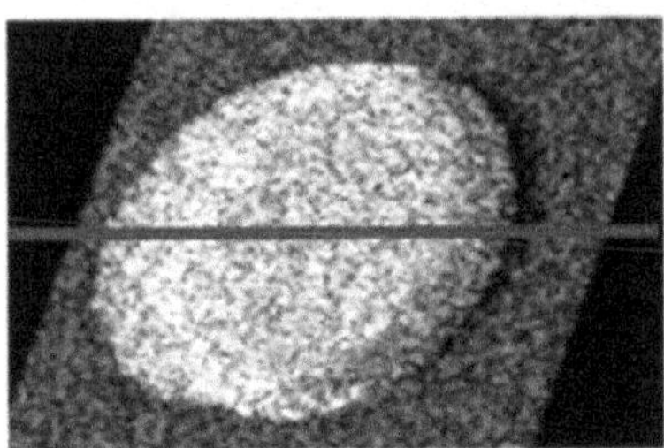

Fig. 3. Pauli coded forest image (Color figure online)

Figure 4 shows the tree forest height estimated by the optimal eigenvalue-based decomposition approach. The tree height obtained is completely consistent with it in a simulated scenario. In this picture, most peaks of the tree height are located at 18 m. At some pixels, tree height is estimated to be higher but still less than 22 m. Therefore, we can conclude that the estimated results from the proposed method are relatively accurate with small error rates.

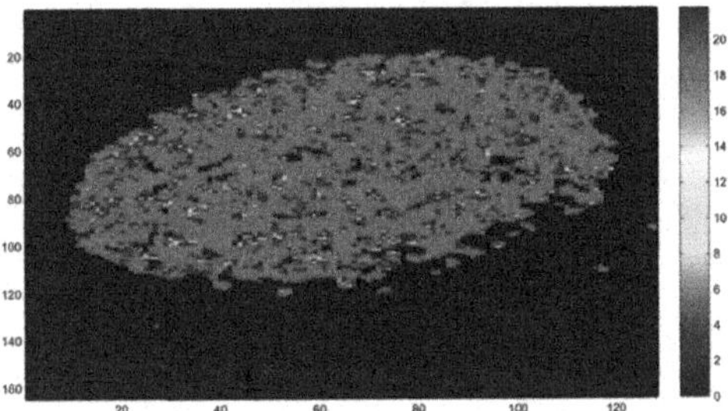

Fig. 4. The tree forest height estimated by the proposed approach

In order to further demonstrate the effectiveness of the optimal eigenvalue-based decomposition method, we implement comparing the results of the optimal eigenvalue-based decomposition approach with those obtained from the 3-stage inversion method [19]. This comparison is conducted in the 86th row of the transect line. The results are detailed in Fig. 5 and Table 1.

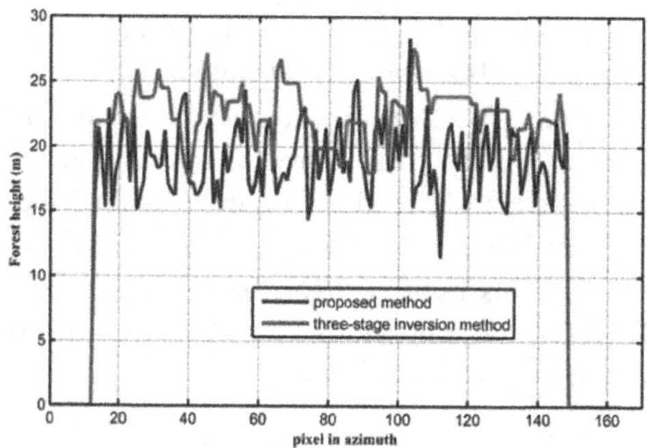

Fig. 5. The comparison of the results obtained from two methods

Table 1. The parameters are estimated by two methods

Parameter	Forest height h_v [m]	ϕ_v [rad]	Average errors [m]	RMSE (h_v) [m]
True	18	−0.157	0	0
Three - stage inversion	19.7325	−0.1218	1.5237	2.7052
Optimal eigenvalue-based decomposition method	17.5339	−0.1425	0.5788	2.5641

As mentioned in [19] by Cloude and Papathanassiou, the 3-stage algorithm performs forest height estimation by using a look-up table (LUT) to compare with the volume coherence. The accuracy of this algorithm depends on model predictions [19]. Therefore, the estimated results using this method are more inaccurate. Based on mathematical principles of eigenvalue analysis, the proposed method determined the optimal parameters for forest vegetation on sloped terrain. Therefore, the estimation results are significantly improved. This improvement is clearly shown by the parameters obtained in Fig. 6 and Table 1.

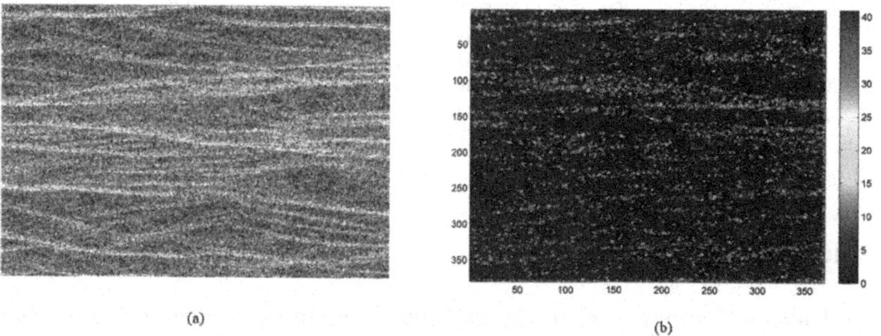

Fig. 6. The survey area: (a) HH polarized image and (b) the result of the proposed method

Next, we apply the proposed method to the satellite data set obtained from the ALOS system/PALSAR on March 12, 2007 for the first time and the second time on April 27, 2007. These include a pair of images of the jungle area at Kalimantan (1150 13′ 19″N, 1° 0′ 57″E), Indonesia. PolInSAR satellite system is operated at 1.3 Ghz frequency with baseline is 330 m and incidence angle is 21.5°. Figure 6a shows the HH polarized image of the survey area which is 375 × 384 pixels in size and the result of the proposed method is shown in Fig. 6b. This figure proves that most tree height on forest areas is approximately 20 m. Table 2 shows the detailed results in the forest parameter estimation using the proposed method for the survey area.

Table 2. Forest parameter estimation for survey area

Average height [m]	Fraction fill canopy r_h	RMSE (h_v) [m]	Volume phase [rad]
19.6942	0.4310	4.3976	0.0079

The amplitude of the ground, double-bounce and volume scattering mechanisms are shown in Fig. 7(a) and 7(b). For VV polarization channel, the volume scattering is the dominant component. For HH cross-correlation polarization channel, the contribution of the double-bound scattering component increased significantly. However, due to the slope of the terrain, this scattering mechanism is still not dominant in the vegetation areas and the contribution of volume scattering is still the largest. These results are completely consistent with the definition of Fresnel coefficients, in which the HH contribution exceeds the VV contribution.

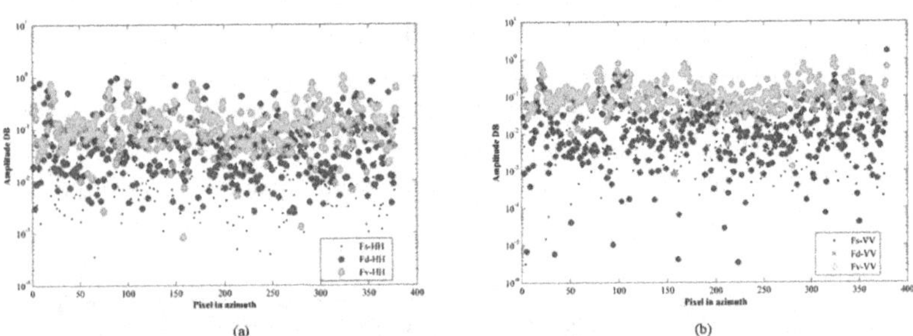

Fig. 7. Amplitude contributions of the three scattering mechanisms to the (a) HH, (b) VV polarized channel

5 Conclusion

A method for estimating vegetation parameters over forest mountain areas based on optimal eigenvalue-based decomposition technique has been proposed in this paper. The proposed decomposition method is implemented by introducing a simplified Neumann scattering model to fit the observed data on sloped terrain. Forest parameters were obtained by the optimal eigenvalue analysis technique. In addition, scattering power and the contribution of scattering components are also estimated. Simulated data and experimental data from satellite system have been used to test this optimal eigenvalue algorithm. Obtained results indicated that vegetation parameters can be extracted directly and accurately. In the future, experiments will continue to be carried out on different terrains to more fully assess the effectiveness of the proposed method.

Acknowledgments. The research was funded by the Vietnam National Foundation for Science and Technology Development (NAFOSTED) under Grant No. 102.01-2017.04.

References

1. Cloude, S.R., Pottier, E.: A review of target decomposition theorems in radar polarimetry. IEEE Trans. Geosci. Remote Sens. **34**(2), 498–518 (1996)
2. Lee, J.-S., Pottier, E.: Polarimetric Radar Imaging, From Basics to Applications. CRC Press, Boca Raton (2009)
3. Cloude, S.: Polarisation Applications in Remote Sensing. Oxford University, London (2009)
4. Chen, S.-W., Li, Y.-Z., Wang, X.-S., Xiao, S.-P., Sato, M.: Modeling and interpretation of scattering mechanisms in polarimetric synthetic aperture radar: advances and perspectives. IEEE Signal Process. Mag. **31**(4), 79–89 (2014)
5. Freeman, A., Durden, S.L.: A three-component scattering model for polarimetric SAR data. IEEE Trans. Geosci. Remote Sens. **36**(3), 963–973 (1998)
6. Xie, Q., Zhu, J., Lopez-Sanchez, J.M., Wang, C., Fu, H.: A modified general polarimetric model-based decomposition method with the simplified Neumann volume scattering model. IEEE Geosci. Remote Sens. Lett. **15**, 1229–1233 (2018)
7. Boerner, W.M., Mott, H., Luneburg, E.: Polarimetry in radar remote sensing: basic and applied concepts. In: Henderson, F.M., Lewis, A.J. (eds.) Manual of Remote Sensing: Principles and Applications of Imaging Radar, New York, NY, USA, vol. 2 (1998)
8. Chen, S.W., Wang, X.S., Xiao, S.P., Sato, M.: General polarimetric model-based decomposition for coherency matrix. IEEE Trans. Geosci. Remote Sens. **52**(3), 1843–1855 (2014)
9. Xie, Q., Ballester-Berman, J.D., Lopez-Sanchez, J.M., Zhu, J., Wang, C.: On the use of generalized volume scattering models for the improvement of general polarimetric model-based decomposition. Remote Sens. **9**(2), 117 (2017)
10. Neumann, M.: Remote sensing of vegetation using multi-baseline polarimetric SAR interferometry: Theoretical modeling and physical parameter retrieval. Ph.D. dissertation, Institute of Electronics and Telecommunications of Rennes, Rennes University, Rennes, France, vol. 1 (January 2009)
11. Neumann, M., Ferro-Famil, L., Reigber, A.: Estimation of forest structure, ground, and canopy layer characteristics from multibaseline polarimetric interferometric SAR data. IEEE Trans. Geosci. Remote Sens. **48**(3), 1086–1104 (2010)
12. Antropov, O., Rauste, Y., Hame, T.: Volume scattering modeling in PolSAR decompositions: Study of ALOS PALSAR data over boreal forest. IEEE Trans. Geosci. Remote Sens. **49**(10), 3838–3848 (2011)
13. Park, S., Moon, W., Pottier, E.: Assessment of scattering mechanism of polarimetric SAR signal from mountain forest areas. IEEE Trans. Geosci. Remote Sens. **50**, 4711–4719 (2012)
14. Hajnsek, I., Pottier, E., Cloude, S.R.: Inversion of surface parameters from polarimetric SAR. IEEE Trans. Geosci. Remote Sens. **41**(7), 727–744 (2003)

15. Cloude, S.R., Pottier, E.: An entropy based classification scheme for land application of polarimetric SAR. IEEE Trans. Geosci. Remote Sens. **35**(1), 68–78 (1997)
16. Yamada, H., Yamazaki, M., Yamaguchi, Y.: On scattering model decomposition of PolSAR and its application to ESPRIT-based PolInSAR. In: Proceeding of 6th European Conference on Synthetic Aperture Radar, Dresden, Germany (May 2006)
17. Zou, B., Lu, D., Cai, H., Zhang, Y.: Ground topography estimation over forests using PolInSar image by means of coherence set. In: 18th IEEE International Conference Image Processing (ICIP) (2011)
18. Williams, M.L.: PolSARproSim: a coherent, polarimetric SAR simulation of forest for PolSARPro (2006). http://earth.eo.esa.int/polsarpro/SimulatedDataSources.html
19. Cloude, S.R., Papathanassiou, K.P.: Three-stage inversion process for polarimetric SAR interferometric. IEEE Proc. Radar Sonar Navig. **150**(3), 125–134 (2003)

An Improved Forest Height Inversion Method Using Dual-Polarization PolInSAR Data

HuuCuong Thieu and MinhNghia Pham(✉)

Faculty of Radio-Electronics, Le Quy Don Technical University, Ha Noi, Viet Nam
nghiapm2018@mta.edu.vn

Abstract. The Extended 3-stage method has somewhat improved the precision of estimating forest parameters of the traditional 3-stage inversion method. However, the topography phase and optimal volume coherence coefficient determined by this method are not really optimal, that lead to the forest parameters estimation of this method is unstable and inaccuracy. Therefore, this paper proposes an improved forest height inversion method by dual-polarization channel PolInSAR data to enhance the precision of forest parameters extraction. In the suggested approach, the surface phase is calculated based on the mean coherence set theory. A comprehensive search method is then proposed to determine the polarization channels corresponding to the optimal polarization channel coherence coefficients for the volume scattering component. The effectiveness of the suggested approach was assessed with simulation data from PolSARprosim 5.2 software. The empirical results show that the suggested approach not only improves the effectiveness of the forest parameters estimation of the extended 3-stage method but also reduces the complexity in the calculation.

Keywords: Dual-polarization PolInSAR · Extended 3-stage inversion method · Forest height · Mean coherence set · Ground phase estimation

1 Introduction

Forest height is one vital parameters to serve forest management, monitoring and protection activities. Along with the strong development of science-technique and technology, many methods have been applied to improve the effectiveness of the mentioned activities related to forest. Currently, the polarimetric interferometry synthetic aperture radar (PolInSAR) system still shows superior advantages in extracting forest parameters [1–3]. The fully polarimetric PolInSAR system provides full information and a more complete description of the vertical structure of the forest. However, it does not have high resolution and its viewing area is not large compared to the dual-polarization PolInSAR system. Hence, the dual-polarization PolInSAR system can meet the requirements of large-scale surveys and global survey. In recent years, there has been a great deal of researches on estimating forest height using dual-polarization PolInSAR images [4–6], which shows a great potential of this technique. Nonetheless, the previously introduced

N.-S. Vo and V.-P. Hoang (Eds.): INISCOM 2020, LNICST 334, pp. 233–242, 2020.
https://doi.org/10.1007/978-3-030-63083-6_18

method mainly performed calibration based on the combination of HH and HV polarization channels without any process of optimal region coherence. Therefore, the region coherence with few polarization channels can results in instability and inaccuracy for the extraction of forest parameters. In 2016, Fu Wenxue suggested the extended 3-stage inversion manner [5] by dual-polarization PolInSAR data. This method somewhat improved on two disadvantages of the previously suggested approaches: (1) an optimization coherence process with the line fit method to estimate the surface phase; (2) optimal search technique is introduced to estimate polarimetric coherence coefficient for the volume scattering component. The extended 3-stage inversion method has significantly improved the effectiveness of estimating forest parameters compared to the 3-stage inversion approach. However, this method still has some drawbacks. Firstly, this method used four polarization channels $\tilde{\gamma}_{opt1}$, $\tilde{\gamma}_{opt2}$, $\tilde{\gamma}_{HH}$ and $\tilde{\gamma}_{HV}$ for ground phase estimation but the accuracy was not high. In addition, the process of determining this parameter is time-consuming and increases the complexity of the algorithm. Secondly, one of the two optimum polarization channels $\left(\tilde{\gamma}_{opt1}, \tilde{\gamma}_{opt2}\right)$ will be selected as the polarization channel for the volume alone scattering component, which causes large errors when calculating forest height.

From the mentioned reasons, this article suggests an advanced forest height conversion method using dual-polarization channels PolInSAR data for accuracy improvement in extracting forest parameters. In the suggested approach, the topography phase is estimated based on the mean coherence set theory. This method not only improves the accuracy of the topography phase estimation, but also minimizes the computation complexity of the suggested approach. After that, a comprehensive search method to identify the polarimetric channels corresponds to the volume scattering component. Finally, the forest height is extracted by comparing the forecast model with the optimal volume coherence factors for the scattering component directly from the canopy. Thus, the suggested approach not only improves the efficiency of the extended 3-stage inversion method, but also lessens the complexity in the calculation.

2 Methodology

2.1 The Complex Interferometry Coherence Coefficient of the Dual-Polarization PolInSAR System

In comparison with the full polarization PolInSAR system, scattering vector of the 2-polarization channels PolInSAR system will not have the VH and VV polarization components. So, its scattering vector is expressed as follow:

$$\begin{aligned} \vec{k}_1 &= \sqrt{2}\left[S_{HH}^1 \; S_{HV}^1 \right]^T \\ \vec{k}_2 &= \sqrt{2}\left[S_{HH}^2 \; S_{HV}^2 \right]^T \end{aligned} \tag{1}$$

Then, the coherence matrix and cross decorrelation matrix of dual Pol - PolInSAR system become level 2 square matrices as follows:

$$[T_4] = \left\langle \vec{k}.\vec{k}^H \right\rangle = \begin{bmatrix} T_{11} & \Omega \\ \Omega^H & T_{22} \end{bmatrix} \quad \text{with} \quad \vec{k} = \begin{bmatrix} \vec{k}_1 \\ \vec{k}_2 \end{bmatrix} \tag{2}$$

The complex interferometry coherence coefficient of the dual-polarization PolInSAR system is presented as Eq. 3.

$$\tilde{\gamma}(\vec{\omega}) = \frac{\vec{\omega}_1^H\ \Omega\ \vec{\omega}_2}{\sqrt{\langle\vec{\omega}_1^H\ T_{11}\ \vec{\omega}_1\rangle\langle\vec{\omega}_2^H\ T_{22}\ \vec{\omega}_2\rangle}} = \frac{\vec{\omega}^H\ \Omega\ \vec{\omega}}{\vec{\omega}^H\ T\ \vec{\omega}} \tag{3}$$

Where $\vec{\omega}_1 = \vec{\omega}_2 = \vec{\omega}$ are 2-dimension complex unitary vectors, the superscript "H" denotes complex conjugation and transposition and $T = (T_{11} + T_{22})/2$.

2.2 Estimating Ground Phase Based on the Mean Coherence Set Theory

Tabb and Flynn are pioneers in applying the numerical range theory of square matrix in analyzing and processing PolInSAR data [8, 9]. The shape of coherence region in the complex plane will then be extracted by using phase density function of the complex interferometry coherence coefficients. Thus, the complex coherence coefficient of dual-polarization PolInSAR data was defined as Eq. 4.

$$\tilde{\gamma}(\vec{\omega}) = \frac{\vec{\omega}^H\ \Omega\ \vec{\omega}}{\vec{\omega}^H\ T\ \vec{\omega}} = \vec{\nu}_i^H.H.\vec{\nu} \tag{4}$$

In which $H = T^{\frac{1}{2}}\ \Omega\ T^{\frac{1}{2}}, \vec{\nu} = \frac{\sqrt{T}.\vec{\omega}}{\vec{\omega}^H T^{\frac{1}{2}}\ \vec{\omega}}$ and $\vec{\nu}_i^H.\vec{\nu} = 1$.

According to the numerical range theory of square matrix, coherence set of all complex interferometry coherence coefficients is defined as below:

$$\Gamma = \left\{\vec{\nu}^H.H.\nu :\ \vec{\nu}^H.\vec{\nu} = 1,\ \vec{\nu} \in \mathbb{C}^2\right\} \tag{5}$$

The Eq. (5) has the similar form with the numerical range theory of a square matrix $[A]\ \left(A \in \mathbb{C}^{2\times2}\right)$. Then, we have the numerical range of A matrix as follow:

$$[A] = \left\{x^n A.x :\ x^H.x = 1,\ x \in \mathbb{C}^2\right\} \tag{6}$$

Therefore, according to the theorem of numerical range of the matrix, the theorem of numerical distance of matrix H is a convex contour of its eigenvalues. Without the loss of generality, we assume that matrix H have two eigenvalues $\lambda_1,\ \lambda_2$ with $(\arg(\lambda_1) < \arg(\lambda_2))$. In fact, λ_2 is very close to the value of $\tilde{\gamma}_{HV}$ [7]. Therefore, when connecting two points $\lambda_1,\ \lambda_2$ we get a straightforward line intersecting the complex unit circle (CuC) at two positions and one of these two positions will be the topography phase. Then the terrain phase will be determined as formula (7).

$$\phi_0 = \arg\{\lambda_1 - \lambda_2(1 - B)\} \tag{7}$$

Wherein, B is defined as follow:

$$a_0B^2 + a_1B + a_2 = 0 \Rightarrow B = \frac{-a_1 - \sqrt{a_1^2 - 4a_0a_2}}{2a_0} \tag{8}$$

With $a_0 = |\lambda_2|^2 - 1,\ a_1 = 2\mathrm{Re}\{(\lambda_1 - \lambda_2)\lambda_2\}$ and $a_2 = |\lambda_1 - \lambda_2|^2$ (9)

2.3 Estimating Forest Parameters by the Polarimetric Channel Comprehensive Search Method

To overcome the drawbacks of the extended 3-stage inversion method, we propose a comprehensive search method to find an optimum polarimetric channel of which there is at least the contribution of surface scattering component.

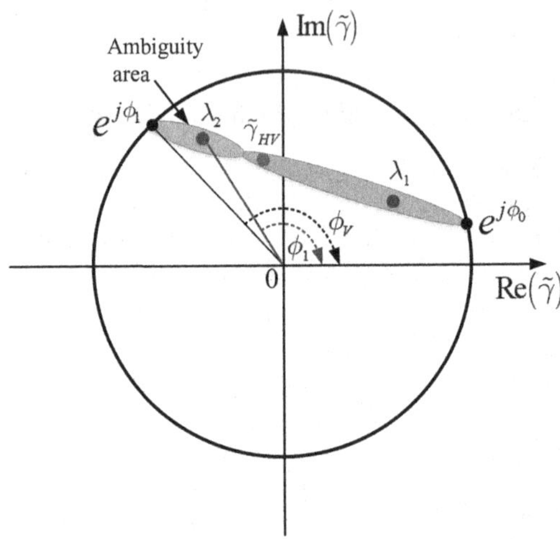

Fig. 1. Schematic representation of the optimization procedure of the mentioned method.

Equation (6) shows that the complex volume coherence factor of the dual-polarization PolInSAR system is a factor, which depends on polarization vectors. In this case, the polarization vector is a 2-dimension complex unitary vector and is represented as follow:

$$\vec{\omega} = \left[\cos\alpha \; \sin\alpha \, e^{j\psi}\right]^T \tag{10}$$

Where α and ψ are real coefficients that satisfy the conditions in Eq. (11).

$$\begin{cases} 0 \leq \alpha \leq \pi/2 \\ 0 \leq \psi \leq 2\pi \end{cases} \tag{11}$$

In this paper, we suggest a comprehensive search method to find a complex interferometry coherence coefficient representing for volume alone scattering component. This mean that, we search for the complex unitary vectors $\vec{\omega}$ to determine all phases of the coherence coefficient on the ambiguity area (the green domain in Fig. 2). For this purpose, we allow values $(\alpha,\ \psi)$ to vary within their value range. After that, a set of complex interferometry coherence coefficients $\tilde{\gamma}_c(\vec{\omega})$ will be determined according to the following condition.

$$\phi_1 > \arg(\tilde{\gamma}_c(\vec{\omega})) > \phi_{HV} \tag{12}$$

In which ϕ_1 is the phase respective to the second position between the coherency straight line and CuC, as is depicted in Fig. 1.

With the set of complex coherence coefficients $\tilde{\gamma}_c(\vec{\omega})$ satisfying the condition (12), we can completely estimate the HV and HH polarization channels from the Eq. 13.

$$\begin{cases} \tilde{\gamma}_{HH\_est} = e^{j\phi_0}\left[\tilde{\gamma}_c(\vec{\omega}) + L_{HH}(\vec{\omega})(1 - \tilde{\gamma}_c(\vec{\omega}))\right] \\ \tilde{\gamma}_{HV\_est} = e^{j\phi_0}\left[\tilde{\gamma}_c(\vec{\omega}) + L_{HV}(\vec{\omega})(1 - \tilde{\gamma}_c(\vec{\omega}))\right] \end{cases} \tag{13}$$

In which, the distance between the estimated channels and the two polarization interferometry channels is as follow:

$$\begin{cases} d_1 = \left|\tilde{\gamma}_{HH} - \tilde{\gamma}_{HH\_est}\right| \\ d_2 = \left|\tilde{\gamma}_{HV} - \tilde{\gamma}_{HV\_est}\right| \end{cases} \tag{14}$$

The coherence coefficient of the optimum polarization channel $\tilde{\gamma}_{c\_opt}(\vec{\omega}_{opt})$ will then be determined according to condition (15).

$$\min_{\alpha,\psi} \left\| \sum_{i=1}^{2} d_i \right\| \tag{15}$$

After that, a look-up table (LUT) for volume only coherence is developed based on formula (3). Finally, forest height and the mean extinction coefficient will be extracted by comparing the coherence coefficient of the optimal polarization channel $\tilde{\gamma}_{c\_opt}(\vec{\omega}_{opt})$ with values in LUT.

3 Experimental Result

The effectiveness of the suggested approach was assessed with simulation data generated from PolSARProSim 5.2 software [10]. Simulation data is received from the PolInSAR

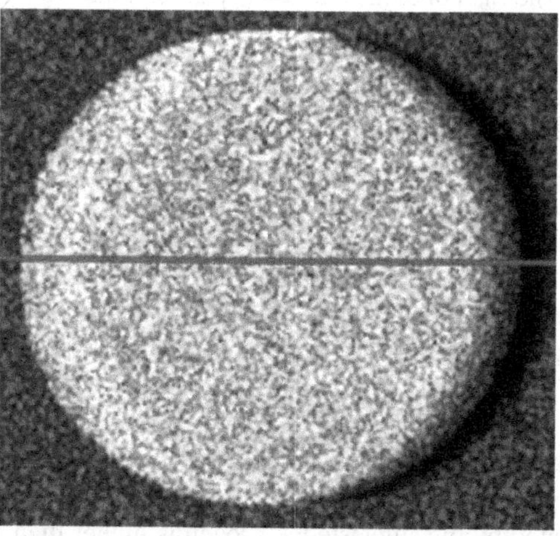

Fig. 2. Pauli image of the surveyed forest area (Color figure online)

band L system at the central frequency of 1.3 GHz with baseline of the 20 m horizontally and the 1.5 m vertically. The surveyed forest area has an average wood height of 20 m over place on a relatively flat terrain. The surveyed forest covers an area of 2,8274 Ha with a distribution of 900 trees per hectare. Figure 2 illustrates a Pauli color image of the observed forest area of 221 $\times$ 259 pixels. The effectiveness of the proposed model is assessed by comparing the results of the suggested approach with the extended 3-stage inversion manner [5].

Figure 3 is a chart comparing the topology phase that is detected by the proposed manner (the black line) and the extended 3-stage inversion algorithm. It can be seen that the estimated ground phase of the mean coherence set method has an average of 0.0982 rad and fluctuates quite close to the true topology phase. Meanwhile, what is estimated by the line fit method has an average of 0.1585 rad and usually oscillates over a relatively wide range and it has a large error compared to the true topology phase. Thus, the ground phase estimated by the suggested approach gives higher precision and reduces the computation time compared to the extended 3-stage inversion method.

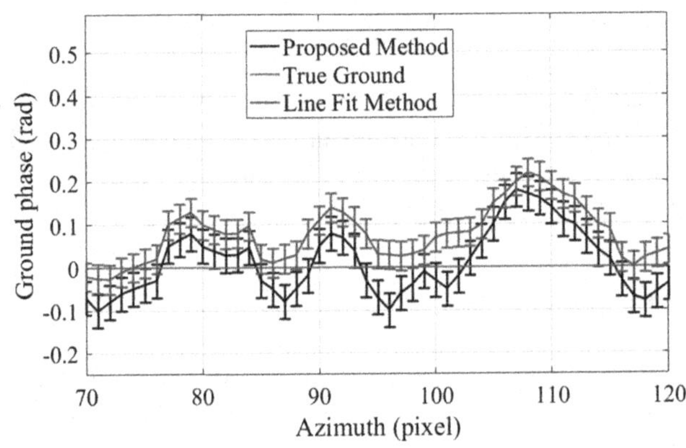

Fig. 3. The chart of estimated ground phases of the two methods.

Figure 4 is a line chart comparing the detected forest heights from the suggested approach (red line) with the extended three-stage inversion procedure (blue line). Figure 4 indicates that the average estimated forest height of the suggested approach fluctuates steadily around the height of 20 m (except for some pixels that exceed 21 m). Meanwhile, the volume height estimated by the extended 3-stage inversion procedure often oscillates strongly in the range of 16 m to 19.5 m (Especially, there are some pixels lower than 12 m). Although this method has significantly improved the accuracy better than the traditional 3-stage method. However, the estimated forest height has not yet been highly effective and stable.

Table 1 shows the forest height estimated by the extended 3-stage inversion manner and suggested approach with average values of 18,696 and 19,629 m, respectively. With the actual forest height in the simulation data of 20 m, we can see that the accuracy of the tree height estimated by the suggested approach is higher than that in the extended 3-stage inversion method of 4.67%. In addition, an important parameter representing the

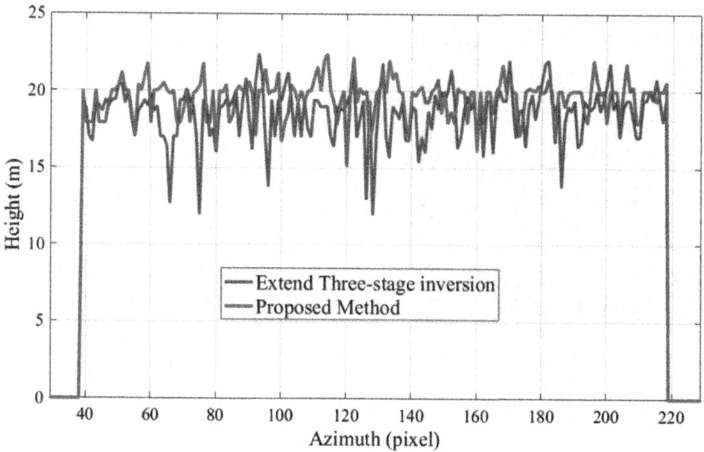

Fig. 4. A graph comparing the height of forest estimated by 2 methods. (Color figure online)

accuracy of the forest height estimated by the two methods is the root mean square error (RMSE). That are of 3,062 m and 1,582 m, respectively. The value of the suggested approach is smaller, meaning that the efficiency of volume height estimation of this method is higher than the extended 3-stage inversion method. Furthermore, the other parameters in Table 1, including the surface phase and the mean extinction coefficient estimated by the suggested approach, also show high accuracy and are close to these values of the system.

Table 1. Forest height estimation for two approaches.

Parameters	True values	Extended 3-stage inversion	Proposed manner
h_V [m]	20	18.696	19.629
ϕ_0 [rad]	0.0875	0.1343	0.0948
Mean extinction σ [dB/m]	0.156	0.264	0.185
RMSE [m]	0	3.062	1.582
Accuracy [%]	100	93.480	98.145
h_V [m]	20	18.696	19.629

Figure 5 (a) is a 2D image and Fig. 5 (b) is a 3D image describing the wood height in the entire surveyed forest detected from the suggested approach. It can be seen that the pixels are shown in 2D and 3D, they are mostly concentrated at 20 m approximately (there are some pixels higher than 20 m but not significant). From the results shown in Fig. 5, the suggested approach is relatively accurate and reliable.

After that, to analyze the influence of tree species on the accurate forest height retrieval of the suggested manner, we apply the proposed manner and the extended 3-stage inversion method with the simulated forest areas having different tree species

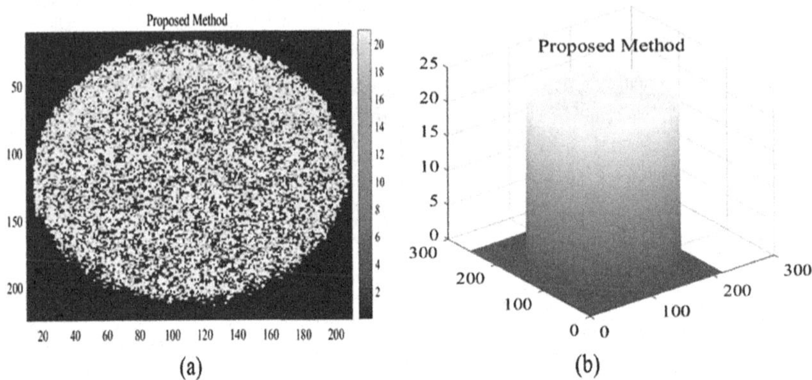

Fig. 5. Forest height is estimated by the suggested approach.

(coniferous forests, mixed forests and broadleaf forests). In the simulation data, we only change the tree types, the other parameters remain unchanged. Figure 6 describes the forest height estimated by two methods with three different tree species. We can see that the forest tree species of the conifer family (Pine) often cause large errors for the estimated results in both methods. The main reason is that with the forest trees belonging to the conifer family, scattering waves easily penetrate to the ground and then the central phase of ground scattering and mass scattering components are relatively close together.

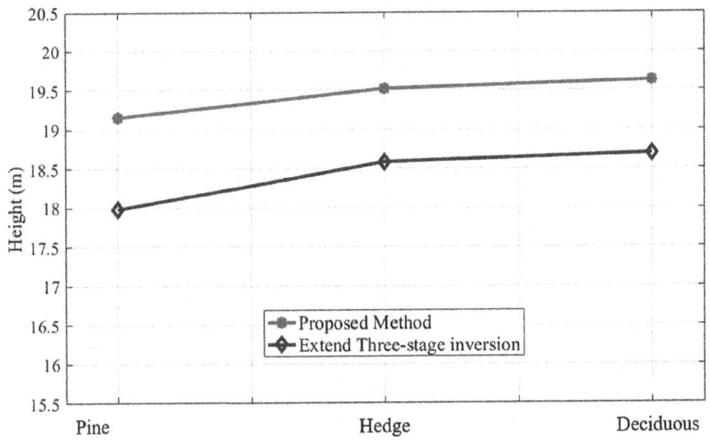

Fig. 6. The chart compares the forest height estimated by two methods with different tree species.

In contrast, with the forest area consisting of mixed trees (Hedge) and broad canopy trees (Deciduous), the radar waves are almost difficult to penetrate the canopy to affect the ground. That mean, the most radar microwaves scatter at the peak of the foliage so the center phase of these two scattering components is relatively far apart. However, in this case, the density of selected forest trees for simulation data is 900 trees/Ha. At that time, radar microwaves will not only be able to penetrate to the surface but also scatter on the canopy. Therefore, the backscattering signals will be received well and the center phase deviation of the volume alone scattering and surface scattering is not too close or too far, and it ensures that the calculation of the forest parameters is correctly. From the above analysis, it can be concluded that the forest tree species is also a parameter that has an effect on the effectiveness of the estimated forest height. Figure 6 presents the results of estimating forest height of the two methods with changes corresponding to the simulation data of each different tree species. However, because the density set for the initial simulation data was 900 trees/Ha, it was relatively favorable with the backscattering of radar waves. Therefore, the estimated forest height results in this case are not much changed. From this figure, it is again shown that the estimated forest height of the proposed algorithm is always more accurate and flexible than the extended 3-stage manner.

4 Conclusion

The paper has researched and developed an accurate method for estimating forest height from dual-polarization PolInSAR images. In the suggested approach, the terrain phase is first determined based on the mean coherence set theory. Then a comprehensive polarization channel search method is proposed to extract forest height and average wave penetration factor. The experiment results indicate that the accuracy of forest height detected by the suggested approach was enhanced by around 4.65% in comparison with the extended 3-state one. In the incoming years, the suggested approach can possibly be applied to different types of data and in different forest areas to optimize the efficiency of suggested approach.

Acknowledgments. The research was funded by the Vietnam National Foundation for Science and Technology Development (NAFOSTED) under Grant No. 102.01-2017.04.

References

1. Mette, T., Papathanassiou, K.P., Hajnsek, I., Pretzsch, H., Biber, P.: Applying a common allometric equation to convert forest height from Pol-InSAR data to forest biomass. In: Proceedings of the IGARSS, Anchorage, AK, USA, pp. 272–276 (2004)
2. Cai, H., Zou, B., Lin, M.: Parameter inversion model base on PolInSAR images. In: Proceedings of APSAR 2007, Huangshan, 5–9 November 2007, pp. 751–754 (2007)
3. Papathanssiou, K.P., Cloud, S.R.: Single baseline polarimetric SAR interferometry. IEEE Trans. Geosci. Remote Sens. **39**(11), 2352–2363 (2001)
4. Cloud, S.R., Papathanssiou, K.P.: Three-stage inversion process for polarimetric SAR interferometry. IEE Proc. - Radar Sonar Navig. **150**(3), 125–134 (2003)
5. Wenxue, F., Huadong, G., Xinwu, L., Bangsen, T., Zhongchang, S.: Extended three-stage polarimetric SAR interferometry algorithm by dual-polarization data. IEEE Trans. Geosci. Remote Sens. **54**(5), 2792–2802 (2016)
6. Xie, Y., Fu, H., Zhu, J., Wang, C., Xie, Q.: A LiDAR-aided multibaseline PolInSAR method for forest height estimation: with emphasis on dual-baseline selection. IEEE Trans. Geosci. Remote Sens. **17**(10), 1807–1811 (2019)

7. Treuhaft, R.N., Cloude, S.R.: The structure of oriented vegetation from polarimetric interferometry. IEEE Trans. Geosci. Remote Sens. **37**(5), 2620–2624 (1999)
8. Managhebi, T., Maghsoudi, Y., Zoej, M.J.V.: A volume optimization method to improve the three-stage inversion algorithm for forest height estimation using PolInSAR data. IEEE Geosci. Remote Sens. Lett. **15**(3), 1214–1218 (2018)
9. Managhebi, T., Maghsoudi, Y., Zoej, M.J.V.: An improved three-stage inversion algorithm in forest height estimation using single-baseline polarimetric SAR interferometry data. IEEE Geosci. Remote Sens. Lett. **15**(6), 887–891 (2018)
10. Williams, M.L.: PolSARproSim: A Coherent, Polarimetric SAR Simulation of Forest for PolSARProSim (2006). http://earth.eo.esa.int/polsarpro/SimulatedDataSources.html

An Attempt to Perform TCP ACK Storm Based DoS Attack on Virtual and Docker Network

Khanh Tran Nam, Thanh Nguyen Kim, and Ta Minh Thanh(✉)

Le Quy Don University, Ha Noi, Viet Nam
noangel0607@gmail.com, thanhcuchp@gmail.com,
thanhtm@mta.edu.vn
http://www.lqdtu.edu.vn/

Abstract. Recently, the server virtualization (hypervisor) market is growing up fast because server virtualization has many benefits. More and more businesses use hypervisors as an alternative solution to a physical server. However, hypervisors are more vulnerable than traditional servers according to recent researches. Therefore, stand on the position of a system administrator, it's necessary to prepare for the worst circumstances, understand clearly, and research for new threats that can break down the virtual system. In this paper, we attempt to perform TCP ACK storm based DoS (Denial of Service) attack on virtual and Docker networks and propose some solutions to prevent them.

Keywords: DoS · Hypervisor · TCP · ACK storm · Virtual network · Docker network

1 Introduction

1.1 Overview

Network security is one important aspect of many aspects that a system administrator is interested in because there are many potential cybersecurity threats to a hypervisor system [9–12]. The DoS/DDoS attack is one of those threats [8]. In 2019, Imperva [1] had reported an SYN DDoS attack in which 500 million packets per second (PPS) in January and another in which 580 million packets per second (PPS) in April. Each of the packets was thought to a median number of 850 bytes per packet. That means that 580 million 850-byte packets would result in about 3944 Gbps of data targeting your network protocol every second to render it unresponsive. Previously, in 2018, the GitHub DDoS Attack was recognized as sustaining a 1.35 Tbps (with 129.6 million PPS) attacks without the help of botnet. "Size" of DDoS attack is increasing year by year and cost businesses thousands to millions of dollars in losses. To prevent and minimize the DoS/DDoS attack's sabotage, analyzing more types of DoS attack

N.-S. Vo and V.-P. Hoang (Eds.): INISCOM 2020, LNICST 334, pp. 243–258, 2020.
https://doi.org/10.1007/978-3-030-63083-6_19

and is necessary. In this paper, we attempt to perform TCP ACK-Storm based DoS attack on virtual and Docker networks. Besides, we propose a new attack method based on ACK-Storm DoS attack with FIN-ACK packets which can make vSwitch/vBridge fall into a state of port-exhaustion in a period of time.

1.2 Our Contributions

In our knowledge, the research on the attacks to the vSwitch/vBridge of hypervisor systems and docker systems is not focused on, especially on the docker. Therefore, real services deployed on hypervisor systems or docker systems are vulnerable to attack via a network. That is the motivation of our paper to research the related network attacks on such systems. In summary, we briefly introduce our contributions in this paper as follows:

1. We propose to attempt DoS attacks using FIN-ACK storm on services deployed on virtual systems so that service providers understand the risks, and security vulnerabilities when deploying the services in a virtual environment. That implies that real applications deployed on virtual systems can be easily attacked by hackers via the Internet.
2. The DoS attacks proposed on the Virtual Machine and Docker systems in this paper is the first attempt that is made to prove feasible when a hacker wants to attack services on a hypervisor system.
3. We propose a new attack method based on ACK-Storm DoS attack with FIN-ACK packet which can make vSwitch/vBridge fall into a state of port-exhaustion in a period of time.
4. Based on these attack experiments, our paper also offers some solutions to prevent and decrease the destruction of these types of attacks in real applications.

1.3 Roadmap

The rest of this paper is organized as follows: In Sect. 2, a brief overview of the related works are presented. We focus on the explanation of the ACK-Storm DoS attack and the FIN-ACK-Storm DoS attack. To illustrate, In Sect. 3 and Sect. 4, the detail of the proposed DoS attacks on the hypervisor systems, VMware and Docker, is explained. Based on our experimental attacks, the simulation results and discussion are shown. Furthermore, In Sect. 5, we assess the feasibility of the Ack-Storm DoS attack on hypervisor systems. Finally, Sect. 6 gives the conclusions of this paper.

2 Related Works

2.1 Transmission Control Protocol - TCP

TCP is a Connection-oriented transmission protocol, it establishes connection channel before transferring data. Through TCP, applications on networked

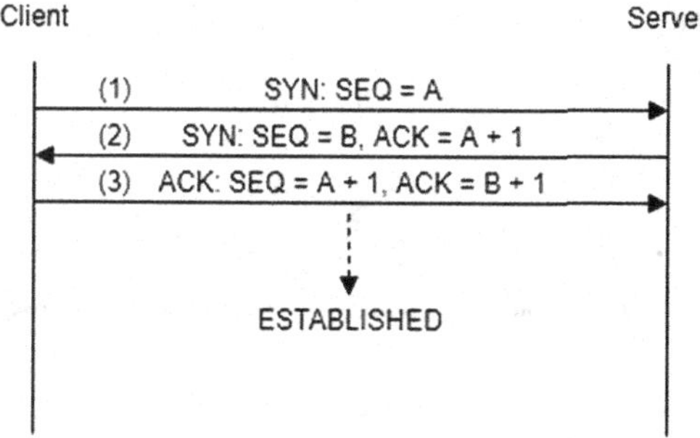

Fig. 1. TCP Three-step Handshake - Establish TCP connection

servers can communicate with each other, through which they can exchange data or packets. This protocol ensures reliable data delivery to the receivers. Moreover, TCP has the function of distinguishing between data of many applications (such as Web services, Email services, and so on) simultaneously running on the same server. The operation of the protocol TCP is described in the RFC-793 [7]. Nowadays, TCP is still widely used in many server systems.

Three-Step Handshake. Three-step handshake, or maybe called a Three-way handshake, is used in TCP to establish a connection. TCP uses passive open, a server bind to and listens to a port before a client tries to connect to the server. A client may start an active open after the passive open is established. The three-step handshake occurring in three steps (see Fig. 1) can be described as follows:

1. The client, who wants to connect to the server, sends a TCP packet to the server with a random value A for the segment's sequence number (SEQ) and a bit SYN set. This packet is called a SYN packet (1).
2. After the server receives the SYN packet, it responds to the client a TCP packet with a random value B for SEQ number, the acknowledgment number is set to one more than the received sequence number (A+1), and two-bit SYN, ACK are set. This packet is called a SYN-ACK packet (2).
3. Finally, the client sends an ACK packet (3), which has the acknowledgment number is set to one more than the received sequence number (B+1), the sequence number is set to the received acknowledgment value (A+1), and only bit ACK set.

Four-Step Handshake. Four-step handshake, or four-way handshake, is used to terminate TCP connection with each side of the connection terminating independently. It occurs as follows:

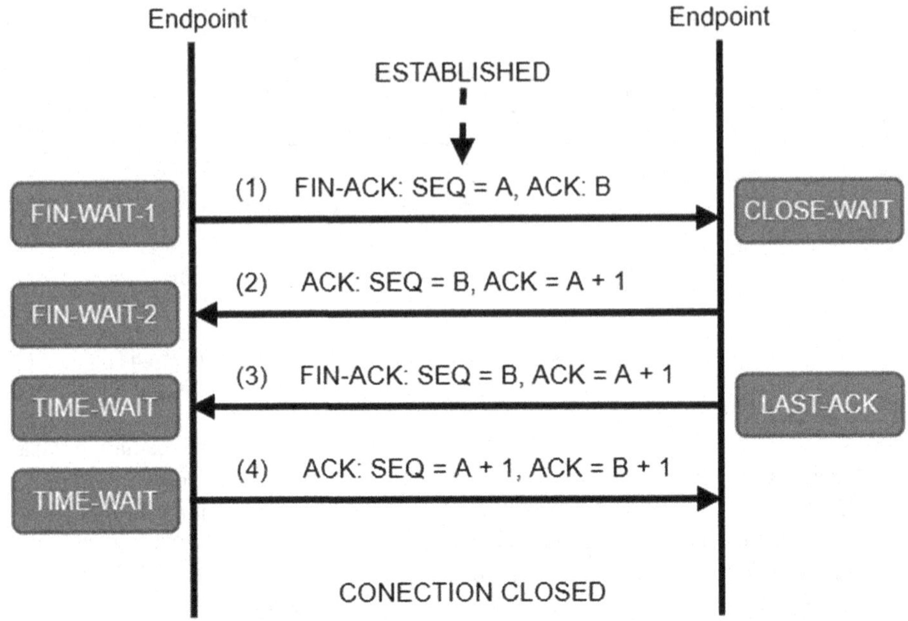

Fig. 2. TCP Four-step Handshake - Terminate TCP connection

1. When one party X of a TCP connection wants to terminate its half of the connection, it sends to the other endpoint a FIN packet, which has FIN bit set, the SEQ number (A), and ACK number (B) depending on the current state of the TCP connection. Then it enters the FIN-WAIT-1 state.
2. The other endpoint Y receives the FIN packet, free up its buffer, and responds an ACK packet, with acknowledgment number is more one than the received sequence number (A+1) and the sequence number is set to the received acknowledgment value (B). It enters the CLOSE-WAIT state. After receiving the ACK packet from endpoint Y, endpoint X enters the FIN-WAIT-2 state from the FIN-WAIT-1 state. The connection from endpoint X to endpoint Y is terminate, but the connection from endpoint Y to endpoint X still opens, this is the half-close connection.
3. Endpoint Y sends a FIN packet to terminate the connection from endpoint Y to endpoint X and wait for an acknowledgment from endpoint X.
4. Finally, when endpoint X receives the FIN packet from endpoint Y, it enters the CLOSED state.

2.2 Vmware Workstation [4]

Virtual Machines (VMs). Virtual machines are software computers that provide the same functionality as physical computers. They are based on computer architectures but they behave as separate computer systems. With them, the users could run different software requesting different environments without conflict at the same time. They bring many benefits for users and businesses: Reduced hardware costs; Faster desktop and server provisioning and deployment; Small footprint and energy saving; Increasing IT operational efficiency, and so on.

Hypervisor. A hypervisor, called a virtual machine monitor (VMM), can be hardware, software, or firmware that provides virtualization capability. A hypervisor allows one host computer, which the hypervisor operates in, to support multiple guest VMs by virtually sharing its resources such as memory and processing, and so on. According to the resources that have been allocated for each virtual machine, the hypervisor gives and manages the scheduling of VM resources against the physical resources. The hypervisor has two types: type 1,"bare metal", run directly on the host's hardware, like an operation system, while type 2, "hosted", run as software on an operating system, as an application. What type of usage is based on the purpose and need of the user and businesses.

Virtual Switch (vSwitch). A virtual switch is a software application that allows communication between virtual machines, between the physical machine and virtual machines. It directs the communication on a network in an intelligent way by ensuring the integrity of the virtual machine's profile, which includes network and security settings checking data packets before moving them to a destination. A virtual switch is completely virtual and can connect to a network interface card (NIC). The vSwitch merges physical switches into a single logical switch. This helps to increase bandwidth and create an active mesh between a server and switches. It also helps in easy deployment and migration of virtual servers, allows network administrators to manage virtual switch deployed through a hypervisor, and easy to roll out new functionality, which can be hardware or firmware related.

Virtual Network - Network Virtualization. Network Virtualization is a method of splitting up the available bandwidth into channels to combine available resources in a network. Each of the channels can be assigned (or reassigned) to a particular server or device in real-time and is independently secured. The main idea of Network virtualization is that virtualization disguises the true complexity of the network by splitting it into manageable parts, like a partitioned hard drive, making it easier to manage files. Every subscriber has shared access to all the resources on the network from a single computer.

2.3 Docker [5]

Docker Engine - Docker Daemon. Docker Engine, which may be called Docker Daemon, is a background-running service that manages everything required to run and interact with Docker containers on the host operating system. It's used to run Docker containers which bundled up all application dependencies inside. Docker Engine enables containerized applications to run anywhere consistently on any infrastructure. Docker Daemon communicates directly with the host operating system and knows how to ration out resources for the running Docker containers. It's also an expert at ensuring each container is isolated from both the host OS and other containers. In simple terms, it replaces the hypervisor.

Docker Container Image. Docker container image, or Docker image, is a lightweight, standalone, executable package of software. It includes code, runtime, system tools, system libraries, and setting files - all things needed to run an application. There are many Docker images available that could be used to rebuild new images or deployed Docker containers.

Docker Container. A Docker container is an instance of a deploying Docker image. But we could modify Docker containers. Multiple containers can run on the same machine at the same time and share the OS kernel. They run as independent processes in userspace. Containers take up less space than VMs, can handle more applications, and require fewer VMs and Operating systems.

Docker Network. The Docker networking philosophy is application-driven. Docker network isolation achieved using Network namespace. Typically, Services gets separate IP and maps to multiple containers. Microservices done as Container puts more emphasis on integrated Service discovery. As the Container scale on a single host can run to hundreds, host networking has to be very scalable.

Virtual Bridge (vBridge). A virtual bridge (vBridge), which may be called Network bridge or Linux bridge, is a piece of software used to unite two or more network segments. It works like a virtual network switch (vSwitch) and working transparently [13,14]. In Docker container built with a Linux image base, the Docker network is managed by vBridge.

2.4 Network and Port Address Translation [6]

Network Address Translation - NAT. Network Address Translation (NAT) is a technique that allows one or more internal IP addresses to be converted to one or more external IP addresses. This technique makes a device in a local/private network could connect to the public network (Internet). NAT is responsible for transmitting packets from one network layer to another in the same network. NAT will make changes to the IP address inside the packet. Then move through routers and network devices. On the contrary, when the packet is transmitted from the internet (public) back to the NAT, NAT performs the task of changing the destination address to the IP address inside the local network and sending it. Moreover, NAT can act as a firewall. It helps users secure computer IP information. Specifically, if the computer is having trouble connecting to the internet, the public IP address (previously configured) is displayed instead of the local network IP address.

Port Address Translation - PAT. Port Address Translation (PAT) is an extension technique of NAT which could help multiple devices on a local/private network connect to the public network by mapping their local IP address to a single public IP address and specific ports. With each set of local-IP:local-port is mapped to a public IP address with a specific port, multiple devices could communicate with the Internet with the corresponding ports provided to them. This technique could conserve IP addresses but the number of ports is not unlimited, only 65,536 ports so there can be a theoretical maximum of 65536 PAT entries at a time for each inside global address. If an attacker can occupy all 65,536 ports, there is a port-exhaustion, and no communications can be made between local devices and the public network.

2.5 Original ACK-Storm DoS Attack

The original idea [2] is suggested by Mr. Abramov and Prof. Herzberg depending on the vulnerable handle exceptions of TCP that described on page 72 of RFC 793 [7] about TCP: when a TCP connection is in the ESTABLISHED state received a packet with not-yet-sent acknowledges data SEG.ACK > SND.NXT (Acknowledgement from the receiving TCP higher than the sequence number of the next byte of data to be sent to the other), the received one handle as follows: Send an ACK (with the last sent SEQ/ACK number) to another party of connection, then drop the segment and return. In particular, ignore the payload in the segment. However, this state has a timeout and stopped when the timer reaches the timeout. For raising the basic ACK-storm DoS attack (Two Packets ACK Storm), the attackers act as the following scenario:

1. Pick up (at least) one packet from a TCP connection between a client and a server (Just need to eavesdrop one packet and do not need any impacts on the connection).
2. Generate two packets, each addressed to one party, and with a sender address of the other party (*i.e.* spoofed). The packets must be inside the TCP windows of both sides. The packets should have content - at least one byte of data or it will not be implemented.
3. Send the packets to the client and the server at the same time. The connection will then enter an infinite loop of sending ACK packets back and forth between both parties.

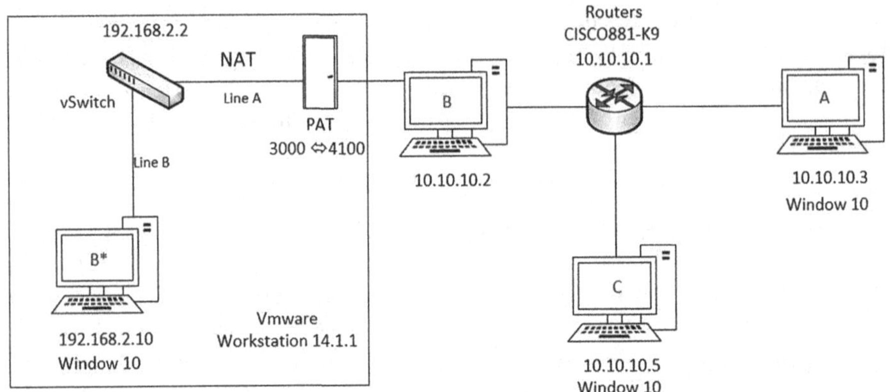

Fig. 3. Experimental model - Three physical computers as above connected with each other by a router Cisco 800 Series Routers CISCO881-K9, the C is attacker which can eavesdrop and inject packets into TCP connection between Host A and Hypervisor B

2.6 ACK-Storm DoS Attack Using FIN-ACK Packet

Depending on Abramov's idea [2], Son proposed another ACK-Storm DoS attack [3] with the same mechanism, but the starting point is sparked by a couple of FIN-ACK packets created by the same way when creating a couple of ACK packets to trigger ACK-Storm DoS attack. According to the description in RFC-793 [7] (p. 73): if a TCP connection is in CLOSE-WAIT state, it does the same processing as for the ESTABLISHED state. That means if attackers force each party running into CLOSE-WAIT, each party waits for an ACK packet (see Fig. 2) never come but not-yet-sent ACK packet instead, then the ACK-Storm DoS raised by two FIN-ACK packets starts. Because no timeout by default for CLOSE-WAIT state, this DoS attack will never stop in theory, and parties of the connection will stay in CLOSE-WAIT state forever.

3 Experiment in VMware Workstation

3.1 Experiment Original ACK-Storm DoS Attack

Environment for Experiment (see Fig. 3)

Hypervisor. We use VMware Workstation 14.1.1 installed Windows 10 64 bit for Virtual Machine server B*.

Physical Computers. Host A and Hypervisor B have the same configuration as follows: Window 10 64 bit, Chip Inter® Core^{TM} i7-6700 CPU @ 3.40 GHz, RAM 8 GB, the network interface is a 100 Mbps Ethernet adapter attached to the PCI-E bus. And the attacker C is Windows 10 installed Python.

Router. We use a router instead of the hub in Son's experimental test because, in reality, businesses use the router for establishing their LAN network or connect to the Internet. The router here is a Cisco 800 Series Routers CISCO881-K9.

Attack Execution. We use a simple TCP connection created with socket python (v3.7.3) scripts. For the experiment, we just let A and B* create a connection with a Three-Way Handshake and pick up the last ACK packet from the connection for SEQ and ACK sequence number. Then, in attacker C, we use scapy[1] to create a couple of fake ACK packets with source IP is one party and destination IP is the other. We sent those ACK packets to each of the respective parties.

In Line A. We captured about 55000 retransmitted ACK packets in 60 s while the ACK storm was occurring. This result is similar to the results in previous experiments of Abramov [2] and Son [3] but the number of packets is less because of smaller Ethernet adapter bandwidth. However, task manager still displayed the bandwidth used by VMware NAT services was 0.6 Mbps. This result is much bigger than 120 bytes (two packets) sent by the attacker C.

In Line B. Only the first fake ACK packet, which we created, is forwarded to virtual machine B* through vSwitch. No retransmitted ACK packet is directed to virtual machine B*.

3.2 Experiment ACK-Storm DoS Attack Using FIN-ACK Packet

Environment for Experiment. We use the same environment with the experiment of ACK-Storm DoS attack described above.

Attack Execution. We do the same action with the experiment ACK-Storm DoS attack but we use a couple FIN-ACK packets instead of a couple of ACK packets. It means that we just turn the bit FIN flag to 1 and keep all conditions as the experiment ACK-Storm DoS attack as above.

Analysis. When each party received fake FIN-ACK packets, they respond with invalid retransmitted ACK packets, and the ACK-Storm was begun.

In Line A. We captured about 86000 ACK retransmitted packets in 60 s (about 1434 packets per second) in Line A while the ACK storm was occurring. This result is similar to the results in previous experiments of Abramov [2] and Son [3] but the number of packets is less because of smaller Ethernet adapter bandwidth. However, the task manager still displayed the bandwidth used by VMware NAT services was 0.6 Mbps. This result is much bigger than 120 bytes (two packets) sent by the attacker C. We keep experiment for 24 h and it's still working. This

[1] https://scapy.net/.

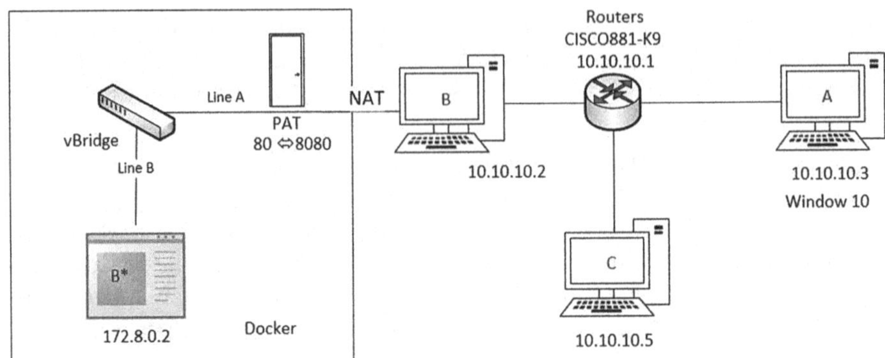

Fig. 4. Experimental model - Three physical computers as above connected by a router Cisco 800 Series Routers CISCO881-K9, computer C is attacker which can eavesdrop and inject packets into TCP connection between Host A and Hypervisor B. Docker container B* use Nginx for server with the local port is 80 and the public port is 8080.

proves that Son's hypothesis [3] is correct. The TCP connection is stuck in the CLOSE-WAIT state while the process established TCP connection keeps running. As the ACK-Storm activated by a couple of FIN-ACK packets can be last forever, the attacker could completely raise DDoS attack to hypervisor B and virtual machine B*. If an attacker could raise a DDoS attack with all available ports, about 65500 ports, he could "play" a DDoS attack which 94 million packets per second (with 40 Gbps), and also cause the port exhaustion.

In Line B. As explained above, in the ACK-Storm DoS attack experiment, only the first fake FIN-ACK packet, which we created, is forwarded to virtual machine B* through vSwicth. Still, no retransmitted ACK packet is directed to virtual machine B*, the same result with the ACK-Storm DoS attack. We assumed that vSwitch responses all retransmitted ACK packets instead of virtual machine B* as long as no RST request is sent between Host A and virtual machine B*. To prove this, we try suspending virtual machine B*, the ACK-Storm attack still going on. We keep experiment for 4 h more and the ACK-Storm attack is no sign of stopping. When we resume virtual machine B*, nothing happens. This is a feature of VMware workstation that prevent all invalid packets to virtual machine inside it.

4 Experiment in Docker

4.1 Experiment Original ACK-Storm DoS Attack

Environment for Experiment

Hypervisor. We use Docker Desktop v2.2.0.5 with Docker Engine v19.03.8, Docker Compose v1.25.4. We build a new image depending on base image Ubuntu 18.04 and install Htop for monitoring the container's activity. We also use TCPdump for monitoring network flow, and use Nginx for server B.

Physical Computers. Host A and Hypervisor B have the same configuration as follows: Window 10 64 bit, Chip Inter® Core^{TM} i7-6700 CPU @ 3.40 GHz, RAM 8 GB, the network interface is a 100 Mbps Ethernet adapter attached to the PCI-E bus. We use Windows 10 for attacker C.

Router We use a router for establishing a LAN network or connect to the Internet. The router here is a Cisco 800 Series Routers CISCO881-K9.

Attack Execution. With TCP connection created by using Chrome to access to the server Nginx in Container B* and do the same as experiment original ACK-Storm DoS attack in VMware Workstation, we pick the last ACK packet in the TCP connection between Host A and Docker container B* for obtaining SEQ and ACK number. Then, we use scapy to create a couple of fake ACK packets with source IP is one party and destination IP is the other. After that, we send those ACK packets to each of the respective parties.

Analysis. When each party received a fake ACK packet, they responded with invalid retransmitted ACK packets, and the ACK-Storm was begun. The ACK-Storm was terminated by Host A (see Fig. 4) after 1 s while TCP connection was timeout after 30 s.

In Line A. We captured about 550 retransmitted ACK packets in 1 s while the ACK storm was occurring. 10 s later, Container B* send an RST-ACK packet. The TCP connection timeout after 30 s. There are still some retransmitted ACK packets generated by the original ACK-Storm attack but the retention time is very short. This is because of Docker's feature (or Nginx's) when the container received many retransmitted ACK packets with the same ACK/SEQ number. This result is not as expected but still proves a flaw in TCP connection, which was discovered by Abramov [2], when receiving not-yet-sent acknowledges data packet.

In Line B. The same result with the experiment ACK-Storm DoS attack, only the first fake ACK packet is forwarded to Docker container B* through vBridge. No retransmitted ACK packet is directed to Docker container B*. It seems like vBridge response all retransmitted ACK packets instead of Docker container B*, same behavior with vSwitch.

4.2 Experiment ACK-Storm DoS Attack Using FIN-ACK Packet

Environment for Experiment. We use the same environment with the experiment ACK-Storm DoS attack in Docker that's described above.

Attack Execution. We do the same with the experiment ACK-Storm DoS attack but we use a couple FIN-ACK packets instead of a couple of ACK packets. It means that we just turn the bit FIN flag to 1 and keep all conditions as the experiment ACK-Storm DoS attack as above.

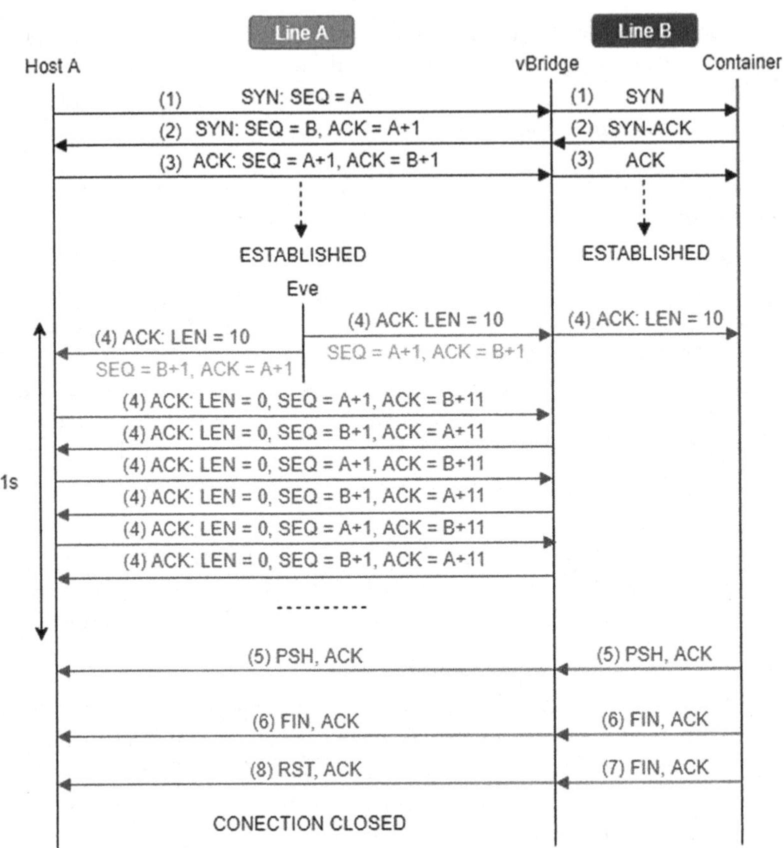

Fig. 5. TCP communication during the experiment in Docker with Ack packet

Analysis. When each party received fake FIN-ACK packets, they responded with invalid retransmitted ACK packets, and the ACK-Storm was begun.

In Line A. We captured a few retransmitted ACK packets between Host A and Container B* before Container B*'s side stopped responding.

In Line B. Same as above, in the experiment ACK-Storm DoS attack, only the first fake ACK packet is forwarded to Container B* through vBridge. No retransmitted ACK packet is directed to Container B*.

Propose. We tried to replace simple ACK packets for attacking by HTTP packets. However, we got the same results as Subsect. 4.1 and 4.2. Because ACK-Storm DoS attack with FIN-ACK [3] packet seem does not work with TCP connection created by browsers like Chrome or Firefox and Nginx, we propose a new attack method based on it: the attacker C create his own fake TCP connection with server B, send a fake FIN-ACK packet to server B after three-ways handshake completed and constantly send fake retransmitted ACK packets of the same format as retransmitted ACK packets of ACK-Storm DoS attack. Server B, more specifically vBridge, will respond to them and stuck in CLOSE-WAIT state. With a simple TCP connection, the CLOSE-WAIT state has no timeout [7], so we assume server B may be stuck in CLOSE-WAIT state forever. The detail of this attack method will be described in Subsect. 4.3.

4.3 Experiment ACK-Storm DoS Attack Using FIN-ACK Packet and Fake Retransmitted ACK Packets from Attacker

Environment for Experiment. We use the same environment with the experiment ACK-Storm DoS attack in Docker that's described above (see Fig. 5) but no Host A.

Attack Execution. We use scapy (version 2.4.3) to create a fake TCP connection with Container B* by Three Steps Handshake as follows:

1. Use scapy create a fake SYN packet manually with the destination is Container B*, the source is attacker C, ACK number is 0, SEQ number is voluntary, and send to Container B*.
2. Capture response SYN/ACK packet from Container B* then create a fake ACK packet depending on the SYN/ACK packet (use ACK and SEQ number).
3. Send the fake ACK packet to Container B* and the TCP connection is established.

After that, we create a fake FIN-ACK packet as same as the previous experiments and create a fake retransmitted ACK packets like attacker C received FIN-ACK packet similar to Container B*'s. Then, we follow these three steps:

1. Send the fake FIN-ACK packet to the Container B*.
2. Send the fake retransmitted ACK packet to the Container B*.
3. Delay about 1 s and repeat step 2.

Analysis. When Container B* received the fake FIN-ACK packet, it switches to CLOSE-WAIT state and responses with invalid retransmitted ACK packets.

In Line A. Whenever vBridge receives the fake retransmitted ACK packet from the attacker C, it responses with invalid retransmitted ACK packets. While the attacker C keeps sending fake retransmitted ACK packet to the Container B*, the Container B* is stuck in CLOSE-WAIT state. But the attack does not last too long as same as the experiment ACK-Storm DoS attack with FIN-ACK packet [3]. After about 10 min, Container B* sends an RST packet to close the TCP connection. In the experiment, we dump some FIN-PSH-ACK packets but there are no signs of disconnection until the Container B* sends RST packet suddenly. This might be a feature of vBridge when received too many retransmitted ACK packets with the same ACK/SEQ number.

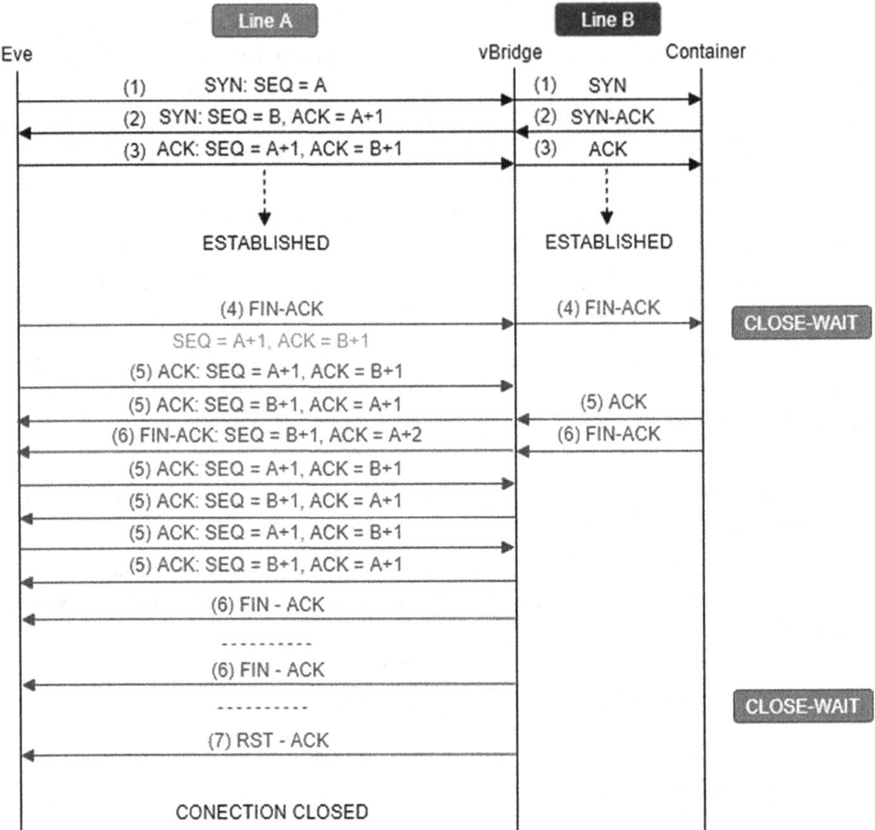

Fig. 6. TCP communication during the experiment using FIN-ACK packet in Docker with fake retransmitted ACK packets from attacker C

In Line B. Only valid packets are directed to Container B*, the same as the result of the experiment ACK-Storm DoS attack. No retransmitted ACK packet is directed to Container B*. It means that vBridge response all retransmitted ACK packets instead of Container B* as long as no RST request is sent between Host A and Container B* (Fig. 6).

5 Discussion

5.1 Feasibility

From the results of the above experiments, we have concluded that the ACK-Storm DoS attack is possible with basic TCP connections in a virtualized environment. Based on the vulnerability discovered by Abramov [2] described in RFC 793 [7] and based on the idea of developing a method of attack with FIN-ACK packet of Son [3], an attacker can perform an attack which is made by creating a fake connection and sending packets just like a normal TCP connection being attacked by ACK-Storm DoS using FIN-ACK packet.

In our experiment, this attack method is feasible even with virtualization environments like Docker with Nginx for the server. However, in the default configuration of Nginx, each state of the connection has a timeout setting so the attack method using ACK packets is blocked and the risk of being attacked by an attacker with the FIN-ACK packet is also limited. However, attacker C can use a simple python script with scapy to create many connections to the server on the Docker container that uses Nginx to occupy ports within 10 min and can cause an issue, which is called port exhaustion, with vBridge when they combine with a botnet. The time, just about 10 min, is not too long, but the damage to businesses will not be small even though the cost of conducting the attack is not very high. With this attack method, attackers do not need to eavesdrop TCP connection but still can occupy port.

5.2 Countermeasures

As we mentioned above, in Sect. 3, the vSwitch (or the vBridge) responds (or ignores) invalid packets instead of the hypervisor (or the container) and only directs valid packets to the hypervisor. Therefore, the hypervisor will receive no packets from the client during the ACK-Storm DoS attack. So, we can create a timer running on another thread to count the time unresponsive from the client and close that connection when "timeout". Besides, we can create a duplicate-ACK-checker to count the number of duplicate retransmitted ACK packets, when the number reaches the maximum, the hypervisor creates its own RST-ACK packet to force the connection stop completely.

5.3 Ethical Considerations

We disconnected the Internet when conducting our experiments and only tested attacks with the lab computers during the process. These computers only connected to each other during the experiment and did not connect even to the university local area network. Therefore, our experiments are completely harmless to the Internet.

6 Conclusions

For well-known software built by professional teams, such as Nginx, this type of attack is difficult to perform. However, its implications for the virtual server system are still evident once it is successfully implemented. This paper emphasizes the proper attention to handling timeout vulnerabilities with CLOSE-WAIT state in TCP connection when building and developing new software using TCP connection. This flaw could lead to an unlimited ACK-Storm DoS which could harm servers. Finally, TCP is still an important protocol and is used in many software, it was built long ago so inevitably there are vulnerabilities, so find the flaws and carefully study the vulnerabilities to find out methods to prevent and fix these gaps are the necessary work of all information security experts. In this article, we have focused on researching and analyzing deeply the ACK-Storm DoS attack models of Abramov [2] and Son [3] as well as proposing a new attack version to capture the network ports of a virtual server. In the future, we will focus on researching other virtualized server systems such as Hyper-V, Oracle, an so on, with these types of attacks as well as proposing new attack measures and effective countermeasures.

References

1. Imperva's DDoS Attack reports. https://www.imperva.com/blog/this-ddos-attack-unleashed-the-most-packets-per-second-ever-heres-why-thats-important. https://www.imperva.com/blog/2019-global-ddos-threat-landscape-report
2. Abramov, R., Herzberg, A.: TCP ACK storm DoS attacks. In: Proceedings of the 26th IFIP TC 11 International Information Security Conference, SEC 2011, pp. 12–27, June 2011. https://www.researchgate.net/publication/225532285_TCP_ack_storm_DoS_attacks
3. Duc, S.N., Mimura, M., Tanaka, H.: An analysis of TCP ACK Storm DoS attack on virtual network. In: 2019 19th International Symposium on Communications and Information Technologies (ISCIT), Ho Chi Minh City, Vietnam, pp. 288–293 (2019). https://ieeexplore.ieee.org/document/8905220
4. VMWare: Workstation for Windows, VMWare Workstation Pro 14. https://www.vmware.com/products/workstation. Accessed 10 Apr 2020
5. Docker: Docker desktop for Windows, Docker Nginx. https://www.docker.com/. Accessed 20 Apr 2020
6. Bansal, A., Goel, P.: Simulation and analysis of network address translation (NAT) & port address translation (PAT) techniques. Int. J. Eng. Res. Appl. **7**(7, Part 2), 50–56 (2017). http://www.ijera.com/papers/Vol7_issue7/Part-2/I0707025056.pdf. ISSN 2248–9622
7. RFC 793 - Transmission Control Protocol, DARPA Internet Program, Protocol Specification, pp. 72–73, September 1981. https://tools.ietf.org/html/rfc793
8. Chelladhurai, J., Chelliah, P.R., Kumar, S.A.: Securing docker containers from Denial of Service (DoS) attacks. In: 2016 IEEE International Conference on Services Computing (SCC), San Francisco, CA, pp. 856–859 (2016)
9. Blenk, A., Basta, A., Reisslein, M., Kellerer, W.: Survey on network virtualization hypervisors for software defined networking. IEEE Commun. Surv. Tutor. **18**(1), 655–685 (2016)
10. Bauman, E., Ayoade, G., Lin, Z.: A survey on hypervisor-based monitoring. ACM Comput. Surv. **48**, 1–33 (2015)
11. Chowdhury, N.M.K., Boutaba, R.: A survey of network virtualization. Comput. Netw. **54**(5), 862–876 (2010)
12. Fischer, A., Botero, J.F., Beck, M.T., de Meer, H., Hesselbach, X.: Virtual network embedding: a survey. IEEE Commun. Surv. Tutor. **15**, 1888–1906 (2013)
13. Arch Linux: Network Bridge. https://wiki.archlinux.org/index.php/Network_bridge
14. Varis, N.: Anatomy of a Linux bridge. In: Proceedings of Seminar on Network Protocols in Operating Systems, p. 58 (2012). https://wiki.aalto.fi/download/attachments/70789083/linux_bridging_final.pdf

Identification of Chicken Diseases Using VGGNet and ResNet Models

Luyl-Da Quach[1](✉), Nghi Pham-Quoc[1], Duc Chung Tran[2], and Mohd. Fadzil Hassan[3]

[1] Information Technology Department, FPT University, Can Tho, Vietnam
luyldaquach@gmail.com, nghipqce140179@fpt.edu.vn
[2] Computing Fundamental Department, FPT University, Hanoi, Vietnam
chungtd6@fe.edu.vn
[3] Department of Computer and Information Sciences, Universiti Teknologi PETRONAS, Seri Iskandar, Perak, Malaysia
mfadzil_hassan@utp.edu.my

Abstract. Nowadays, food security is essential in human life, especially for poultry meat. Therefore, the poultry raising is growing over years. This leads to the development of diseases on poultry, resulting in potentially great harm to human and the surrounding environment. It is estimated that when the diseases spread, the economic and environmental damages are relatively large. In addition, small-scale animal husbandry and an automated process to identify diseased chickens are essential. Therefore, this work presents an application of machine learning algorithms for automatic poultry disease identification. Here, the deep convolutional neural networks (CNNs) namely VGGNet and ResNet are used. The algorithms can identify four common diseases in chickens namely Avian Pox, Infectious Laryngotracheitis, Newscalte, and Marek against healthy ones. The obtained experimental results indicate that the highest achievable accuracies are 74.1% and 66.91% for VGGNet-16 and ResNet-50 respectively… The initial results showed positive results, serving the needs of the building and improving the model to achieve higher results.

Keywords: Chicken · Disease · Pox · VGGNet · ResNet

1 Introduction

The population is growing in significant numbers, which boosts the demand for food and requires a large amount of protein to support the whole of humanity [1, 2]. Estimates of demand for poultry - representing a relatively healthy and efficient source of protein to feed the world's population will grow to more than 9 billion by 2050. This leads to expected consumption demand that the world will consume 40% more eggs [3]. Vietnam is no exception, according to Vietnam Livestock magazine [4], the total meat output reaches over 1 million tons of poultry, 11 billion eggs, thus poultry farming is increasing to on average over 6% annually.

N.-S. Vo and V.-P. Hoang (Eds.): INISCOM 2020, LNICST 334, pp. 259–269, 2020.
https://doi.org/10.1007/978-3-030-63083-6_20

In only 3 years 2016–2018, the total poultry population increased very high by 6.33%. In 2018, 317 million chickens accounted for 77.6%, while laying hens accounted for 22.44%. In Table 1, the number of broilers increased to 7.24% and the number of laying hens also increased by 5.88% per year. This led to an increase in meat output to 6.46% and egg count to 13.30%. This shows that the demand for chicken products of the world and Vietnam is huge. Therefore, the development of the poultry industry is necessary. However, for the livestock industry to develop, it is necessary to pay attention to the factors of nutrition, environment, seed quality, water quality, and, importantly, the disease issue.

Table 1. Total number of samples of sick and normal chickens collected.

Label	# Species
Normal chicken	130
Chicken with Avian Pox	145
Chicken with Infectious Laryngotracheitis	111
Chicken with Newscalte	65
Chicken with Marek	96
Total	547

The development of poultry industry has led to the development of a disease affecting humans and poultry. In the world, the H5N2 pandemic on Turkey farms in the US in 2015 destroyed more than 47 million chickens in 21 states of the United States [5], followed by the appearance of the influenza virus H7N9 in China [6]. In particular, in Vietnam, according to the Department of Animal Health (Ministry of Agriculture and Rural Development) [7] avian influenza A/H5N1 epidemic took place in Vietnam from 2014 to March 2019, each year nearly 90,000 poultry were culled, with 127 people infected with bird flu and 65 people died from 2003 to April 2019, the amount spent on prevention is about VND 180 billion per year. Besides, there are losses of other diseases such as Newscalte, Marek, ILT, IC, etc. This shows that the need for predicting infectious diseases in poultry is becoming feasible when the new technologies and algorithms can build predictive models.

The detection of disease in chickens is a very urgent issue, but recent studies have focused on solving detailed problems of the disease and its origin. In particular, the study of Jake Astill et al. [8] to detect and predict new disease in poultry is based on new technology of big data, but only focus on avian influenza virus. In this study, the author Jake Astill focused on analysis and showed the need for a disease detection and prediction system based on gathering information from big data and analysis. In addition, big data is accompanied by data that changes during the prediction process based on time-varying activities.

Unlike Jake Astill's research, Hemalatha's team [9] focused on the identification of poultry diseases with machine learning methods such as Support Vector Machine (SVM) with Gaussian Radial Basis Function (GRBF)) and Extreme Learning Machine

(ELM). This study uses gray-level matrices to calculate mean values, standard deviations, noise, etc. They used Matrix Mean Square Error method to evaluate the performance of the Support Vector Machine model with Gaussian Radial Basis Function. However, Hemalatha's research stopped at detecting chicken pox in chickens. A new method to mention is the identification of chicken disease based on its sound, Muhammad Rizwan and et al. [10] used the extreme machine learning method extreme learning machine and support vector machine to detect rales, it helps to detect healthy or sick chickens. To conduct healthy and sick chicken stratification based on 20 min of recorded and labeled data for 25 consecutive days, the results of the study show the potential for automatic sound monitoring health of herd of chicken.

From the above analysis, it is feasible to identify disease in chickens with the development of machine learning algorithms, including deep convolution neural networks (CNN). Deep CNN [11] is a prediction algorithm using neural networks which is widely used in data stratification, it consists of a CNN. When building the network, CNN was developed from basic neural networks from 1 to 100 players, from there, it evolved into many different architectures.

Therefore, the construction of a prediction system can be based on changes in depth and width such as ResNet [12] or VGGNet [13] to classify images. Both algorithms are easier to use in training deeper networks than previously used networks. Thus, it is used in image classification studies [14–20]. In terms of the structure of VGGNet and ResNet, there are relatively similar architectures with many stack layers, making the model learn more deeply, but ResNet easily training with hundreds of layers.

Based on these important insights, we propose design recommendations to identify disease in chickens. The contributions of this research are:

- Firstly, we have collected samples from chickens for use in the study, which can be considered as a standard data set for comparison.
- Secondly, we propose a combination of deep CNN, namely ResNet and VGGNet, to formulate ideas and foundations for building a toolkit that has never been done before.
- Thirdly, we obtained the results after making the comparison, assessing the advantages and disadvantages of the two models, thereby strengthening the theory about them.

2 The Proposed Methodology

2.1 ResNet Model

Deep residual network (ResNet) [12] is a network of many "Residual Units" stacked on top of each other. Each unit can be expressed in general form as [21]:

$$y_l = h(x_l) + F(x_l, W_l), \tag{1}$$

$$x_l + 1 = f(y_l) \tag{2}$$

where, x_l *và* $x_l + 1$are input and output of the ith units, and F is residual function. In [2], $h(x_l) = x_l$ is identity map and f is ReLU function [22].

The main idea of ResNet is to find the residual function F related to $h(x_l)$, with the main option to use mapping $h(x_l) = x_l$. The basic idea of ResNet is to use a uniform off connection to cross one or more layers. Such block is called a residual block as shown in Fig. 1.

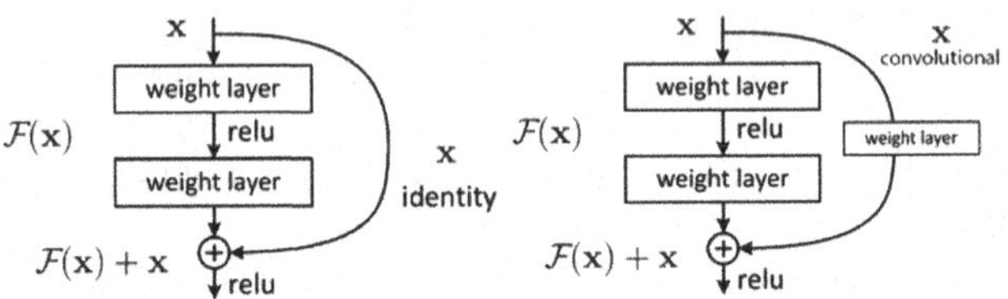

Fig. 1. Residual learning: a building Identity block and convolutional block. Left: identity block; Right: convolutional blocks.

Identity Block is the standard block used in ResNet and corresponds to the input trigger case ($a^{|i|}$) having the same size with trigger output ($a^{|i+2|}$). One Identity Block is defined as:

$$y_l = W_s x_l + F(x_l, W_l) \tag{3}$$

Here, x_l and y_l are the input and output vectors of the considered layers. $F(x_l, W_l)$ function represents the rest of the learned model. I.e., in Fig. 1, there are two layers, $\mathrm{F} = W_2\sigma(W_1 x_l)$ when σ represents rectified linear unit (ReLU) and errors are ignored to simplify symbols. The expression $\mathrm{F} + x_l$ are shown briefly connected and supplemented for the elements. We adopt the second non-linear properties after addition. In Fig. 1, Convolutional block is another block of ResNet. In general, Convolutional Block and Identity Block are relatively similar. However, Convolutional Block has 1 CONV2D block in shortcut path by applying linear projection W_s so that the dimensions match. CONV2D in the shortcut path is used to change the input size to another size so that the dimensions match the last addition before the shortcut path returns to the main path.

Based on that idea, the ResNet-50 model [16] is built with Convolution Block as shown. In particular, the Convolutional Block is used to handle when the input and output sizes do not match (Fig. 2). The ResNet block is divided into 5 stages corresponding to the increasing order of the size of the set filters from the 2nd stage: [64, 64, 256], [128, 128, 512], [256, 256, 1024] and [512, 512, 2048].

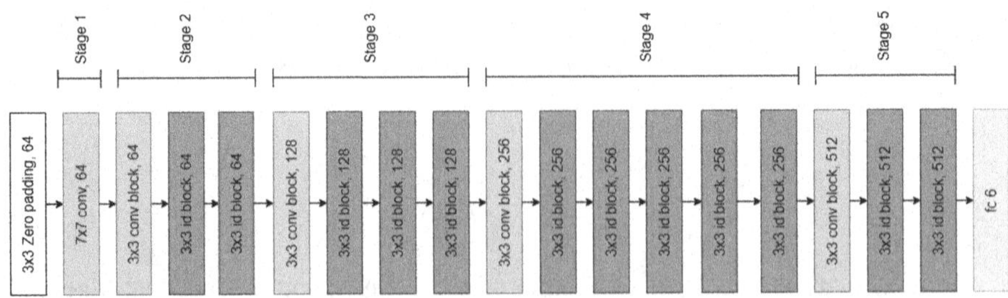

Fig. 2. Building ResNet with 50 Players.

2.2 VGGNet

The VGG network architecture (VGGNet) was introduced by Simonyan and Zisserman [18]. VGGNet architecture (see Fig. 3) uses 3 × 3 convolutional layers stacked on top of each other in increasing depth, which helps to minimize processing size. The two classes are fully connected, each with 4,096 nodes which are then followed by a softmax classifier, consisting of 138 million parameters. The design principles of VGGNets are generally very simple: 2 or 3 layers Convolution (Conv) and followed by a Max Pooling 2D layer. Immediately after the last Conv is a Flatten layer to convert the 4-dimensional matrix of the Conv layer to a 2-dimensional matrix. Following are the Fully-connected layers and 1 Softmax layer. Because VGGNet is trained on the ImageNet data [28] set of 1000 classes, the final Fully-connected layer will have 1000 units (Fig. 3).

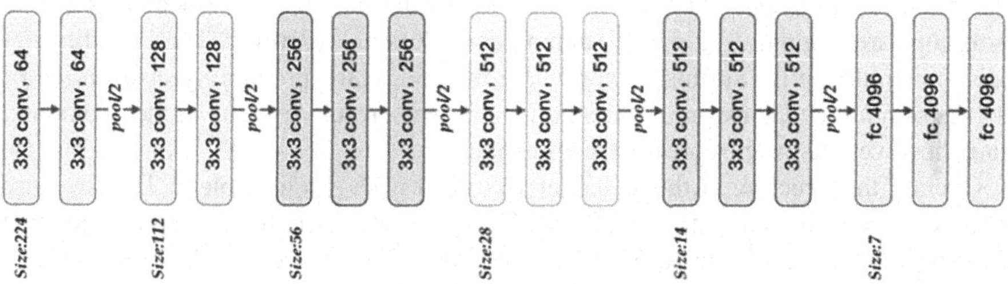

Fig. 3. Illustration of the VGGNet architecture.

3 Experimental Setup

3.1 Data Collection

This research, with the support of animal experts at FVET Vietnam Co., Ltd., has collected 600 picture samples of chicken diseases from 30 to 90 days old. The images were randomly collected in the natural environment like in a barn, natural stocking, etc. The photos have the same caption from the experts. At the end of the collection process, we obtained a total of 547 samples with 5 layers, of which there were 4 disease classes and 1 distinguished normal class which is shown in Table 1.

In these 4 diseases, signs to identify diseases are shown in chickens as follows:

Chicken with Avian pox disease:

- Chicken with Infectious Laryngotracheitis disease
- Chicken with Newscalte disease
- Chicken with Marek disease
- Normal chicken: Chickens do not show any disease and are different from the above.

3.2 Implementation

In our tests, we put the necessary hyper-parameters as follows. The learning rate is {0.05}. The number of epochs is {1500}. We have used different learning rates from 0.01 to 0.1 on 2 training models VGGNet-16 and ResNet-50. The training process on the 2 models has a big difference in time. To test the run time difference of the two models, we have installed a running timer on both models for epochs {10}, the result is VGGNet takes about 200.5 s, to train and ResNet. It took about 821.3 s for our dataset, thus the ResNet model ran about 4.2 times slower. In addition, we also tried training on 3 different epochs: {1000, 1500, 2000}.

Here we retrained the final layers of the model using our dataset. We randomly divided the dataset into a training set, a test set in a 75/25 ratio. Because both the VGGNet and ResNet models have a fixed set size filter at the stages, we set the input size of the model to 64 × 64 × 3 for height, width and channel respectively. These resolutions are common settings of running convolutional networks. The classification decision is made at the softmax layer where its input is the probability distribution of 5 labels of adult chickens. In each combination, we carefully run the model again and again, however the exact score does not change.

Overall, the structure of the VGGNet-16 model is shown in Table 2. The structure of the ResNet-50 model is presented with 50 layers. Our experiments were conducted on a workstation Intel Xeon X5675 with 3.07 GHz clock speed, 16 GB of RAM. The GPU Quad Quad K2200 with 4 GB of GDDR5 is activated by default.

Table 2. VGGNet-16 architecture.

Layers (type)	Output shape	Param #
conv2d_19 (Conv2D)	(None, 64, 64, 32)	896
activation_25 (Activation)	(None, 64, 64, 32)	0
batch_normalization_22 (Batch)	(None, 64, 64, 32)	128
max_pooling2d_10 (MaxPooling)	(None, 32, 32, 32)	0
dropout_13 (Dropout)	(None, 32, 32, 32)	0
conv2d_20 (Conv2D)	(None, 32, 32, 64)	18496
activation_26 (Activation)	(None, 32, 32, 64)	0
batch_normalization_23 (Batch)	(None, 32, 32, 64)	256
conv2d_21 (Conv2D)	(None, 32, 32, 64)	36928
activation_27 (Activation)	(None, 32, 32, 64)	0
batch_normalization_24 (Batch)	(None, 32, 32, 64)	256
max_pooling2d_11 (MaxPooling)	(None, 16, 16, 64)	0
dropout_14 (Dropout)	(None, 16, 16, 64)	0
conv2d_22 (Conv2D)	(None, 16, 16, 128)	73856
activation_28 (Activation)	(None, 16, 16, 128)	0

(*continued*)

Table 2. (*continued*)

Layers (type)	Output shape	Param #
batch_normalization_25 (Batch)	(None, 16, 16, 128)	512
conv2d_23 (Conv2D)	(None, 16, 16, 128)	147584
activation_29 (Activation)	(None, 16, 16, 128)	0
batch_normalization_26 (Batch)	(None, 16, 16, 128)	512
conv2d_24 (Conv2D)	(None, 16, 16, 128)	147584
activation_30 (Activation)	(None, 16, 16, 128)	0
batch_normalization_27 (Batch)	(None, 16, 16, 128)	512
max_pooling2d_12 (MaxPooling)	(None, 8, 8, 128)	0
dropout_15 (Dropout)	(None, 8, 8, 128)	0
flatten_4 (Flatten)	(None, 8192)	0
dense_7 (Dense)	(None, 512)	4194816
activation_31 (Activation)	(None, 512)	0
batch_normalization_28 (Batch)	(None, 512)	2048
dropout_16 (Dropout)	(None, 512)	0
dense_8 (Dense)	(None, 5)	2565
activation_32 (Activation)	(None, 5)	0

4 Results

Figure 4 presents the samples of sick and normal chickens. From left to right, the labels of the chickens are: normal chicken, chicken with head disease, and chicken with ITL, Newcastle, and Marek diseases.

Fig. 4. Samples of sick and normal chickens. From left to right, the labels are normal chickens, chickens with chicken head disease, and chickens with ILT, Newscalte, and Marek diseases.

In this work, the training and test accuracies of VGGNet and ResNet are shown in Fig. 5 and Fig. 6 respectively. By looking at the figures, one observes that the results of accuracies are not much different between the two models. Both achieves approximately 70% of accuracy after only 200 training epochs. Overall, the model converges between epoch {60} and epoch {80}.

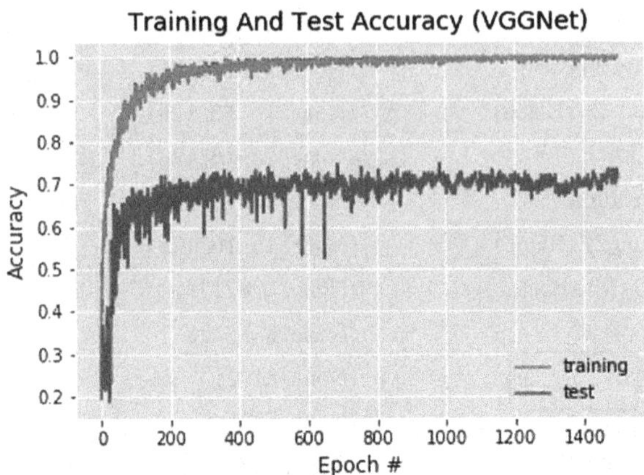

Fig. 5. The classification accuracy on the training and test sets in case of model VGGNet, learning rate = 0.05, epochs = 1500 and input size = 64 × 64.

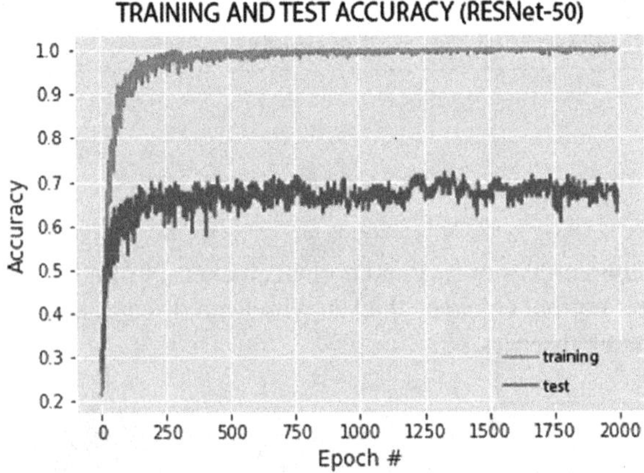

Fig. 6. The classification accuracy on the training and test sets in case of model VGGNet, learning rate = 0.02, epochs = 2000 and input size = 64 × 64.

The complete performance on 20 different combinations of our models is presented in Table 3. Here, it is seen that the best achievement of our tuned model architectures happens at learning rate equals to 0.05. For VGGNet-16 the best classification accuracy is 74.1% while the minimum one is 65.47% happened at learning rates of 0.05 and 0.04 respectively. Meanwhile, for ResNet-50, the highest achieved accuracy is 66.91% and the least achieved one is 61.15% happened for at learning rates of 0.05 and 0.06 correspondingly. In general, VGGNet-16 model performs better than ResNet-50. At close to 75% of accuracy, with further fine-tuning steps, VGGNet-16 will be suitable for used at poultry plant.

Table 3. Classification accuracies of VGGNet-16 and ResNet (%).

Learning rate\models	0.01	0.02	0.03	0.04	0.05	0.06	0.07	0.08	0.09	0.1
VGGNet-16	68.35	70.05	72.66	**65.47**	**74.10**	71.22	69.06	66.91	69.78	69.06
ResNet-50	65.47	62.59	64.03	62.59	**66.91**	**61.15**	63.31	63.31	61.87	62.59

5 Conclusions

With limited resources, the use of state-of-the-art deep CNN greatly reduces the time it takes to perform. As seen from the presented work, VGGNet-16 is more time-efficient and more accurate compared to 50-layer ResNet model. However, VGGNet requires more parameters used for learning deeper the data. The achievable accuracy of approximately 75% by using VGGNet-16 shows the positive results. In future, we will further improve the model's accuracy and deploy the developed application for field trial at poultry plant.

Acknowledgment. We acknowledge the support of staff and management board of FVET LLC Vietnam to support us during the duration of data collection and classification.

References

1. Godfray, H.C.J., et al.: Food security: the challenge of feeding 9 billion people. Science **327**, 812–818 (2010). https://doi.org/10.1126/science.1185383
2. Fraser, E., et al.: Biotechnology or organic? Extensive or intensive? Global or local? A critical review of potential pathways to resolve the global food crisis. Trends Food Sci. Technol. **48**, 78–87. https://doi.org/10.1016/J.TIFS.2015.11.006
3. Smith, D., Lyle, S., Berry, A., Manning, N., Zaki, M., Neely, A.: Internet of animal health things opportunities and challenges data and analytics. Internet of Animal Health Things (2015)
4. Tổng quan về chăn nuôi gà. http://nhachannuoi.vn/tong-quan-ve-nganh-chan-nuoi-gia-cam-cua-viet-nam-p1-tong-dan-va-san-pham-giai-doan-2016-2018/
5. Reuters TIMELINE-Tracing the Bird Flu Outbreak in N. American Poultry Flocks (2015). https://www.reuters.com/article/health-birdflu-usa-timeline/timeline-tracing-the-bird-flu-outbreak-in-n-american-poultry-flocks-idUSL1N0Y334G20150612
6. Gilbert, M., Xiao, X., Robinson, T.P.: Intensifying poultry production systems and the emergence of avian influenza in China: a "One Health/Ecohealth" epitome. Arch. Public Health **75**, 1–7 (2017). https://doi.org/10.1186/s13690-017-0218-4
7. Cục Thú y (Bộ Nông nghiệp và phát triển nông thôn), Kế hoạch quốc gia phòng chống bệnh cúm gia cầm giai đoạn 2019–2025, Số: 172/QĐ-TTg, Thư tư ́ơng Chính phu

8. Astill, J., Dara, R.A., Fraser, E., Sharif, S.: Detecting and predicting emerging disease in poultry with the implementation of new technologies and big data: a focus on Avian Influenza Virus. Front. Vet. Sci. **5**, 263 (2018). https://doi.org/10.3389/fvets.2018.00263
9. Hemalatha, Muruganand, S., Maheswaran, R.: Recognition of poultry disease in real time using extreme learning machine. In: Proceedings of the International Conference on Inter Disciplinary Research in Engineering & Technology, ICIDRET 2014, pp. 44–50 (2014)
10. Rizwan, M., et al.: Identifying rale sounds in chickens using audio signals for early disease detection in poultry. In: 2016 IEEE Global Conference on Signal and Information Processing (GlobalSIP), USA, pp. 55–59 (2016). https://doi.org/10.1109/globalsip.2016.7905802
11. Krizhevsky, A., Sutskever, I., Hinton, G.E.: ImageNet classification with deep convolutional neural networks. In: Advances in Neural Information Processing Systems, pp. 1097–1105 (2012)
12. He, K., Zhang, X., Ren, S., Sun, J.: Deep residual learning for image recognition. In: Proceedings of the IEEE Conference on Computer Vision and Pattern Recognition, pp. 770–778 (2016)
13. Simonyan, K., Zisserman, A.: Very deep convolutional networks for large-scale image recognition. In: ICLR (2015)
14. Habibzadeh, M., Jannesari, M., Rezaei, Z., Baharvand, H., Totonchi, M.: Automatic white blood cell classification using pre-trained deep learning models: ResNet and Inception. In: Tenth International Conference on Machine Vision, ICMV 2017, vol. 10696, p. 1069612. International Society for Optics and Photonics, April 2018
15. Akbar, S., Peikari, M., Salama, S., Nofech-Mozes, S., Martel, A.L.: Determining tumor cellularity in digital slides using ResNet. In: Medical Imaging 2018: Digital Pathology, vol. 10581, p. 105810U. International Society for Optics and Photonics, March 2018
16. Lin, B., XIe, J., Li, C., Qu, Y.: DeepTongue: tongue segmentation via ResNet. In: 2018 IEEE International Conference on Acoustics, Speech and Signal Processing (ICASSP), pp. 1035–1039, April 2018
17. Reddy, A.S.B., Juliet, D.S.: Transfer learning with ResNet-50 for malaria cell-image classification. In: 2019 International Conference on Communication and Signal Processing (ICCSP), pp. 0945–0949, April 2019
18. Ke, H., Chen, D., Li, X., Tang, Y., Shah, T., Ranjan, R.: Towards brain big data classification: epileptic EEG identification with a lightweight VGGNet on global MIC. IEEE Access **6**, 14722–14733 (2018)
19. Singh, V.K., et al.: Classification of breast cancer molecular subtypes from their micro-texture in mammograms using a VGGNet-based convolutional neural network. In: CCIA, pp. 76–85, October 2017
20. Muhammad, U., Wang, W., Chattha, S.P., Ali, S.: Pre-trained VGGNet architecture for remote-sensing image scene classification. In: 2018 24th International Conference on Pattern Recognition (ICPR), pp. 1622–1627, August 2018
21. Nair, V., Hinton, G.E.: Rectified linear units improve restricted Boltzmann machines. In: Proceedings of the 27th International Conference on Machine Learning, ICML 2010, pp. 807–814 (2010)
22. Nair, V., Hinton, G.E.: Rectified linear units improve restricted Boltzmann machines. In: ICML (2010)
23. Maggiori, E., Tarabalka, Y., Charpiat, G., Alliez, P.: Convolutional neural networks for large-scale remote-sensing image classification. IEEE Trans. Geosci. Remote Sens. **55**(2), 645–657 (2017)
24. Nair, V., Hinton, G.E.: Rectified linear units improve restricted Boltzmann machines. In: Proceedings of the International Conference on Machine Learning (ICML), pp. 807–814 (2010)

25. Schuster, M., et al.: TensorFlow: Large-Scale Machine Learning on Heterogeneous Systems (2015). http://tensorflow.org/
26. Xie, S., Girshick, R., Dollár, P., Tu, Z., He, K.: Aggregated residual transformations for deep neural networks. In: Proceedings of the IEEE Conference on Computer Vision and Pattern Recognition, pp. 1492–1500 (2017)
27. Simonyan, K., Zisserman, A.: Very Deep Convolutional Networks for Large-Scale Image Recognition. arXiv preprint arXiv:1409.1556 (2014) (2015)
28. Deng, J., Dong, W., Socher, R., Li, L.J., Li, K., Fei-Fei, L.: ImageNet: a large-scale hierarchical image database. In: 2009 IEEE Conference on Computer Vision and Pattern Recognition, pp. 248–255, June 2009

Design and Evaluation of the Grid-Connected Solar Power System at the Stage of DC BUS with Optimization of Modulation Frequency for Performance Improvement

Nguyen Duc Minh[1](✉), Quach Duc-Cuong[2], Nguyen Quang Ninh[1], Y Nhu Do[3], and Trinh Trong Chuong[2]

[1] Institute of Energy Science, Vietnam Academy of Science and Technology, Cau Giay, Vietnam
minhnguyenduc.ies@gmail.com
[2] Hanoi University of Industry, Hanoi, Vietnam
[3] Hanoi University of Mining and Geology, Hanoi, Vietnam

Abstract. In grid-connected solar panel systems, the power converters play a very im-portant role in control systems, because the characteristics of solar panel system are that the generation power is constantly changing due to dependence on weath-er conditions. This article presents the research results of the application of power electronic converter in grid-connected solar power system. In particular, we focus on building an algorithm to control and simulate the grid-connected solar power system at the DC-BUS stage by setting an optimal set of SVPWM (Space Vector Pulse Width Modulation) modulation frequencies when the pulse width values are different. The simulation results on Matlab/Simulink show that the system op-erates stably, ensures the requirements of DC-BUS grid integration. The set-up system operates safely, has a simple control structure and algorithm, is easy to calibrate and makes it ready for practical applications.

Keywords: Maximum power point tracking · Space Vector Pulse Width Modulation · Incremental conductance · Grid-connected solar power system

1 Introduction

The power converters are commonly used in distributed energy sources: wind power, solar power, ... and play a very important role in control systems [1]. In a solar panel system, the commonly used DC/DC converter is the implementation mechanism of the MPPT maximum power sticking algorithm of the solar panel array. MPPT uses a DC/DC converter to adjust the input voltage from the solar panel, convert it and provide the load so that the output from the solar panel is maximum. However, the use of a DC/DC converter can increase the power loss in the system [2] and may lead to a reduction in the energy conversion performance of the entire solar power system. This raises the increase of DC/DC converter's conversion performance. In fact, the performance of the DC/DC

N.-S. Vo and V.-P. Hoang (Eds.): INISCOM 2020, LNICST 334, pp. 270–288, 2020.
https://doi.org/10.1007/978-3-030-63083-6_21

converter is not constant, but rather depends on the power transmitted through it [3]. Typically, the performance of the DC/DC converter reaches its maximum value within 50%–60% of the design power and decreases rapidly if the power through it becomes smaller [4]. However, in solar panels, the output power is not fixed, the power reaches the rated value at the time of noon and the output power is small in the morning and afternoon [5], the period in which the power is smaller than 40% can reach several hours a day, not to mention shade and sunny days. Thus, in this case, the power through the DC/DC converter will be quite small (less than 40%), therefore, the performance of the DC/DC converter is very low and most of the power is consumed in the converter. Therefore, the design of a high-performance DC/DC converter is essential. Many authors have proposed the structure of DC/DC units with high performance [6–8]. Several other studies have suggested solutions to reduce the loss in the converter to improve the performance [9, 10]. However, its performance still depends on the power passing through it. That means that, during the period when the solar panel output power is very low, the DC/DC converter performance is still very low and there is not a reliable enough dataset in a sufficient range of the modulation frequencies to determine the relationship between the hashing frequency and the modulation pulse width to the performance of the DC/DC Boost unit.

This article presents the research results of the application of power electronic converter in grid-connected solar power system. In particular, we focus on building a algorithm to control and simulate the grid-connected solar power system at the DC-BUS stage by setting an optimal set of SVPWM modulation frequencies when the control value is different (width D is different). The simulation results on Matlab/Simulink show that the system operates stably, ensures the requirements of DC-BUS-side grid integration.

2 Structure of Power Blocks in the System

Figure 1 illustrates the structure of the system integrated into the solar power source. The DC voltage from solar power sources is usually low voltage, therefore, step-up DC/DC converters are necessary. Meanwhile, the grid power source is the AC power source with fixed frequency and amplitude. The DC bus power is directly supplied to local DC loads. From the DC bus, through the DC voltage inverter, it changes to AC at the AC bus. The power at the AC bus is fed to the AC load and connected to the grid through

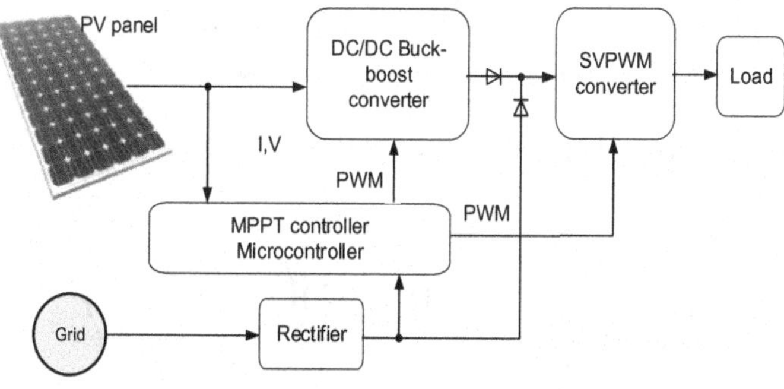

Fig. 1. Block structure of the system.

the transformer. Currently, the inverter technology to generate sinusoidal voltage mainly uses SVPWM modulation technology.

2.1 Boost DC/DC Block

The DC/DC converters are usually divided into 2 types with isolation and non-isolation. The isolation type uses small size high frequency electrically isolated transformer to isolate the input DC power from the DC output power supply and to increase or decrease the voltage by adjusting the transformer factor. This type is often used for DC power supplies using electronic keys. The most common is the bridge, half bridge and flyback circuits. In fact, many grid-working systems often use electrically isolated types for many safety reasons. Non-isolated DC/DC type do not use the isolation transformer. Common types of DC/DC converters used in PV systems include:

- Buck unit
- Boost unit
- Buck-boost unit

The selection of which type of DC/DC to use in the PV system depends on the system requirements and the load on the output voltage of the solar panel array [11–13]. The Buck unit is able to determine the optimum power operating point whenever the input voltage exceeds the output voltage of the converter, which is less likely when the radiation intensity of the light is low. The Boost unit can set the optimal working point even with low light intensity. The system that works with the grid uses the Boost unit to increase the output voltage for the load before putting it into the DC/AC converter. If the pulse hash frequency of the DC/DC voltage hashing stage is sufficient, and the system operates in continuous current mode (relating to parameters such as modulation pulse width, modulation frequency, inductance and capacitance of the DC/DC stage), the Boost DC/DC unit model on Fig. 2 in dynamic mode is described by Eq. (1) [14].

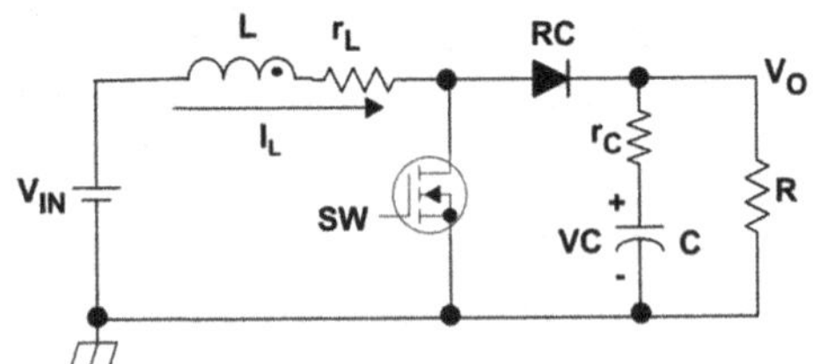

Fig. 2. Boost DC/DC circuit diagram

$$G_{dv}(s) = \frac{\hat{V}(s)}{\hat{D}(s)} \approx G_{d0}\frac{(1 + r_c Cs)\left(1 - \frac{L}{R}\left(\frac{V_O}{V_{IN}}\right)^2 s\right)}{\frac{1}{\omega_0^2}s^2 + \frac{1}{\omega_0 Q}s + 1} \quad (1)$$

In which:
$G_{dv}(s)$ is the transfer function of the Boost unit;
$\hat{V}(s)$ is the Laplace image of the output voltage;
$\hat{D}(s)$ is the Laplace image of the control signal which is the modulation pulse width;
G_{do} is a transfer function in a set mode, calculated by:

$$G_{d0} = \frac{V_{IN}}{(1-D)^2} = \frac{V_O^2}{V_{IN}} \tag{2}$$

Frequency:

$$\omega_0 \approx \frac{1}{\sqrt{LC}}\sqrt{\frac{r_L + 1 - D^2 R}{R}} \approx \frac{1}{\sqrt{LC}} \cdot \frac{V_{IN}}{V_0} \tag{3}$$

and:

$$Q \approx \frac{\omega_0}{\frac{r_L}{L} + \frac{1}{C(R+r_c)}} \approx \frac{\omega_0}{RC} = \frac{RC}{\sqrt{LC}} \times \frac{V_{IN}}{V_O} = R\sqrt{\frac{C}{L}} \times \frac{V_{IN}}{V_O} \tag{4}$$

Through the above model, it can be seen that the transfer function of Boost DC/DC stage depends on the working status in the system's set mode (depending on D and load R). To design PID controller for Boost object, we can use methods of pattern suppression, Bode diagrams, experiments … These are traditional methods. These traditional control methods are suitable for systems operating in a state of low fluctuations of load as well as input/output variable values of voltage. For systems operating in the state of large fluctuations (D and R with large variation), it is necessary to have appropriate adaptive control solutions. In the case of using the pattern suppression method, we can choose PID controller with the form:

$$PID(s) = \frac{D(s)}{E(s)} = K\frac{\frac{1}{\omega_0^2}s^2 + \frac{1}{\omega_0 Q}s + 1}{s} \tag{5}$$

Then:

$$K_p = \frac{K}{\omega_0 Q};\ K_i = K;\ K_d = \frac{K}{\omega_0^2} \tag{6}$$

The open loop transfer function of Boost DC/DC stage when using the PID controller follows the pattern suppression method as in formula (7).

$$G_H(s) = G_{do}K\frac{(1 + r_c Cs)\left(1 - \frac{L}{R}\left(\frac{V_O}{V_{IN}}\right)^2 s\right)}{s} \tag{7}$$

The closed loop transfer function has the following formula (8):

$$G_k(s) = \frac{G_{do}K(1 + r_c Cs)\left(1 - \frac{L}{R}\left(\frac{V_O}{V_{IN}}\right)^2 s\right)}{s + G_{do}K(1 + r_c Cs)\left(1 - \frac{L}{R}\left(\frac{V_O}{V_{IN}}\right)^2 s\right)} \tag{8}$$

K can be selected so that the closed system has the pole point at the desired position. When the closed system have have un-desired zero points, we can also use sensible filters to eliminate these zero points.

2.2 Single-Phase SVPWM Inverter

The SVPWM inverter [2] acts as stage to convert the power from DC to AC. From the control point of view, it can be considered that the amplification stage has a factor of 1 (under voltage perspective) when the pulse hashing frequency is large enough. Technologically, this is the unit that performs the inverter vector modulation, power amplification, control circuit isolation and dynamic circuit.

The diagram of single-phase bridge voltage inverter is shown in Fig. 3 including 4 fully control valves V1, V2, V3, V4 and negative Diodes D1, D2, D3, D4. The negative diodes of the voltage inverter diagrams help the process of exchanging reactive power between the load and the source. The DC input is a voltage source with a characteristic of capacitor C with a value sufficiently large. Capacitor C serves both as a voltage balancing filter capacitor in case E source is a rectifier, and also has a role of reactive power storage exchanged with the load through negative Diodes.

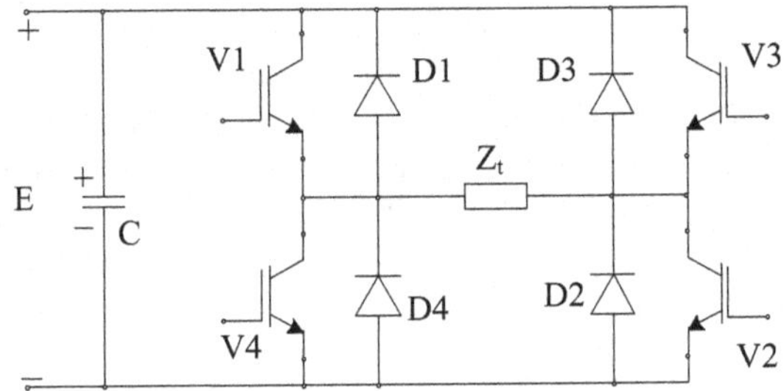

Fig. 3. Half-bridge voltage source inverter

3 Maximum Power Sticking Control Algorithm

Maximum Power Point Tracker (MPPT) is responsible for finding the working point of the panel array so that the maximum power received is corresponding to each given temperature and light intensity (MPPT point). To determine the MPPT point, two methods can be used: INC (Incremental Conductance); or P&O (Perturbation and Observe) [15, 16].

The P&O method is a simple and heavily used method thanks to the algorithm's simplicity and ease of implementation. However, when weather conditions change, this algorithm will become slower to follow the MPPT point. This method also has the disadvantage of not accurately identifying MPPT when the weather changes quickly, not suitable for frequent and sudden changes in weather conditions.

For the INC method [17]:

INC method is a type of MPPT algorithm. This method utilizes the incremental conductance (dI/dV) of the photovoltaic array to compute the sign of the change in power with respect to voltage (dP/dV). INC method provides rapid MPP tracking even in rapidly changing irradiation conditions with higher accuracy than the Perturb and observe method.

The input signal is the signal of voltage and current of solar panel array V_PV, I_PV; The output signal is the control signal D that controls the opening angle of the IGBT (Insulated Gate Bipolar Transistor) to get the maximum power. This method uses the total conductive power of the panel array to find the maximum power point as shown in Fig. 4. This method is based on the feature: The slope of the panel characteristic line is zero at the MPPT point, this slope is positive when on left of MPPT point, negative when on right of MPMP point.

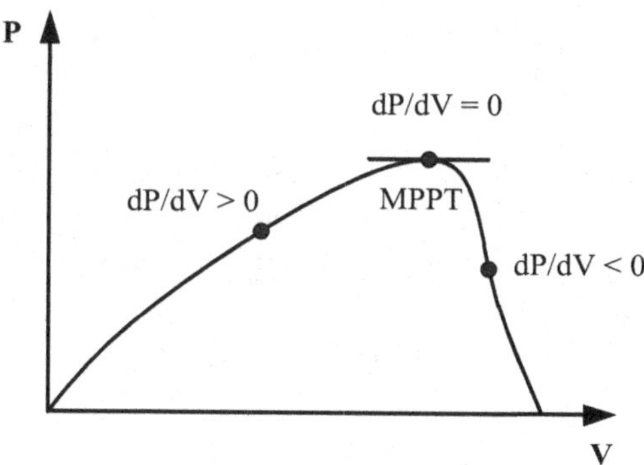

Fig. 4. Incremental Conductance method

Since dP/dV = d(IV)/dV = I + V*(ΔI/ΔV):
ΔI/ΔV = - (I/V): at MPPT point
ΔI/ΔV > - (I/V): on the left of the MPPT point
ΔI/ΔV < - (I/V): on the right of MPPT point

By comparing the instantaneous conductive value (I/V) with the incremental inductance value (ΔI/ΔV), this algorithm will find the maximum power operating point. At the MPPT point, the standard voltage $V_{ref} = V_{MPPT}$. Every time the MPPT point is detected, panel operation is maintained at this point unless there is a change in the current ΔI, a change of current ΔI indicates a change of condition of weather and MPPT point. However, when the incremental inductance is too large, it will cause the system to operate incorrectly at the MPPT point and will be oscillated. The INC algorithm controlled via the reference voltage V_{ref} is shown in Fig. 5.

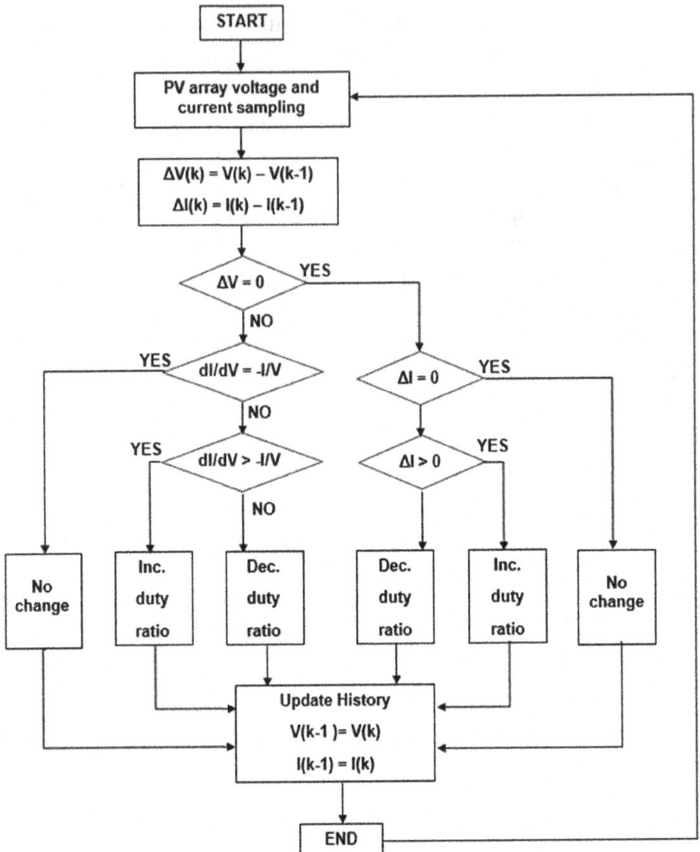

Fig. 5. INC algorithm controlled via V_{ref} reference voltage [12]

4 Setting Up Simulation Model

4.1 Solar Panel Model

Simulation model of grid-connected solar power system in the form of DC bus is implemented on Matlab software. The power of the solar power calculated for the simulation is 1 kW. The output voltage of the system is 220 V and the frequency is 50 Hz. Solar panel model is set up as shown in Fig. 6.

The panel's input is light intensity and temperature. The output is voltage, direct current. The structure of a panel model is to use a function to control the current source. The function is written on M-file and embedded in the S-Function block.

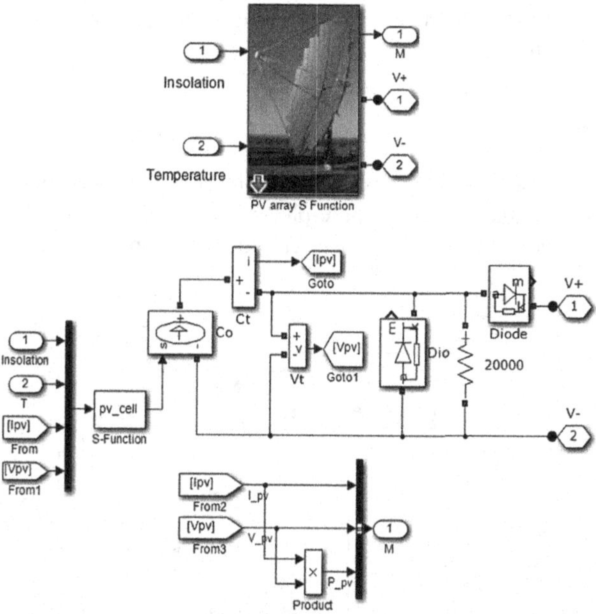

Fig. 6. Model of solar panel

4.2 DC/DC Power Converter Model

The model of DC/DC power converter is essentially a voltage BOOST unit as shown in Fig. 7. The model is built based on Matlab's powersim library. Parameters of the model are used to simulate: inductance L = 3 mH; 100 μF capacitor; power switching circuit using IGBT power semiconductor valve. Selected modulation frequency is 5 kHz. At the

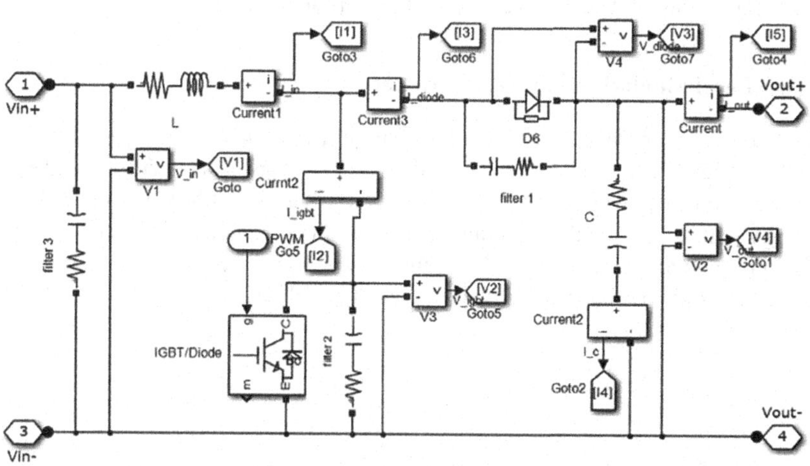

Fig. 7. Model of DC/DC voltage converter

points on the power circuit, both current and voltage sensor models are used to observe the time diagram of the current and voltage in the circuit.

4.3 Rectifier Stage

The rectifier stage uses a single-phase bridge rectifier circuit and does not use a filter capacitor because the quality of the single-phase bridge rectifier voltage is quite good. The connection between the two DC power circuits is shown in Fig. 8.

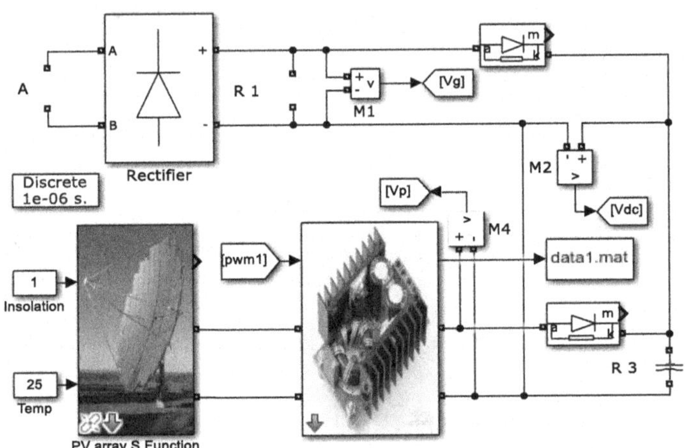

Fig. 8. DC power connection diagram

The principle of DC control ensures 2 requirements: 1) control the PV system to its maximum power; 2) the control ensures that the DC voltage of the PV power stage is always greater than the DC voltage at the output of the rectifier.

4.4 Two-Level Single-Phase Inverter Model

Single-phase inverter stage uses a two-level single-phase inverter with 4 IGBT valves as shown in Fig. 9. The sensor stage is for the purpose of measuring current and single-phase voltage on the power circuit. The modulation part uses SVPWM vector modulation technology with a 5 kHz pulse hashing frequency. The parameters used for modulation are Us amplitude and phase angle. Amplitude Us = 1 (reaches the maximum value to minimize harmonics). Phase angle is with cycle T = 20 ms.

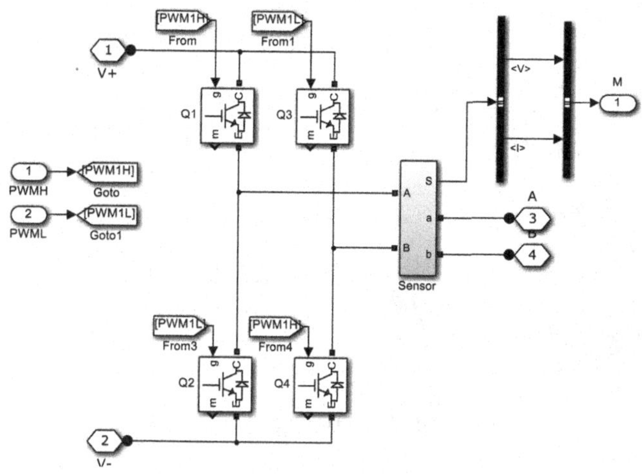

Fig. 9. H bridge single-phase inverter model

4.5 LCL Filter Model

The LC filter has a high-order harmonic removal function (described in Fig. 10). L1 and L2 inductors both select the value 1mH and filter capacitors have the value 2.2 μF.

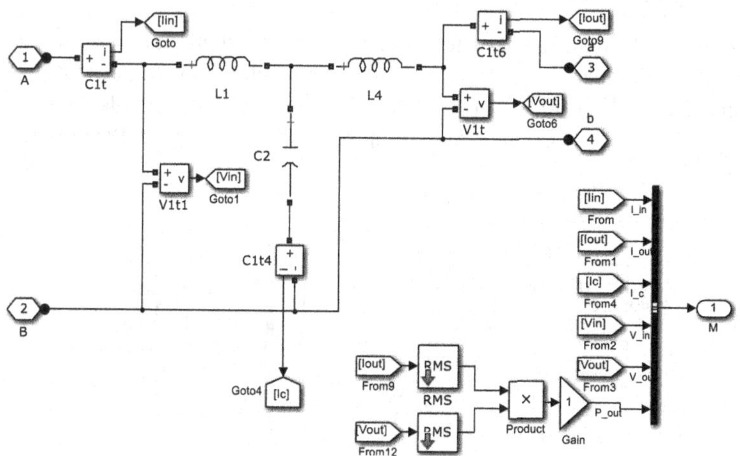

Fig. 10. LCL power filter diagram

4.6 Controller

PID unit model is as shown in Fig. 11.

PID controller is with signal set as the DC voltage supplied by the network plus an amount dV = 5 V and the feedback signal is a single-phase bridge rectifier voltage on

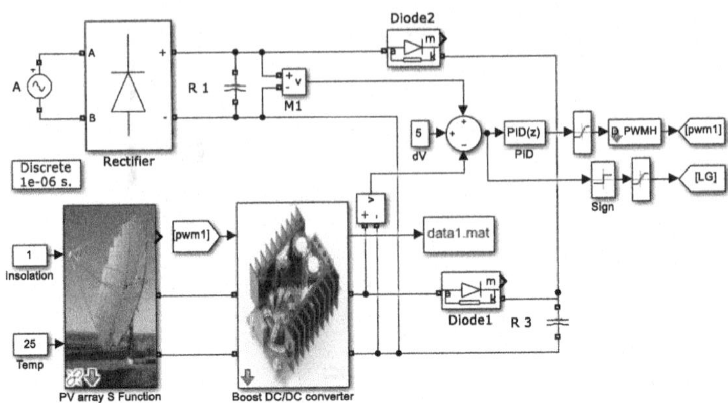

Fig. 11. PID controller

the load. In practice, this stage will use a single-phase voltage reduction transformer or specialized sensors to measure high DC voltage directly (not a voltage reduction transformer).

5 Simulation Results

5.1 Simulation of PV Panel Properties

The simulation of PV matrix properties from the library built on Matlab/Simulink applies to solar panels with the following specifications: Maximum power $P_{max} = 110$ W; Open circuit voltage $V_{OC} = 21{,}99$ V; Short-circuit current $I_{SC} = 6{,}72$ A; Voltage at maximum power $V_{mp} = 17{,}53$ V; Current at maximum power $I_{mp} = 6{,}28$ A; Panel performance 15.4% (Fig. 12).

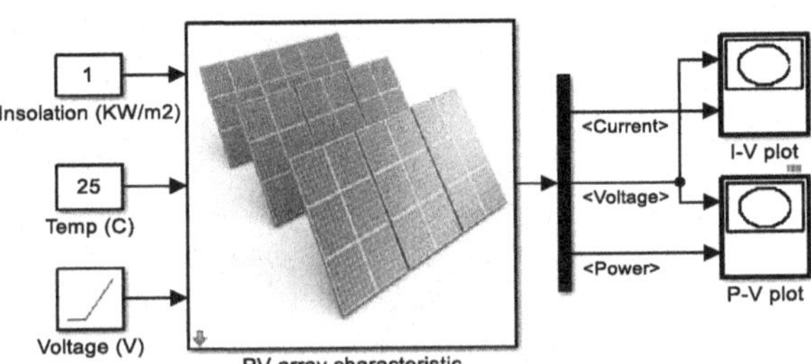

Fig. 12. Simulation diagram of PV panel

Characteristics IV and PV depend on the temperature and light intensity shown in Figs. 13, 14, 15 and 16. From the above characteristics, it can be seen that the working point with maximum power is always moving and depending on temperature as well as light intensity. At the high temperature, the power will decrease and at low temperature, the power will increase. For light intensity when the intensity is large, the power will be large and vice versa. Comparing the shape of the IV and PV characteristics obtained from the construction library with the characteristics I-V and P-V in documents [1–5], it is found that the results of building the PV matrix library above are completely appropriate.

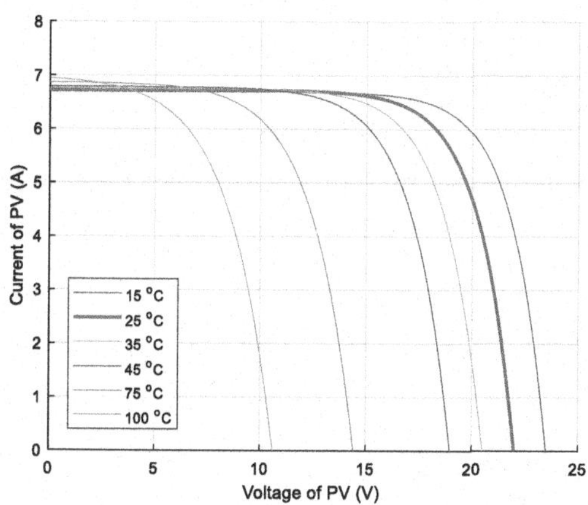

Fig. 13. I-V characteristics dependent on temperature at 1 kW/m^2 light intensity

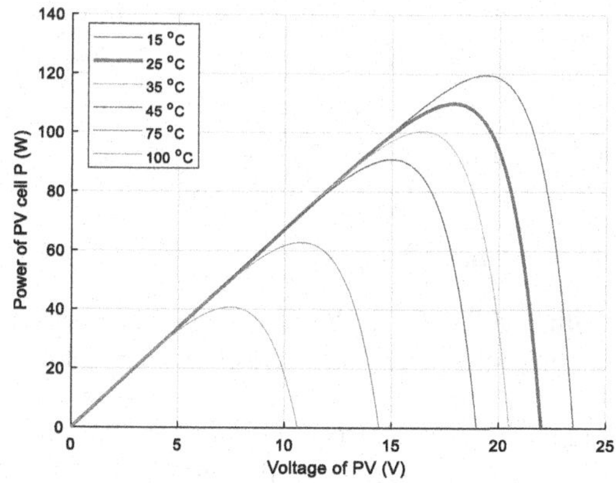

Fig. 14. P-V characteristic depending on temperature at 1 kW/m^2 light intensity

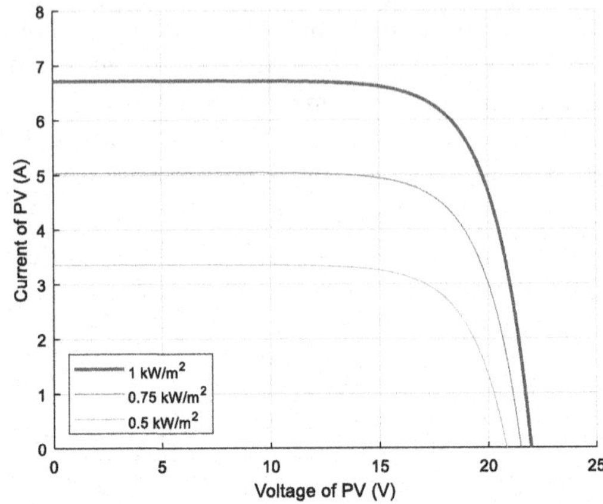

Fig. 15. I-V characteristic according to light intensity at 25 °C

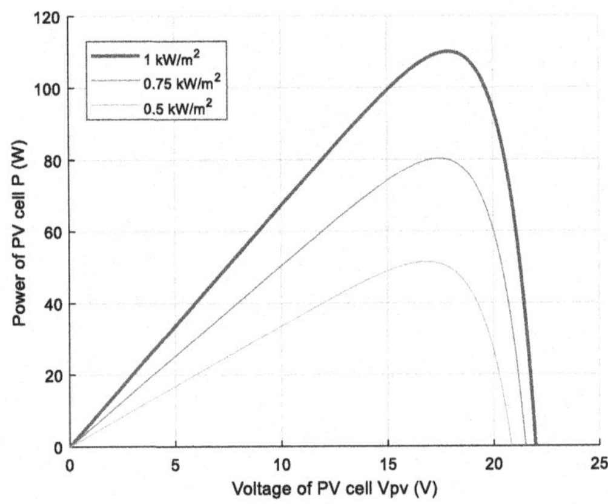

Fig. 16. P-V characteristic depending on the light intensity at 25 °C

5.2 Simulation of System Characteristics

Simulation system is with load R = 50 Ω. When the light intensity is as shown in Fig. 17a and the temperature is constant 25 °C. The DC response at DC bus grid points is given in Fig. 17b.

From Fig. 17, it can be seen: 1) the DC voltage on the PV side (V_{DC}) is controlled according to the DC side voltage (V_g); 2) when the PV power source is not sufficient for the load, the DC side voltage is connected to the DC BUS system and the DC voltage on the PV side falls into non-conduction state. This is the mechanism and principle of connecting DC-BUS grid in the system.

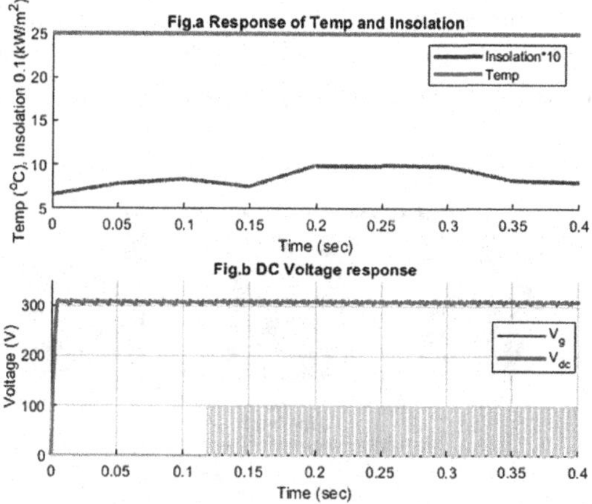

Fig. 17. DC voltage response at the stages

A. Simulation of Boost DC/DC Unit's Characteristics The current and voltage characteristics of the Boost DC/DC stage are shown in Fig. 18a.

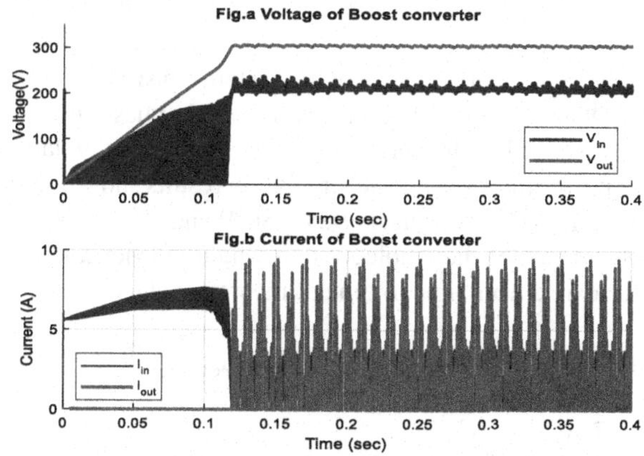

Fig. 18. Input/output current characteristics of Boost DC/DC stage

It can be seen that the time response (current and voltage) of Boost DC/DC stage has a pulse shape with the same frequency as the SVPWM modulation frequency at the control part. The interruption of the Boost DC/DC output current in Fig. 18b shows the DC-BUS separation of the Boost part. At this time, the energy from the rectifier will be fed to the inverter.

B. Investigation of Phase-to-Phase Inverter Characteristics SVPWM The phase voltage before filtering at the output of the SVPWM inverter is shown in Fig. 19a. To reduce harmonics, an additional LC or LCL power filter is required. Figure 19b is the wire voltage diagram of the inverter after the power filter. Properly calculating to select the filter and the pulse hash frequency will produce a high quality sinusoidal signal while reducing the power valve switching losses. Figure 19b shows the wire voltage when the modulation frequency is 5 kHz, resistive load Y, value 50Ω, LCL filter with 1 mH, 2.2 μF and 1 mH values.

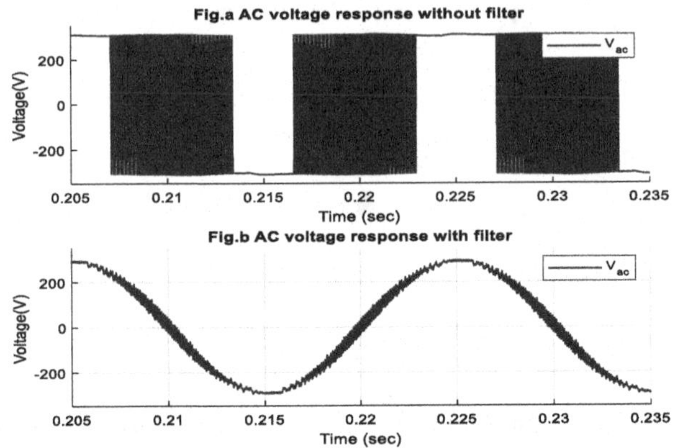

Fig. 19. Voltage diagram of the inverter before filtering

C. System Performance The performance of the system consists of 3 parts: performance of the rectifier, the DC/DC Boost DC unit and the single phase inverter. Note that the performance of the inverter has the performance of SVPWM modulation stage and the performance of the power filter stage. The simulation results show that the performance of rectifiers, boost units and inverters is 98%, 86% and 95%, respectively. Thereby leading to the performance of the whole system which reaches about 80%. Figure 20 simulates the performance from time to time of the power stages and the whole system.

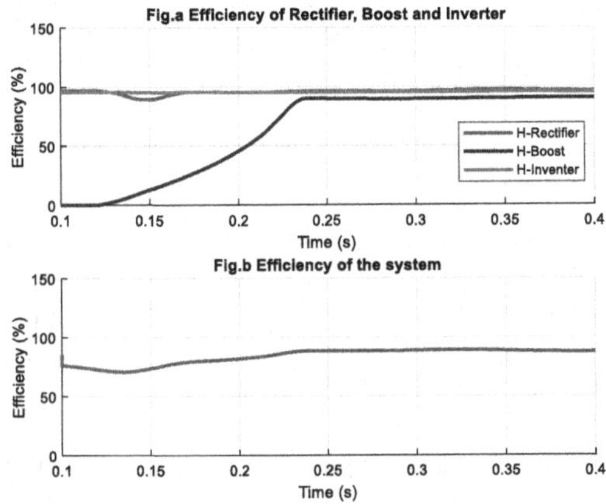

Fig. 20. System performance

We see that before 0.126 s, the performance of the Boost DC/DC unit is 0%. The reason is that this time the energy from PV is in the charging stage of Boost DC/DC stage and therefore, the PV source is not supplying to DC-BUS. In order to improve the performance of the system, it is necessary to focus on improving the performance of the Boost DC/DC stage, this is the stage with the lowest performance and the biggest impact on the performance of the system.

D. Impacts of Modulation Frequency on the Performance of the Variable Stages
To investigate the impacts of the pulse hashing frequency and the pulse width modulation on the performance of the Boost DC/DC, set the system with the following parameters: Light intensity: 1 kW/m^2; Temperature on PV: 25 °C; Voltage hashing frequency at Boost DC/DC stage (f-kHz): 2 kHz; 4 kHz; 6 kHz; 8 kHz; 10 kHz; 12 kHz; 14 Hz; 16 Hz; 18 Hz; 20 Hz; 22 Hz; 24 Hz; 26 Hz; 28 Hz; load value: 150Ω. Simulation results are shown in Table 1 and Fig. 21.

Table 1. BOOST DC/DC unit performance

F (kHz)	The pulse width modulation D								
	0.1	0.2	0.3	0.4	0.5	0.6	0.7	0.8	0.9
6	63.35	57.67	51.09	50.11	49.32	48.37	48.47	46.63	41.50
8	78.57	66.34	64.79	60.75	57.73	56.57	52.84	51.13	47.16
10	88.77	75.43	67.88	62.17	58.10	57.14	52.47	53.93	48.26
12	92.39	83.95	79.34	74.36	63.40	58.73	53.30	54.99	51.32
14	93.06	87.06	83.76	79.46	67.44	61.45	54.85	54.23	55.46
16	**93.12**	88.49	85.84	83.81	71.71	67.44	55.91	55.69	56.85
18	92.70	**88.96**	86.77	85.30	74.58	67.89	56.72	57.52	56.04
20	92.12	88.88	86.90	**85.78**	76.82	67.92	57.18	58.42	56.50
22	91.21	88.46	**86.92**	85.68	77.07	67.63	57.99	60.29	57.30
24	90.31	87.86	86.29	85.28	77.27	68.12	64.56	59.97	58.31
26	89.34	87.10	85.63	84.64	**79.90**	**76.42**	68.19	63.44	60.93
28	88.15	86.27	84.96	83.93	79.64	75.62	**70.83**	**67.64**	**61.82**

From the results in Fig. 21, it can be seen: Boost stage performance depends greatly on the pulse hashing frequency and the width of the voltage hashing pulse. Increasing the width of the voltage hashing pulse width will reduce the system performance. The pulse hashing frequency also needs to be selected appropriately. If this value is too large, it will cause losses on the IGBT protection RC circuits and the valve ON/OFF valve power will increase, resulting in a decrease in performance. However, if the pulse hash frequency is not large enough, the quality of the DC voltage in the circuit will not be high. Therefore, the design process needs to select the frequency of the voltage hashing pulse and the scope of adjusting the pulse width appropriately for the system to achieve high performance.

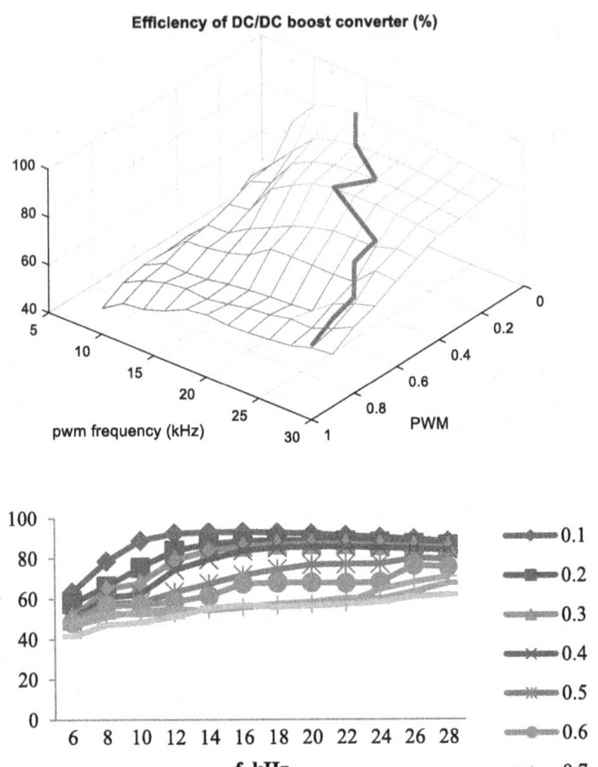

Fig. 21. System performance depending on the frequency and width of the voltage hashing pulse width

With the simulation results, it is recommended to select the working area D = 0–0.4 and the pulse hash frequency range 16–22 kHz to achieve the highest possible performance. In the case where a large modulation range (D with a large range of values) must be used, a functional structure that automatically adjusts the pulse hashing frequency so that the system will achieve the highest possible performance.

6 Conclusion

From the results of model building, control algorithm and simulation of the grid-connected solar power system at the DC-BUS stage, the system operates stably, ensures the requirements of DC-BUS grid. For the above system, we can realize some advantages as follows:

- The system operates safely because it does not have to handle the oasis phenomenon as in the AC grid system.
- Control structure and algorithm are simple, easy to adjust and execute

- The modulation part of SVPWM does not need to implement the algorithm to stick to the grid.

In order to improve converter performance further, in the coming time, we will:

1. Research, develop and test the real system based on embedded system platform to get the most accurate evaluation results on grid-connected solar power system at DC-BUS stage.
2. Implement a self-tuning mechanism of voltage hash frequency based on the modulation pulse width in order to maximize the system performance.

Acknowledgements. The authors wish to thank the Institute of Energy Science (IES) and Vietnam Academy of Science and Technology (VAST) for their support to the research activity of Project "Research on completing the technology, constructing and transferring the model of exploiting and proper using solar and wind energy for production and daily living in Central Highland", TN17/C03.

References

1. Chuong, T.T.: Voltage stability analysis of grids connected wind generators. In: The International Conference on Electrical Engineering, No. O-054 (2008). Article No. 1
2. Minh, N.D., Van Huy, B., Quan, N.T., Ninh, N.Q., Chuong, T.T.: Research and Design of Grid-Connected Inverter in Photovoltaic System With Svpwm Technique. Int. J. Eng. Technol. Manag. Res. 6(11), 18–31 (2020)
3. Amaral, G., et al.: Control in Power Electronics, vol. 369, no. 1. (2013)
4. Patil, U., Kolte, D.M.: Simulation of DC-DC Boost Converter for SPVM (2015)
5. de Cesare, G., Caputo, D., Nascetti, A.: Maximum power point tracker for portable photovoltaic systems with resistive-like load. Sol. Energy **80**(8), 982–988 (2006)
6. Arathy, M., Sreekala, P.: Design and implementation of A PV powered five level inverter using multilevel differential boost converter. Int. J. Adv. Res. Electr. Electron. Instrum. Eng. **3**(2320 – 3765), 10315–10320 (2014)
7. Pal, S., Dalapati, S.: Digital simulation of two level inverter based on space vector pulse width modulation. Indian J. Sci. Technol. **5**(4), 2557–2568 (2012)
8. Anand, R., Gnanambal, I.: Modeling and analysis of cascade multilevel DC-DC boost converter topologies based on H-bridge switched inductor. Res. J. Appl. Sci. Eng. Technol. **9**(3), 145–157 (2015)
9. Rasheed, M., Omar, R., Sabari, A., Sulaiman, M.: Validation of a three-phase cascaded multilevel inverter based on Newton Raphson (N.R.). Indian J. Sci. Technol. **9**(20), 1–13 (2016)
10. Omar, R., Rasheed, M., Sulaiman, M., Tamijis, M.R.: Modeling and simulation of a three phase multilevel inverter for harmonic reduction based on modified space vector pulse width modulation (SVPWM). J. Theor. Appl. Inf. Technol. **77**(2), 178–189 (2015)
11. Stepins, D., Huang, J.: Effects of switching frequency modulation on input power quality of boost power factor correction converter. Int. J. Power Electron. Drive Syst. **8**(2), 882–899 (2017)
12. Eichhorn, T.: Boost converter efficiency through accurate calculations. Power Electron. Technol. **34**(9), 30–35 (2008)

13. Hauke, B.: Basic Calculation of a Boost Converter's Power Stage. Texas Instruments. Application Report November November 2009, pp. 1–9 (2009)
14. Quang, N.P.,Gmbh, E.: Inverter control with space vector modulation. Power Syst. **20**, 17–59 (2008)
15. Ahmed, J., Salam, Z.: An improved perturb and observe (P&O) maximum power point tracking (MPPT) algorithm for higher efficiency. Appl. Energy **150**, 97–108 (2015)
16. Kumar, A.: Overview of genetic algorithm technique for maximum power point tracking (MPPT) of solar PV system. no. Cognition, pp. 21–24 (2015)
17. Liu, F., Duan, S., Liu, F., Liu, B., Kang, Y.: A variable step size INC MPPT method for PV systems. IEEE Trans. Ind. Electron. 55(7), 2622–2628 (2008)

Security and Privacy

An Efficient Side Channel Attack Technique with Improved Correlation Power Analysis

Ngoc-Tuan Do and Van-Phuc Hoang(✉)

Le Quy Don Technical University, 236 Hoang Quoc Viet, Ha Noi, Viet Nam
phuchv@lqdtu.edu.vn

Abstract. Correlation Power Analysis (CPA) is an efficient way to recover the secret key of the target device. CPA technique exploits the linear relationship between the power model and the real power consumption of an encryption device. In theory, we only need fewer power traces to recover secret key bytes successfully. However, due to the impact of noise, we need a larger number of power traces in order to extract the secret key. Therefore, the computation time becomes a serious problem for performing this attack. This paper introduces a new method to reduce the computation time for CPA method with the technique of finding points of interest which was used for template attack. The experimental results have clarified the efficiency of the proposed method.

Keywords: Correlation power analysis · Side channel attack · AES

1 Introduction

Electronic cryptographic devices are widely used today for securing the secret information. However, there are emerging issues about side channel attack. In the scope of the statistical power analysis attack on cryptographic systems, two efficient techniques were proposed. The first one is well known Difference Power Analysis (DPA) introduced by Paul Kocher [1, 2] and formalized by Thomas Messerges et al. [3]. It uses statistical tools to find out the information correlates to confidential key. The second one is Correlation Power Analysis (CPA) that proposed by Brier et al. [4]. It exploits the correlation between the power model and real power consumption, in order to leak the secret key. These techniques and measurements are carefully taken into account by embedded system designers to know an attacker can measure the power consumption or electromagnetic emanations, which are two of the main physical quantities used for non-invasive attack.

In this paper, we focus on CPA in order to evaluate the security of the AES cryptography because it is an efficient technique to recover the secret key with a simple power consumption model. In fact, we realize that when the cryptographic devices run in low frequency without or with very small noise, CPA only needs a hundred of power traces to successfully extract all secret key. However, for the device running in high frequency and high noise, CPA needs a huge amount of power traces to recover the whole key.

N.-S. Vo and V.-P. Hoang (Eds.): INISCOM 2020, LNICST 334, pp. 291–300, 2020.
https://doi.org/10.1007/978-3-030-63083-6_22

Besides, to evaluate the protection method for a secure device, the designers need to consider to the real scenarios. Side channel analysis with real devices usually leads to the high level of noise, then the efficient attacks may require the large numbers of power traces. Therefore, finding an efficient way to analyze those power traces is critical.

The rest of this paper is organized as follows. The original CPA and some related works are described in Sect. 2. In Sect. 3, we will introduce a new CPA, which is an efficient CPA technique that allows to recover secret key byte at least 2× faster than the original CPA. Then, in Sect. 4, we show and compare the analysis results on cryptographic devices, which is an implementation of AES running on XMEGA MCU (on ChipWhisperer board). Finally, we conclude the paper in Sect. 5.

2 Original CPA and Related Works

2.1 Original CPA

CPA was first proposed by Brier et al. in [4]. It exploits the correlation between the real power consumption and the power consumption model. It is based on a power consumption model of the running device at some certain points of time which must depend on the fixed secret key and the plain text changed for each trace. For the Advanced Encryption Standard (AES) attack, attackers mostly used the performing after the first Sbox function to analyze because it has strongly correlation between the real power consumption and power consumption model. The most widely used consumption model is the Hamming distance between two relevant values in the same register [4], or simply the Hamming weight of the particular value [5, 6]. The correlation between the power model and actual power consumption is calculated through Pearson's correlation coefficient by evaluating the linear relationship. The formula of this correlation coefficient between a_i and b_i is given by:

$$\rho = \frac{\mathrm{cov}(A, B)}{\sigma_A \sigma_B} \tag{1}$$

Next, we will discuss the analysis technique for AES algorithm. AES is a cryptographic algorithm executing on one byte separately. Therefore, this paper will focus on analyzing the key bytes independently. Let's denote that p is Pearson's correlation coefficient, N power traces of length L, $t_{n,i}$ is the consumption value of point i in trace n (with $1 \leq i \leq L$, $1 \leq n \leq N$). We have K possible values of a subkey (in this paper $K = 256$), we note $h_{n,k}$ the power consumption model of key k, in trace n and calculated by following formula:

$$h_{n,k} = HW(Sbox(Plaintext_n \oplus k)) \tag{2}$$

where HW denotes the Hamming weight of the Sbox output. With this data, we can see how well the power model and actual power consumption correlation is for each guess key byte k at each time i, by the following formula:

$$\rho_{k,i} = \frac{\sum_{n=1}^{N} (h_{n,k} - \bar{h}_k)(t_{n,i} - \bar{t}_i)}{\sqrt{\sum_{n=1}^{N} (h_{n,k} - \bar{h}_k)^2 \sum_{n=1}^{N} (t_{n,i} - \bar{t}_i)^2}} \tag{3}$$

where $\bar{h}_k$ and $\bar{t}_i$ are the average of the power consumption model and real power consumption respectively at instant i, respectively.

It can be seen that, by taking the maximum of $\rho_{k,i}$ among all values for i and k, we can decide which power consumption model of the key is most correlated with the actual power consumption. In (3), i is the value of number of samples that we acquired in a trace. Fortunately, we don't need to use all of samples in a power trace, we only focus on a byte that has a leakage position as illustrated in Fig. 1. We use N random plaintexts corresponding to N traces in which each trace has L samples. Note that $t_{i,j}$ is the value of jth sample in the trace number i^{th} ($1 < j < L$, $1 < i < N$), $d_{i,B}$ is the byte value of byte B ($B \in (1; 16)$) in the plaintext number i^{th}. In order to compute the correlation coefficient based on (3), Algorithm 1 as shown below is used. From this algorithm, it is clear that we can easily calculate the mean value of power consumption model and the real power consumption. However, the problem is the iteration of algorithm, it can be computed by $16 \times 256 \times L \times N$. It leads to a huge number of iterations that directly impact on the execution time. Therefore, the values of L and N need to be reduced for improving computing effectiveness. Hereafter, some related works will be discussed.

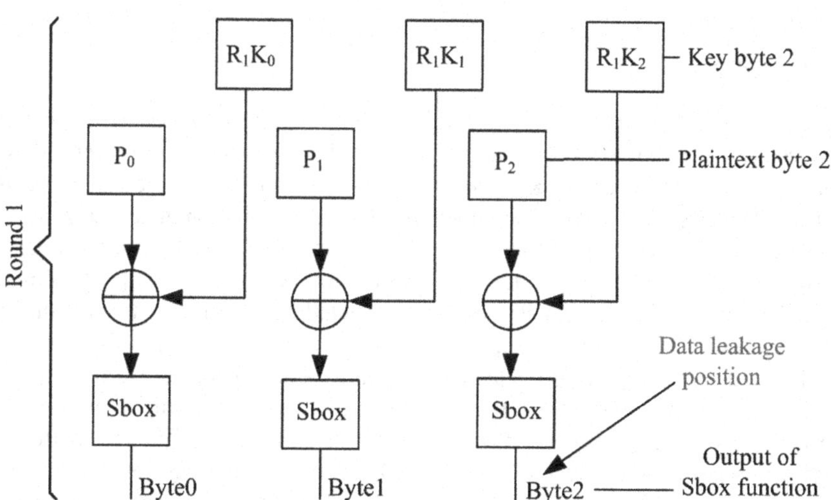

Fig. 1. Data leakage position of AES.

Algorithm 1: Algorithm for computing the correlation coefficient based on the conventional CPA technique.

Input: $d_{i,B}$, $t_{i,j\ (1 \le j \le L)}$, L, N
for $B \in (1;16)$ **do**
  for $KEY \in (0;255)$ **do**
    $h_{1<i<N} = HW(SubByte(d_{i,B} \oplus Key))$
    $mean_h = \sum_{i=1}^{N} h_i, mean_{t_j} = \sum_{i=1}^{N} t_{(i,j)}$
   for $j \in (1;N)$ **do**
      $hdiff = h_i - mean_h$
      $tdiff = t_{i,j} - mean_t$
      $sum = sum + hdiff * tdiff$
      $sum1 = sum1 + hdiff * hdiff$
      $sum2 = sum2 + tdiff.*tdiff$
    end for
   $\rho_{KEY,L} = \text{abs}(sum \ ./\ sqrt(sum1.*sum2));$
  end for
   $\rho_{KEY} = \arg\max(\rho_{KEY,L})$
   $Subbyte_B = KEY$
end for

2.2 Related Works

Recently, CPA technique focuses on the computing effectiveness. In [7], the authors introduced a selection method that can be used to reduce the number of power traces (it means reducing N) to reveal the keys. This is a pre-processing technique which consists selecting a biased subset of the power traces have more information than average traces. Their method improves the performance of the original analysis but the authors does not show the computation analysis in terms of execution time. The work in [8] proposed a technique for CPA enhancement. The authors presented the Partitioning Power Analysis (PPA) technique to combine the techniques of DPA, multi-bit DPA and CPA in a single form. The notion of class was introduced by grouping the power consumption. Each class contains the messages which have the same Hamming distance as the reference state. After that, the correlation is calculated on the classes which use difference coefficients for the difference classes. According to [8], CPA is a particular case of PPA. Despite enhancement of CPA, the computation time is not presented in this method. Another technique is presented in [9], the authors show a new method to recover secret keys with less power consumption traces than the standard CPA. This improvement is done by selecting appropriate plaintexts, namely non-adaptive and adaptive. The authors choose a set of messages with the most pairwise decorrelated subkeys consumption model. This technique reduces the number of power traces for CPA, but the author does not give any information of computation time.

Most recently, in [10], the authors proposed an algorithm that is similar to the original CPA. This technique is based on the idea of creating, at each point in time, a vector of consumption values indexed by the plaintext byte value. This method also refers to L and N and has two phases. In phase 1, the vector consumption is created in a single pass, requiring $(N \times L)$ iterations. Phase 2 uses these correlated vectors and corresponding power model with Hamming weight. This phase takes $L \times K^2$ $(K = 256)$ iterations. Consequently, this technique uses $(N \times L + L \times K^2)$ iterations for computing one byte key. Despite reducing execution time significantly, this method only works well if the number of power traces is large enough.

This work also employs the idea to reduce the iterations. As mentioned previously, the original CPA requires many power traces and the number of iterations is $16 \times 256 \times L \times N$. Therefore, to enhance the effectiveness of CPA, especially the computation time, in this paper, we aim to propose a technique that reduces the value of L for calculating the final result in CPA technique.

3 Proposed CPA Technique

In this Section, we describe the process of our proposed method and explain how to reduce the number of samples of a power trace in computing. CPA exploits the correlation of power consumption model. In this work, we use the Hamming weight with the data leakage positions as describing in Sect. 2.

In Algorithm 1, we need to calculate the mean of modeled power consumption and the real power consumption. The mean value of power trace needs to use all of the samples of all power traces. Therefore, the number of iterations in computing is large number. If we use a small amount of power traces, it means that we can reduce the number of iterations. Moreover, not every points is important in calculating the correlation, the subkeys only influence the power consumption at a few critical times, and corresponding samples are called Points of interest (POI). If we can pick POI, we can ignore most of the samples. Therefore, using the POI will reduce the number of iterations dramatically. For these reasons, we propose a method that can take the POI base on conventional CPA, and then they are used to calculate Pearson correlation coefficient for all of power traces with whole possible key.

The POI mentioned above is similar to the template attack, a simple technique was used is sum of differences method. In our proposed, the plaintext is random, and the secret key is fixed. We cannot use sum of differences, so the Pearson correlation coefficient is chosen instead. Since a power trace has L samples, we will have L values of correlation, after each iteration, if i^{th} sample has the linear relationship, it will steadily increase or reduce, by contrast it will flexible like noise. Figure 2 illustrates the correlation of 5,000 samples after computing 50 power traces. It is clear that some samples of power trace are much higher than the rest. One of these samples, which is correlated to correct byte key, then constantly goes up. A question is posed, what happen if we only compute a part of power traces instead of all traces. In this case, we suppose that all correlation values will reduce but these POI still higher than the rest. This is an important point to extract the POI from the large sample. To implement this hypothesis, we propose Algorithm 2.

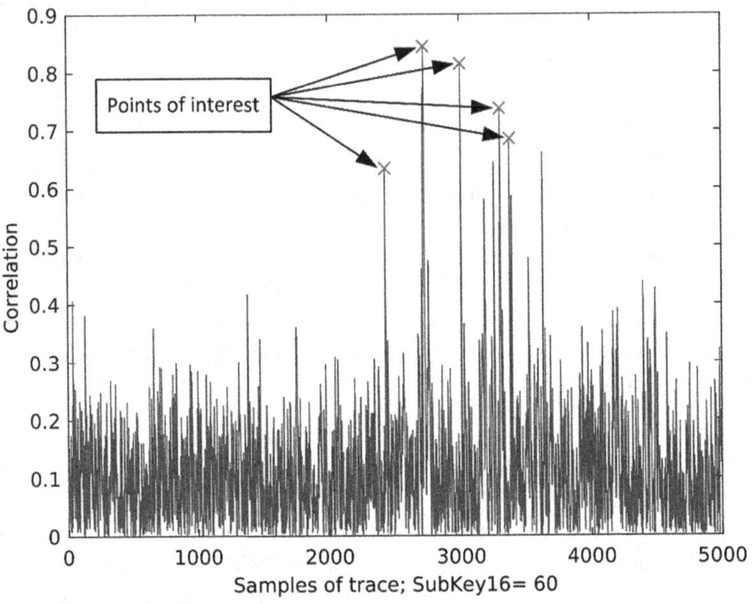

Fig. 2. Points of interest for one byte of secret key.

Inspired by the conventional CPA, our proposed algorithm is also based on the correlation coefficients to find out the highest value. However, the novelty of our proposed approach is the two-phase CPA technique with optimal chosen parameters. In the first phase, a smaller number of power traces is used and denoted as N_1 ($N_1 \leq \frac{1}{2}N$). After that, the Eq. (3) is applied to calculate α highest samples from all guess keys. The value of α is much smaller than *L*. In the last phase, this equation is used again to compute the correlation for all hypothesis keys again. However, only α samples are calculated instead of using *L* samples as the first phase. Next, we present the proposed algorithm in detail (Algorithm 2). The notation *N, L,* $d_{i,B}$ of Algorithm 1 are reused and N_1 denotes for the power traces using in phase 1.

Algorithm 2: Proposed algorithm for computing the correlation coefficient in two phases

Input: $d_{i,B}, t_{i,j\,(1\le i\le N,\, 1\le j\le L)}$, L, N_1, N ($N_1 \le \frac{1}{2}N$)

for $B \in (1;16)$ **do**

//Phase 1: taking α sample which have highest correlation value

for $KEY \in (0;255)$ **do**

$h_{1<i<N} = HW(SubByte(d_{i,B} \oplus KEY))$

$mean_h = \sum_{i=1}^{N_1} h_i, mean_{t_j} = \sum_{i=1}^{N_1} t_{i,j}$

for $i \in (1;N_1)$ **do**

$hdif1 = h_i - mean_h$

$tdif1 = t_{i,j} - mean_t$

$sum1\_1 = sum1\_1 + hdif1 * tdif1$

$sum1\_2 = sum1\_2 + hdif1 * hdif1$

$sum1\_3 = sum1\_3 + tdif1.* tdif1$

end for

$\rho_{KEY,\bar{L}} = \text{abs}(sum1\_1 ./\ sqrt(sum1\_2.* sum1\_3));$

end for

for $i \in (1;\alpha)$ **do**

$\rho_{\bar{L}'} = \arg\max(\rho_{KEY,\bar{L}} - \rho_{\bar{L}'-1})$

$\bar{N}_{i,\bar{L}'} = N_{i,\bar{L}'}$

end for

//Phase 2: Compute the correlation values with α samples found in phase 1

for $KEY \in (0;255)$ **do**

$mean_h = \sum_{i=1}^{\bar{N}} h_i, mean_{t_{\bar{L}'}} = \sum_{i=1}^{\bar{N}} t_{i,\bar{L}'}$

for $i \in (1:\bar{N})$ **do**

$hdif2 = h_i - mean_h$

$tdif2 = t_{i,\bar{L}'} - mean_{t_{\bar{L}'}}$

$sum2\_1 = sum2\_1 + hdif2 * tdif2$

$sum2\_2 = sumden2\_2 + hdif2 * hdif2$

$sum2\_3 = sumden2\_3 + tdif2.* tdif2$

end for

$\rho_{KEY,\bar{L}'} = \text{abs}(sum2\_1 ./\ sqrt(sum2\_2.* sum2\_3));$

end for

$\rho_{KEY} = \arg\max(\rho_{KEY,\bar{L}'})$

$Subyte_B = KEY$

end for

4 Experimental Results

In this section, we evaluate the proposed method on unprotected AES implementation compared to the conventional correlation coefficient used for CPA. We will show the impact of N_1 in computing time, then present two criteria including the accuracy and the execution time for recovering whole key. To prove the practical effectiveness of the proposed technique, we use the synchronized and low noise power traces of low frequency AES implementation acquired on ChipWhisperer board.

To illustrate the accuracy of algorithm, an AES implementation on XMEGA MCU is used. The number of correct key bytes will be used as the criteria to assess. In this work, 10,000 power traces are acquired from ChipWhisperer board and then added Gaussian noise ranging from 0.1 to 0.2 with steps of 0.02. As mentioned above, we will show the impact of N_1 in the proposed algorithm. In this case, the fixed value of $\alpha = 50$ is chosen. All experiments are implemented in MATLAB software.

In Fig. 3, it is clear that the more noise is added, the higher of N_1 is needed for extracting secret key successfully. However, the value of N_1 directly affects the computation time, therefore we cannot choose the too large value of N_1. In the experiments, with the value of N_1 = [1/3 N, 1/4 N, 1/5 N], we find that the value N_1 = 1/3 N will leads to the better results than other ones. Hence, this value is used for evaluating the execution time with the detail results in the next section.

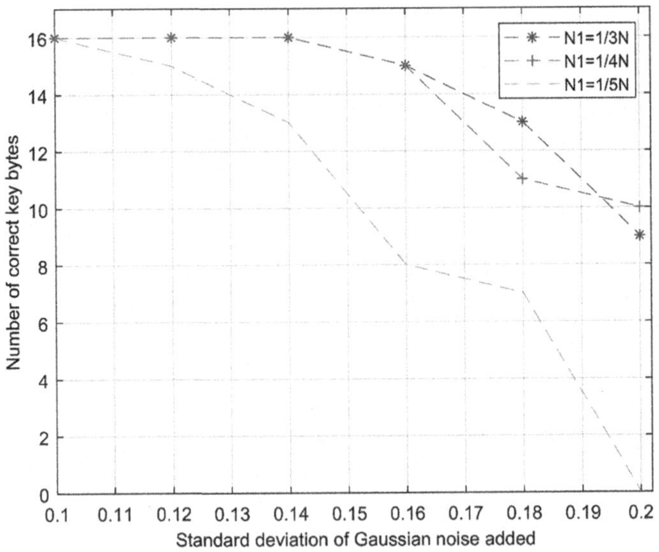

Fig. 3. The average of the correct number of key bytes corresponds to $N1$ and the standard deviation increases.

To demonstrate the effectiveness of timing execution, the proposed algorithm is implemented with the same configurations of the conventional CPA algorithm. In this experiment, we used a number of power traces in the range from 1,000 to 10,000 corresponding to the standard deviation of noise ranging from 0.06 to 0.15 and with $N_1 =$ 1/3 N. These parameters are listed in Table 1.

Table 1. Parameters of CPA attack on the whole AES key for computation time measurements.

Experiment no.	1	2	3	4	5	6	7	8	9	10
N	1000	2000	3000	4000	5000	6000	7000	8000	9000	10000
N_1	330	667	1000	1333	1667	2000	2333	2667	3000	3300
Noise deviation	0.06	0.07	0.08	0.09	0.10	0.11	0.12	0.13	0.14	0.15

Figure 4 shows the results of experiment. It can be seen that the proposed technique is much faster than the conventional CPA. As expected, the value of N_1 directly affects the calculation time and it is effective with the broad range of number of power of trace, not only for the large number as in [10]. When the number of power traces is larger, with the appropriately chosen value of N_1, the higher efficiency of the execution time can be achieved. The smaller N_1 is chosen, the lower computation time is achieved. In this case, we only use the value of $N_1 = 1/3\,N$, however, depending on the standard deviation value of noise, N_1 could be chosen smaller. According to Fig. 3, if the standard deviation is 0.14, the value of $N_1 = 1/4\,N$ can be applied. Hence, the efficiency of execution time will be higher.

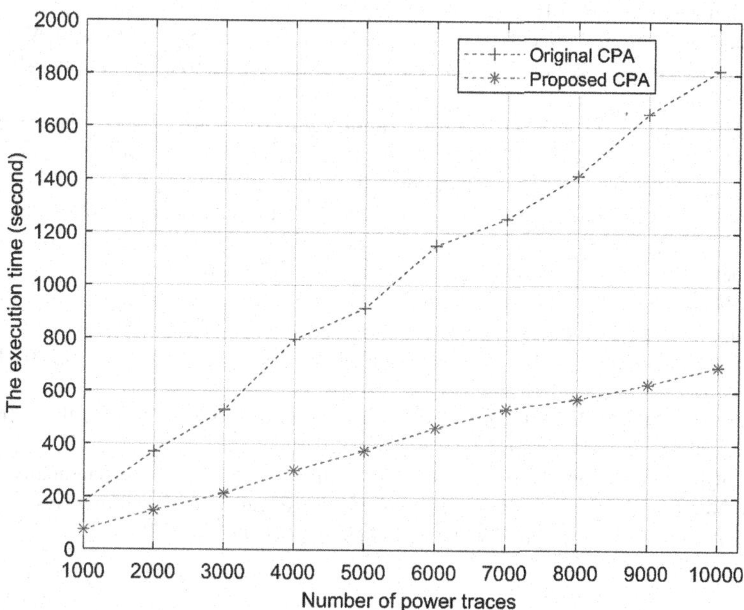

Fig. 4. Results of computation time for conventional and proposed CPA techniques, for power traces containing 5000 samples.

5 Conclusion and Future Work

In this paper, we have proposed a new CPA method in order to recover the secret key. This technique contains two phases, in which the first phase calculate the number of points of interest of all possible key bytes (very small compare to number of power trace samples). Then, the points of interest are used to compute the highest correlation value of all guess key bytes. The effectiveness of execution time is demonstrated in experiments with various number of power traces which have clarified the efficiency of the proposed method. In the future work, a mathematical representation and analysis will be investigated in detail.

Acknowledgment. This work is funded by Ministry of Science and Technology (MOST), Vietnam, under the grant number HNQT/TKCG/04.20.

References

1. Kocher, P., Jaffe, J., Jun, B.: Differential power analysis. In: Wiener, M. (ed.) CRYPTO 1999. LNCS, vol. 1666, pp. 388–397. Springer, Heidelberg (1999). https://doi.org/10.1007/3-540-48405-1_25
2. Kocher, P., Jaffe, J., Jun, B.: Introduction to differential power analysis and related attacks (1998). http://www.cryptography.com
3. Messerges, T., Dabbish, E., Sloan, R.: Investigation of power analysis attacks on smartcards. In: Usenix Workshop on Smartcard Technology (1999). http://www.usenix.org
4. Brier, E., Clavier, C., Olivier, F.: Correlation power analysis with a leakage model. In: Joye, M., Quisquater, J.-J. (eds.) CHES 2004. LNCS, vol. 3156, pp. 16–29. Springer, Heidelberg (2004). https://doi.org/10.1007/978-3-540-28632-5_2
5. Coron, J.-S., Kocher, P., Naccache, D.: Statistics and secret leakage. In: Frankel, Y. (ed.) FC 2000. LNCS, vol. 1962, pp. 157–173. Springer, Heidelberg (2001). https://doi.org/10.1007/3-540-45472-1_12
6. Mayer-Sommer, R.: Smartly analyzing the simplicity and the power of simple power analysis on smartcards. In: Koç, Çetin K., Paar, C. (eds.) CHES 2000. LNCS, vol. 1965, pp. 78–92. Springer, Heidelberg (2000). https://doi.org/10.1007/3-540-44499-8_6
7. Kim, Y., Sugawara, T., Homma, N., Aoki, T.: Akashi Satoh: Biasing power traces to improve correlation in power analysis attacks. In: First International Workshop on Constructive Side-Channel Analysis and Secure Design, Citeseer, pp. 77–80 (2010)
8. Le, T.-H., Clédière, J., Canovas, C., Robisson, B., Servière, C., Lacoume, J.-L.: A proposition for correlation power analysis enhancement. In: Goubin, L., Matsui, M. (eds.) CHES 2006. LNCS, vol. 4249, pp. 174–186. Springer, Heidelberg (2006). https://doi.org/10.1007/11894063_14
9. Ouladj, M., Guillot, P., Mokrane, F.: Chosen message strategy to improve the correlation power analysis. IET Inf. Secur. **13**(4), 304–310 (2019)
10. Quentin L. Meunier. FastCPA: Efficient Correlation Power Analysis Computation with a Large Number of Traces. In: 6th Cryptography and Security in Computing Systems (CS2 2019), https://doi.org/10.1145/3304080.3304082. hal-02172200 (2019)

An Optimal Packet Assignment Algorithm for Multi-level Network Intrusion Detection Systems

Dao Thi-Nga[1(✉)], Chi Hieu Ta[1], Van Son Vu[1], and Duc Van Le[2]

[1] Le Quy Don Technical University, Hanoi 10000, Vietnam
daothinga.mta@gmail.com, hieunda@gmail.com, sontlc246@gmail.com
[2] Nanyang Technological University, Singapore 639798, Singapore
vdle@ntu.edu.sg

Abstract. With the outbreaks of recent cyber-attacks, a network intrusion detection system (NIDS) which can detect and classify abnormal traffic data has drawn a lot of attention. Although detection time and accuracy are important factors, there is no work considering both contrastive objectives in an NIDS. In order to quickly and accurately respond to network threats, intrusion detection algorithms should be implemented on both fog and cloud devices, which have different levels of computing capacity and detection time, in a collaborative manner. Therefore, this work proposes a packet assignment algorithm that assigns detection and classification tasks for appropriate processing devices. Specifically, we formulate a novel optimization problem that minimizes detection time while achieving accuracy performance and computational constraints. Then, an optimal packet assignment algorithm that allocates as many packets as possible to fog devices in order to shorten the detection time is proposed. The experimental results on a state-of-the-art network dataset (UNSW-NB15) show that the proposed packet assignment algorithm produces similar performance to the optimal solution with regard to the detection time and accuracy.

Keywords: Network intrusion detection · Packet assignment · Internet of things

1 Introduction

The advanced development of communication devices and networking technologies has enabled the explosive growth in data exchange and associated services in computer networks [11], but also introduced extra challenges for network security due to the increase of emerging cyber-attacks. A recent report by Forbes has shown that the measured traffic of network attacks increases by three-fold to more than 2.9 billion events worldwide in 2019 [3]. Therefore, it is essential to develop a robust network intrusion detection system (NIDS) that can quickly detect network intrusions with a high detection accuracy.

N.-S. Vo and V.-P. Hoang (Eds.): INISCOM 2020, LNICST 334, pp. 301–313, 2020.
https://doi.org/10.1007/978-3-030-63083-6_23

A number of previous studies [1,4,5,8,12,15] have proposed various network intrusion detection (NID) schemes which aim at detecting the network intrusions based on advanced machine learning (ML) techniques. For instance, Moustafa et al. [8] proposed an ensemble NID algorithm which combines different ML models, including the decision tree, Naïve Bayes, and neural network to detect abnormal packets generated by the network intrusions.

In addition, the study in [5] proposed a two-phase deep learning model which can detect multiple types of network attacks. Specifically, their model first calculates a probability that the network is attacked by network threats based on the collected network data. Then, the network attack probability and other network features are fed to a multi-label classifier for detecting types of the network attacks.

With the primary focus on obtaining a high intrusion detection accuracy, those previous schemes [1,4,5,8,12,15] are computationally expensive in general. Thus, those schemes may not be implemented on real networking devices (e.g., gateways and switches) that often have limited computing and memory capabilities. In those studies, the collected network traffic data are often forwarded to an external computing node (e.g., a fog computing node or a centralized server) with high computing performance CPU/GPUs and large memory, in which the data are processed to detect the network intrusions. As a result, those schemes generally have long detection latencies due to the long communication delay of transmitting the collected data to the external computing devices.

To achieve low NID latency, few existing studies [6,9] have developed low computational complexity NID algorithms which can be implemented in resource-constrained networking devices (e.g., an FPGA-based gateway). For instance, the authors in [6] developed and implemented an entropy-based NID scheme on the data plane of the programmable network devices using the P4 programming language. More specifically, source and destination IP addresses are collected and processed for detecting the DDoS attacks in real time. By processing the collected data at the networking devices rather than sending the data to the external stations, those studies can significantly reduce the NID latency. However, they often suffer from the low detection accuracy due to the limited computational complexity.

Generally, existing algorithms are suitable for a specific type of processing devices and as a result only one factor (detection time or accuracy) can be achieved. There were some initial works [10,14] on multi-level collaborative NIDS which leverages the use of both the fog and the cloud computing levels. However, there is no study which considers both detection time and accuracy factors. In order to address the limitation of existing works, we propose a packet assignment algorithm which can balance between two contrastive requirements (i.e., low detection time and high accuracy). In the network model, we assume that two types of processing devices participate in the intrusion detection tasks in a collaborative manner. Since gateways can detect abnormal data traffic within a short time, we should allocate as many packets as feasible to gateways until the accuracy and memory constraints are not satisfied. The remaining packets are transmitted to servers for inspection. By doing that, we can make use of benefits from gateways and servers.

We summarize the main contributions of our paper as follows.

- We first formulate the novel packet assignment optimization problem with the objective of minimizing detection time and constraints on detection and classification accuracy as well as the computational capacity.
- We analyze and compare the detection and classification performance of two different types of processing devices using state-of-the-art network traces.
- A novel and optimal packet assignment algorithm consisting of two phases (independent assignment and collaborative assignment phases) is proposed where each gateway makes a decision on which packets should be inspected. More specifically, the independent assignment phase allocates packets for gateways and servers based on the performance constraints and only the closest gateway to a sensor is responsible for attack detection of that sensor. Then, the collaborative assignment phase performs fine-tuning packet al.location by considering the computational capacity limitation of gateways. To achieve short detection time, packet assignment which exceeds the capacity of a gateway should be forwarded to nearby nodes.
- We design practical network scenarios and conduct the experiments with different parameters to verify the effectiveness of the proposed packet assignment algorithm.
- The experiments with different network parameters have been conducted and results show that our proposed method is able to obtain a close solution to the optimal one.

The rest of the paper is organized as follows. Section 2 introduces the network model and assumption considered in this study. Section 3 presents the problem formulation. Section 4 describes the proposed packet assignment algorithm. Section 5 analyzes the evaluation setting and results. Section 6 concludes the paper.

2 Network Model

We consider a network which consists of k sensors (or data sources), networking devices (e.g., gateways, switch), and external processing devices (e.g., server or cloud computers) as shown in Fig. 1. In this work, nodes are used to indicate networking devices and external processing devices. Two kinds of nodes are able to detect network attacks with different detection time and accuracy. Let us define a set of sensors $\mathbb{S} = \{1, 2, ..., k\}$ and a set of nodes $\mathbb{N} = \{1, 2, ..., n\}$. Sensor i generates and sends packets to end devices in the network for data storage with data rate r_i. Each gateway is connected to one or several external processing devices. Each node has specific features, e.g., the number of incoming packets per second, the maximal number of packets to be inspected per second, attack detection (task 1) capacity, attack classification (task 2) capacity denoted by $n_j^{in}, n_j^{max}, a_{j,1}, a_{j,2}$, respectively. We define v_j as the node type, e.g., $v_j = 1$ if node j is a gateway while $v_j = 2$ if node j is a server. Since there are two types of nodes, we define $\mathbb{N}^1$ as a set of gateway nodes (type 1) and $\mathbb{N}^2$ as a set of server nodes (type 2).

A routing algorithm (e.g., hop count-based) is used to forward packets to destination addresses, which means routing paths for packets are determined in advance. We use $G(s_i)$ to indicate the closest gateway to sensor i which is in charge of forwarding packets of sensor i according to the routing algorithm. For example, $G(s_1) = \{2\}$ indicates that data collected by sensor i is first sent to gateway 2 before arriving at a destination device. In addition, t_{ij} denotes the detection time if node j is assigned to conduct the attack prediction task for data from sensor i ($t_{ij} > 0$ with $\forall j \in G(s_i)$). For gateway j, we define $S(n_j)$ and $G(n_j)$ as a set of sensors which have direct connection to node j and a set of servers which are affiliated with node j.

NIDS has two related tasks: attack detection (or anomaly detection) and attack classification. Nodes including gateways and servers are responsible for conducting the two tasks. However, we assume that gateway nodes can only perform the attack detection task due to the limitation of memory resources and computational capacity. Meanwhile, thanks to the large memory size and powerful processing units, server nodes can perform both detection and classification tasks. For better understanding of performance difference between gateways and servers, the evaluation section will compare the detection and classification accuracy using two different algorithms which are suitable for two types of processing nodes.

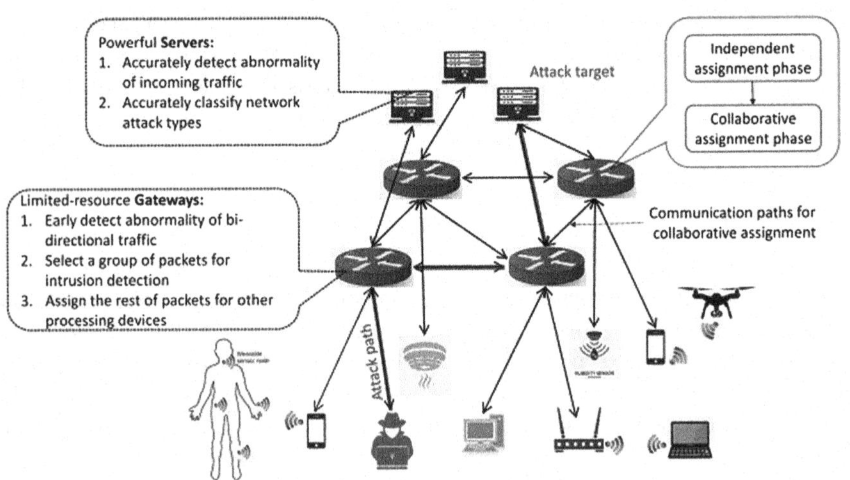

Fig. 1. Collaborative NIDS for interconnected networks with homogeneous devices

3 Problem Formulation

In this section, we present the formulation of an optimization problem for attack detection and classification with the given network model. The optimization problem is to minimize the detection time with performance and resource constraints of processing nodes. We first define decision variables as follows.

$$y_{ij} = \begin{cases} 1, & \text{if node } j \text{ is assigned to detect abnormal packets from } i^{th} \text{ sensor} \\ 0, & \text{otherwise} \end{cases} \tag{1}$$

$$z_{ij} = \begin{cases} 1, & \text{if node } j \text{ is assigned to classify an attack from } i^{th} \text{ sensor} \\ 0, & \text{otherwise} \end{cases} \tag{2}$$

Table 1 summarizes definitions of notations used in the problem formulation.

Table 1. Model parameters

Notation	Description
k	The number of data generated devices
n	The number of processing nodes (gateways and servers) for NIDS
$\mathbb{S}$	A set of data generated devices
$\mathbb{N}$	A set of processing nodes (gateways and servers)
$\mathbb{N}^1$	A set of gateways nodes
$\mathbb{N}^2$	A set of servers nodes
r_i	Data rate of sensor i
n_j^{in}	The number of incoming packets per second at node j
n_j^{max}	The maximal number of packets to be inspected per second at node j
$a_{j,1}$	Attack detection capacity of node j
$a_{j,2}$	Attack classification capacity of node j
v_j	Node type
$G(s_i)$	The closest gateway to sensor i
$S(n_j)$	A set of sensors with direct connection to node j
$G(n_j)$	A set of gateways associated with node j
h_{ij}	The number of hop counts from sensor i to node j
t_{ij}	Detection time of node j for packets of sensor i

Based on the given routing paths in the network, we define h_{ij} as the number of hop counts from sensor i to node j. Assume that average hop-to-hop propagation time is t_0. The processing time at a hop is relatively small and can be ignored. Therefore, detection time of sensor i's packets is estimated as $t(s_i) = t_0 \sum_{j=1}^{n} h_{ij} \times y_{ij}$ and the average detection time of the whole network is $\frac{t_0}{k} \sum_{i=1}^{k} \sum_{j=1}^{n} h_{ij} \times y_{ij}$. The task assignment problem can be formulated as follows:

$$\min \quad \frac{t_0}{k} \sum_{i=1}^{k} \sum_{j=1}^{k} h_{ij} \times y_{ij} \tag{3}$$

subject to

$$\frac{1}{k} \sum_{i=1}^{k} \sum_{j=1}^{n} y_{ij} a_{j,1} \geq a_d \tag{4}$$

$$\frac{1}{k} \sum_{i=1}^{k} \sum_{j=1}^{n} z_{ij} a_{j,2} \geq a_c \tag{5}$$

$$\sum_{i=1}^{k} (y_{ij} + z_{ij}) r_i \leq n_j^{max}, \quad \forall j \in N \tag{6}$$

$$\sum_{j=1}^{n} y_{ij} = 1, \quad \forall i \in S \tag{7}$$

$$\sum_{j=1}^{n} z_{ij} = 1, \quad \forall i \in S \tag{8}$$

$$z_{ij} = 0, \quad \forall i \in S, j \in \mathbb{N}^1 \tag{9}$$

$$z_{ij} \geq y_{ij}, \quad \forall i \in S, j \in \mathbb{N}^2 \tag{10}$$

Equation (3) represents the objective function which aims to minimize the average detection time of all packets in the considered network. Inequalities (4) and (5) make sure that the average detection and classification accuracy should be at least the accuracy constraints a_d and a_c, respectively. Meanwhile, constraint (6) indicates that packet assignment should not violate the computational capacity of nodes. Equations (7) and (8) ensure that only one node is assigned for a specific task. Equation (9) guarantees that the intrusion classification task is not assigned to gateway nodes. At the same time, inequality (10) makes sure that a server node should perform both tasks if this node has already assigned to task 1. If $y_{ij} = 1$, z_{ij} should be set to 1. Inequality (10) is included in the optimization problem since we aim to minimize the packet transmission cost.

4 An Optimal Packet Assignment Algorithm

In this section, we will analyze the motivation of the proposed algorithm and describe the two-phase algorithm procedure in details as follows. First, based on the required detection accuracy, gateway j separately divides the incoming packets into two groups: group 1 processed by gateway nodes and group 2 inspected by server nodes. Let p_j denote the proportion of packets assigned for nodes in group 1 and then $(1-p_j)$ is the proportion of packets for nodes in group 2. Assume that gateway and server nodes have detection performance a_1 and a_2, respectively ($a_{j,1} = a_1$ and $a_{j,2} = a_2, \forall j \in \mathbb{N}$). The average detection performance a_d can be estimated as $a_d = p_j a_1 + (1-p_j)a_2$. Therefore, to achieve the required detection accuracy, we can determine packet grouping for gateway j as below.

$$p_j = \frac{a_2 - a_d}{a_2 - a_1} \tag{11}$$

Note that gateway j has limited processing capacity, n_j^{max}, which means $p_j n_j^{in} \leq n_j^{max}$. Then $p_j = \min(\frac{a_2 - a_d}{a_2 - a_1}, \frac{n_j^{max}}{n_j^{in}})$.

Based on the analysis of accuracy and bandwidth constraints, we design the optimal packet assignment algorithm which consists of two consecutive phases: independent assignment and collaborative assignment phases. As can be seen in Algorithm 1, in the first phase, gateway j assigns packets in a distributed manner. The proportion of packets inspected by nodes in group 1 p_j is estimated by using Eq. 11. Then, gateway j randomly selects traffic packets for intrusion detection with the probability p_j and allocates the detection task of the remaining packets to a server which is connected to gateway j. Since there can be multiple servers ($G(n_j)$) associated with gateway j, we assign a relatively similar load to these servers, i.e., $y_{ik} \sim B\left(1, \frac{1-p_j}{|G(n_j)|}\right), \forall i \in S(n_j), k \in G(n_j)$. Note that notation $x \sim B(1,p)$ is used to indicate that random variable x follows the Bernoulli distribution with mean p. At the end of phase 1, detection tasks are assigned to gateways and servers based on the accuracy requirement. However, there may exist some gateways who could not perform all allocated tasks due to limited computing resource. Therefore, the proposed algorithm contains the second phase where the limited-resources gateways decide which tasks should be offloaded to nearby devices.

In the second phase, each gateway needs to check whether constraint on bandwidth is achieved or not. If $\sum_{i=1}^{k}(y_{ij} + z_{ij})r_i \leq n_j^{max}$, gateway j could not conduct intrusion detection on some packets, called uncompleted packets. Then, gateway j needs to find neighboring nodes which can perform the detection task on uncompleted packets. we use the hop count metric to measure the distance between two nodes. If there is a nearby gateway l who is willing to help gateway j and is closer to gateway j than the server connected to node j, we change task assignment of packet i from gateway j to gateway l, (i.e., $y_{ij} = 0$ and $y_{il} = 1$). Otherwise, the server k which is dedicated to gateway j is selected to perform the detection task on uncompleted packets.

```
Input: Sensor set 𝕊, node set ℕ
Output: Assignment of two tasks to nodes: y_ij, z_ij, ∀i ∈ 𝕊, ∀j ∈ ℕ
Initialize assignment:
y_ij ← 0, z_ij ← 0, ∀i ∈ 𝕊, ∀j ∈ ℕ
Independent assignment phase:
for j ∈ ℕ^1 do
    p_j ← (a_2 − a_d)/(a_2 − a_1)   /* compute p_j using constraints on accuracy */
    y_ij ~ B(1, p_j),  ∀i ∈ S(n_j)   /* assignment to gateway j
        follows Bernoulli distribution with mean p_j */
    y_ik ~ B(1, (1 − p_j)/|G(n_j)|),  ∀i ∈ S(n_j), ∀k ∈ G(n_j)   /* assignment
        to server k has Bernoulli distribution with
        probability (1 − p_j)/|G(n_j)| */
end
Collaborative assignment phase:
for j ∈ ℕ^1 do
    i ← 1
    if (y_ij = 1) & (i ≤ k) & (∑_{i=1}^{k} (y_ij + z_ij) r_i ≥ n_j^max) then
        y_ij ← 0   /* remove task assignment for gateway j */
        for l ∈ ℕ^1 (l ≠ j) do
            if h_il ≤ h_ik with k ∈ G(n_j) then
                y_il ← 1   /* l is a close gateway to node j */
            end
            else
                y_ik ← 1   /* k is a server associated with node j
                    */
            end
        end
    end
    i ← i + 1
end
```

Algorithm 1: An Optimal Packet Assignment Algorithm

5 Performance Evaluation

5.1 Experimental Setup

In this subsection, we describe a practical network scenario and network traces for evaluation of the proposed algorithm. We consider a heterogeneous network including sensors from different applications (e.g., smart home, smart healthcare, smart city, smart agriculture). In each application, sensors are deployed and connected in a sub-network, e.g.., 28 different sensors scattered around a smart home [2,7]. Assume that each gateway is responsible to forward data of several sub-networks. The parameters of network scenarios are summarized in Table 1. For edge devices, we consider Virtex-7 FPGA-based gateways since the Virtex-7 FPGA board [13] with 64 Mb RAM is the most advanced experimental kit manufactured by Xilinx. Each gateway can have 4, 8 Ethernet ports where a half of them is used for incoming traffic and the remaining ports are dedicated for outgoing data. Assume that each port of gateways can support data rate at 100 Mbps.

In this work, the UNSW-NB15 dataset is selected to evaluate the performance of the proposed assignment algorithm. The UNSW-NB15 dataset includes real normal and synthetic abnormal network traffic traces during a 16-hour experimental period and consists of 9 attack classes. For our evaluation, the full dataset is divided into a training data set of 175,341 samples and a test data set of 82,332 samples, and one third of the training data set is used as the validation dataset. The validation set is used to find the best hyper-parameters for the prediction model. In this traffic trace, around 68% of samples are normal while the remaining data belongs to packets associated with one of 9 network attacks. The UNSW-NB15 dataset provides 48 traffic features for each data connection. We convert these features into 196 numeric attributes for the intrusion detection model.

Table 2. Network scenarios

Parameters	Values
The number of gateways	5
The number of servers	2
The number of sub-networks	{15, 21}
Processing capacity of gateways	0.5
The number of available ports at each gateway	{4, 6 }
Data rate of each sub-network	100 Mbps
Intrusion detection accuracy requirement	73%

5.2 Performance Comparison of Processing Nodes

We select and compare performance of detection algorithms which are suitable for gateways and servers based on their memory size and computing capacity. Due to the resource limitation, a logistic regression-based intrusion detection method is constructed for gateway nodes. Meanwhile, in order to gain improvement over the logistic regression-based model, we use an autoencoder model, which learns the latent representation of traffic packet, followed by a fully connected neural network-based intrusion detection model for cloud nodes. The performance of two constructed models are summarized in Table 3. We compare two models in terms of the number of training parameters and detection accuracy. The logistic regression-based detection model which consists of a significantly smaller number of training parameters and produces less accuracy (with roughly 16% difference) compared the neural network-based attack detection model. For the evaluation of the proposed packet assignment algorithm in the next subsection, we use the collected accuracy of the logistic regression-based and neural network-based models as the detection accuracy of gateways and servers.

Table 3. Performance of prediction models

Models	No. of input features	No. of parameters	Accuracy (%)
LR-based detection	196	197	70.62
NN-based detection	196	25,791	86.3
NN-based classification	196	25,980	73.37

5.3 Evaluation of the Proposed Packet Assignment Algorithm

In this subsection, we describe a small example to show how the proposed packet assignment algorithm updates the solution at different phases. Then, using a larger network scenario as shown in Table 2, the proposed algorithm is verified and compared to the optimal solution in terms of detection accuracy and time.

Figure 2 demonstrates the packet assignment when the number of sub-networks and nodes are set to 10 and 5, respectively. Among 5 processing nodes, the first three nodes are gateways and the last two nodes are servers. Gateways 1 and 2 have connected to four sub-networks while gateway 3 is responsible for two sub-networks. The processing capacity of gateways is set to 0.5, i.e., each gateway should only perform network intrusion detection on 50% of sub-networks. Assume that each gateway consists of 4 networking ports then only 2 sub-network flows can be inspected in each gateway. We use a binary matrix consisting of ten rows and five columns to represent the assignment solution. The value at the ith row and j^{th} column represents the assignment of node j to sub-network i, e.g., value 1 means assignment and 0 indicates no assignment. At the initial phase, all values of assignment matrix are set to 0, i.e., there is no assignment for processing nodes.

Then, in the next phase, the proportion of packets which are inspected by each gateway is estimated. In this example, $p_j = \dfrac{a_2 - a_d}{a_2 - a_1} = \dfrac{86.3 - 73}{86.3 - 70.62} = 0.84$. Then, the number of sub-networks which are inspected by gateways 1 and 2 is $\lfloor 0.84 * 4 \rfloor = 3$. Similarly, gateway 3 should perform anomaly detection on one sub-network in order to satisfy the detection accuracy requirement. Since there are four sub-networks which forward data packets to gateway 1 and gateway 1 should detect intrusion for 3 sub-networks due to the accuracy constraint, gateway 1 simply selects the first three sub-networks for intrusion detection and sends data from the fourth sub-network to node 4 (server 1). After the independent assignment phase has completed, anomaly detection for packets of ten sub-networks are assigned to five nodes considering the accuracy requirement as shown in Fig. 2(b).

Finally, the collaborative assignment phase implements fine-tuning the current assignment based on the computing limitation of gateways in order to minimize the detection time. Assume that each gateway can perform the intrusion detection function for two sub-networks due to the limited memory and computing capacity. As a result, gateway 1 needs to ask the neighboring node to conduct the detection task for one sub-network. As can be seen in Fig. 2(c), gateway 1 selects and forwards traffic data of the third sub-network to node 4 (server 1). Similarly, gateway 2 transmits packets of the seventh sub-network to node 5 (server 2) for intrusion detection.

In order to evaluate the effectiveness of our proposed algorithm, we made performance comparison between the proposed algorithm and the optimal solution. Note that the optimal solution is obtained by using the brute-force search while the proposed algorithm determines a packet assignment solution based on a distributed manner. As can be seen in Table 4, the proposed assignment method produces similar detection time to the optimal solution in both cases of 15 and

Table 4. Performance comparison with the optimal solution

	Proposed algorithm	Optimal brute-force-based algorithm
Average detection time (15 sub-networks)	$6.93t_0$	$6.86t_0$
Detection accuracy (15 sub-networks)(15 sub-networks)	75.84%	75.84%
Average detection time (21 sub-networks)	$6.67t_0$	$6.67t_0$
Detection accuracy (21 sub-networks)	75.1%	75.1%

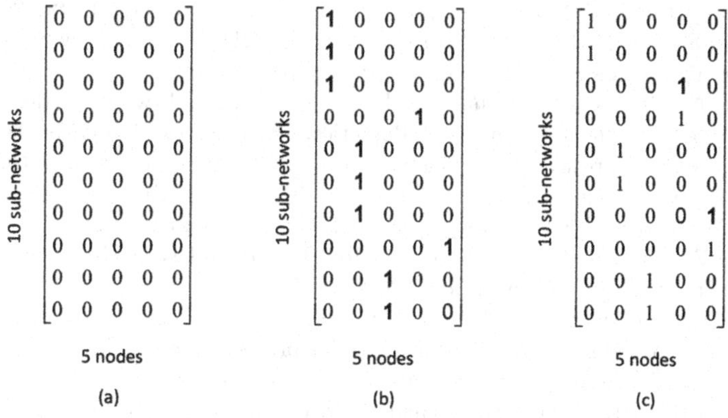

Fig. 2. Packet assignment at different phases of the proposed algorithm, (a): Initial assignment, (b): Independent assignment, and (c): Collaborative assignment

21 sub-networks. For instance, if there are 15 sub-networks, average detection time of our method and the optimal solution is $6.93t_0$ and $6.86t_0$, respectively (t_0 is the one-hop propagation time).

6 Conclusion

In this paper, we address the problem of balancing between detection time and detection accuracy in a multi-level network intrusion detection system. In this system, both fog and cloud devices are assigned tasks of anomaly detection and classification in order to shorten the detection time while satisfying the accuracy requirement. We define the novel optimization problem which aims to minimize the detection time with constraints on accuracy and computational capacity of computing nodes. Then, a heuristic packet assignment method is proposed to find a solution for the optimization problem. The proposed method is evaluated by using the UNSW-NB15 dataset and the results show promising performance in terms of detection time and accuracy. As a future work, we plan to design an intelligent and self-adaptive packet assignment algorithm which is able to learn optimal solutions given the change of network conditions (e.g., the number of sub-networks and the number of neighboring gateways).

Acknowledgment. This work is funded by the Le Quy Don Technical University.

References

1. Cao, V.L., Nicolau, M., McDermott, J.: Learning neural representations for network anomaly detection. IEEE Trans. Cybern. **49**(8), 3074–3087 (2019)
2. Clark, M., Dutta, P.: The haunted house: Networking smart homes to enable casual long-distance social interactions. In: IoT-App 2015 (2015)
3. Doffman, Z.: Cyberattacks on IOT devices surge 300% in 2019, 'measured in billions', report claims (2019). https://bit.ly/35uPCI7. Accessed 04 May 2020
4. Hosseini, S., Azizi, M.: The hybrid technique for DDoS detection with supervised learning algorithms. Comput. Netw. **158**, 35–45 (2019)
5. Khan, F.A., Gumaei, A., Derhab, A., Hussain, A.: A novel two-stage deep learning model for efficient network intrusion detection. IEEE Access **7**, 30373–30385 (2019)
6. Lapolli, A.C., Marques, J.A., Gaspary, L.P.: Offloading real-time DDoS attack detection to programmable data planes. In: 2019 IFIP/IEEE Symposium on Integrated Network and Service Management (IM), pp. 19–27 (2019)
7. Morais, C., Sadok, D., Kelner, J.: An IoT sensor and scenario survey for data researchers. J. Braz. Comput. Soc. **25**, 4 (2019)
8. Moustafa, N., Turnbull, B., Choo, K.R.: An ensemble intrusion detection technique based on proposed statistical flow features for protecting network traffic of internet of things. IEEE Internet Things J. **6**(3), 4815–4830 (2019)
9. Carvalho, R.N., Bordim, J.L., Alchieri, E.A.P: Entropy-based dos attack identification in SDN. In: 2019 IEEE International Parallel and Distributed Processing Symposium Workshops (IPDPSW), pp. 627–634 (2019)

10. Nguyen, T.G., Phan, T.V., Nguyen, B.T., So-In, C., Baig, Z.A., Sanguanpong, S.: Search: a collaborative and intelligent NIDS architecture for SDN-based cloud IoT networks. IEEE Access **7**, 107678–107694 (2019)
11. Systems, C.: Cisco Annual Internet Report (2018–2023) White Paper. Technical Report Cisco Systems (2020)
12. Vu, L., Cao, V.L., Uy, N.Q., Nguyen, D.N., Hoang, D.T., Dutkiewicz, E.: Learning latent distribution for distinguishing network traffic in intrusion detection system, pp. 1–6 (2019)
13. Xilinx: Xilinx Virtex-7 FPGA VC707 Evaluation Kit. Tech. rep., Xilinx
14. Yan, Q., Huang, W., Luo, X., Gong, Q., Yu, F.R.: A multi-level DDos mitigation framework for the industrial internet of things. IEEE Commun. Mag. **56**(2), 30–36 (2018)
15. Yang, Y., Zheng, K., Wu, C., Yang, Y.: Improving the classification effectiveness of intrusion detection by using improved conditional variational autoencoder and deep neural network. Sensors **19**(11), 2528 (2019)

Privacy-Preserving for Web Hosting

Tam T. Huynh[1(✉)], Thuc D. Nguyen[2], Nhung T.H. Nguyen[2], and Hanh Tan[1]

[1] Posts and Telecommunications Institute, Ho Chi Minh City, Vietnam
{tamht,tanhanh}@ptithcm.edu.vn
[2] University of Science, Ho Chi Minh City, Vietnam
{ndthuc,nthnhung}@fit.hcmus.edu.vn

Abstract. Privacy-preserving for data is one of the crucial responsibilities of service providers. For the web hosting service, customers' information and source codes of websites are considered sensitive data that need to protect. The solution decentralized for web hosting is a new technology trend, provides an effective mechanism for storing and accessing websites. This paper addresses some problems related to privacy concerns of the decentralized web hosting service. Based on the blockchain technology, cryptography, and interplanetary file system (IPFS) platform, we propose a protocol for web hosting that provides three features. The first one ensures anonymity for information of customers. The second feature provides confidentiality and authentication for transmitting source codes of websites between customers and the service provider (SP). And the last one is responsible for securing the source code of websites from other nodes on the public IPFS network. The experiments demonstrate that the proposed solution efficiently protects privacy for the decentralized web hosting service.

Keywords: IPFS · Blockchain · Privacy · Web hosting

1 Introduction

Nowadays, most websites are hosted based on the client-server model [1], where, the centralized servers are responsible for storing source codes and data of websites and handling queries of users, and users can use their web browsers to send requests to these web servers. Some challenges for web hosting based on this model: (1) with websites provide data sharing features, when the number of requests to large files increases significantly, the network traffic becomes more congested and the hosting servers become slower in responding. To improve data retrieval speed, content delivery networks can be used to cache and deliver contents to users based on their geographic location [2–4]. However, these solutions may leak information about users [5], and have some challenges concerning scalability, quality of service and flexibility [6], moreover, these entire systems are still the centralized model. (2) If the data is not well protected and centralized systems stop working due to overload, denial-of-service (DoS) or distributed denial-of-service (DDoS) attacks, system errors, etc., users cannot access the websites at that time.

To solve these challenges, the authors in [7] used IPFS to build distributed web applications. In [8] we proposed a decentralized solution for web hosting, which was a

N.-S. Vo and V.-P. Hoang (Eds.): INISCOM 2020, LNICST 334, pp. 314–323, 2020.
https://doi.org/10.1007/978-3-030-63083-6_24

combination of the IPFS protocol and the blockchain technology to form a decentralized platform for the web hosting service. IPFS, was proposed by Juan Benet in 2014 [9], is a distributed file system, provides efficient methods for data storage and retrieval. It is also considered the distributed and permanent website. Files on IPFS are segmented inside objects of 256 kbyte, identified by the SHA-256 hash function. Nodes on the IPFS network connect and transfer objects to each other over a peer-to-peer network. Each node owns a distributed hash table to locate the nodes that store certain objects being requested. Nodes can use the IPFS pinning feature to keep the objects available to the community, and can activate the interplanetary name space (IPNS) service for creating and updating mutable links to contents of websites [9, 10]. Meanwhile, the blockchain technology provides anonymity, transparency, decentralization, auditability [11–13], is used to conduct transactions between customers and the SP. All valid transactions are mined and stored on the shared ledger. To easily deploy electronic contracts between customers and the SP, we select the ethereum blockchain which is a new distributed blockchain network [14].

In this paper, we present three models of web hosting, and illustrate the pros and cons of them. We also propose an overall protocol used for web hosting that provides three privacy-preserving features as follows: (1) ensure anonymity of customers' information; (2) secure data sharing between customers and the SP over a peer-to-peer network; (3) protect source codes of websites from illegal copying. Our experimental results also demonstrate the improvement of privacy-preserving for web hosting, and it is a suitable and necessary protocol for the web hosting service provider based on the distributed platform.

The rest of the paper is organized as follows: Sect. 2 describes the three deployment models for web hosting and analyzes the privacy requirements for the system. Section 3 presents our proposed protocol. Section 4 reports our experimental results, and finally we conclude with Sect. 5.

2 Models and Privacy Requirements

2.1 Models

The decentralized solution for web hosting can be implemented on the three models as shown in Figs. 1 and 2. The SP can build a private IPFS network as shown in Fig. 1a, the size of the network is the number of nodes on the network. In this model, the SP uses some nodes having high performance (especially large storage capacity) to run the IPFS cluster service which is used for data replication between cluster nodes. All websites of customers are stored long-term and replicated on these nodes. Other nodes, called miner nodes, must connect to cluster nodes directly or indirectly, and act as gateways of the network. Websites are segmented and put into objects, and each miner node on the network can cache some objects or all objects of some websites depend on its storage capacity. Objects are stored on miner nodes temporarily and removed periodically by the garbage collection feature or when the cache memory is full. In order to access websites hosted on the SP's network, there are two ways that users can select. One is to use web browsers to access websites through one of the gateway nodes published by the SP, this connection model can be considered the sub-client-server models for the private IPFS network, with each server is a gateway node. The gateway nodes can be congested when receiving a lot of requests from users or due to DoS/DDoS attacks; the other is that each user uses a device to join the network as a node of the network, and use a web browser to access websites via the local address of the joined device (such as http://127.0.0.1:8080/ipfs/<website>). This way is recommended for users because it reduces the network congestion at gateway nodes and users can access websites faster. In general, the model in Fig. 1a will not operate more effectively than the traditional model if the size of the network is small.

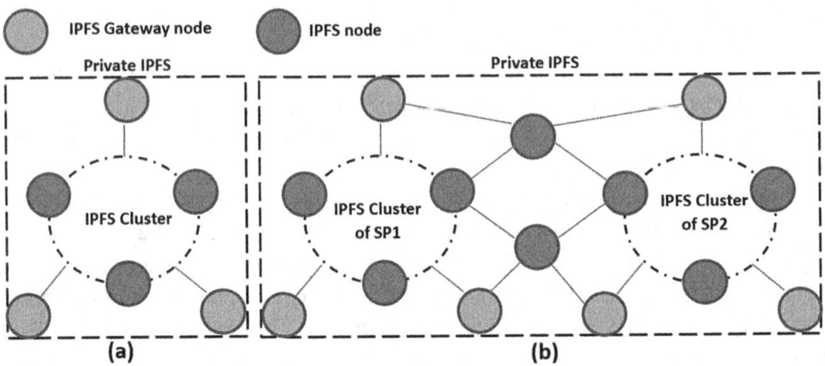

Fig. 1. The private IPFS network for web hosting.

As shown in Fig. 1b, the model is the combination of the two private IPFS networks of the two SPs to form a new private IPFS network. In fact, the more networks join, the more effective the IPFS network operates. However, the models in Fig. 1 can only operate effectively within areas that the network covers because many nodes of the SPs are always in active state to serve users. Users in other regions do not always join the network to access or cache data for other nodes in the regions, hence the stability of the IPFS network will not be guaranteed. Therefore, the model in Fig. 1b is suitable for SPs whose websites only serve users in a specific geographical area such as a region or a country.

In order to serve a lot of users around the world effectively. The SP's network needs to join the public IPFS network. Currently, the *ipfs.io* is one of the largest public IPFS networks, is built by the Protocol Labs [15]. Nodes of the SP can easily join this network to form a private IPFS network and operate based on the public network. In Fig. 2, nodes 1, 2, 3, and 4 of the SP join the public IPFS network. These nodes only store long-term the websites of the SP's customers, other data on the network are cached in a certain period. This is the suitable model for web hosting with lower deployment costs than the two previous models, it is also selected for the experiment section.

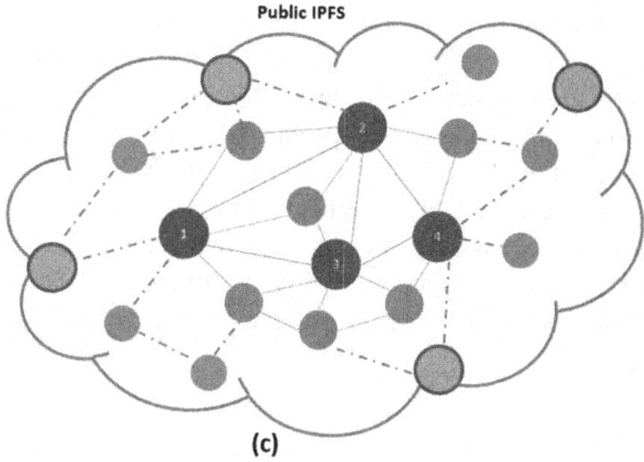

Fig. 2. The private IPFS network based on the public IPFS network.

2.2 Privacy Requirements

In order to implement this solution into practice, it is necessary to consider aspects concerning security and privacy for the service. We indicate three requirements that need to solve for the decentralized web hosting service including: protect customers' information, guarantee the security and privacy of websites' source codes when transmitting from customers to the SP for hosting, and protect websites from illegal copying.

Concerning customers' information, with the traditional method, in order to register the web hosting service, customers are required to send certain information like full name, address, phone number, credit card information, etc. This information is used for making the contract and invoice. Although a non-disclosure agreement between the customer and the SP may be signed, the SP can secretly share this sensitive data with third parties or the SP's database is compromised by hackers. In our solution, the anonymity is guaranteed by the ethereum blockchain, we write smart contracts containing a set of rules under which the parties agree to interact with each other and automatically executed once the conditions of the agreement are met. In this way, customers may create accounts on the ethereum blockchain network with anonymous information and easily conduct transactions to the SP. Transactions are validated by miners and stored in the shared public ledger of the blockchain network. In addition, the service payment can be done by Ethereum coin.

Concerning transmission of the source codes of websites from customers to the SP, our protocol also conducts this work over the public IPFS network. However, IPFS is currently not provided any solutions for privacy-preserving data storage. Hence, nodes on the IPFS network can view data through hash values that they own. To ensure this property, source codes of websites must be encrypted by a symmetric algorithm before uploading on the public IPFS network, and the secret key is sent to the SP for decryption through a secure channel.

Concerning protection for source codes of websites from illegal copying, by default hosted websites on the IPFS network are accessed from normal web browsers like Chrome, Firefox, etc. and source codes of websites are easily downloaded by any users. To overcome this problem, the SP needs to select a symmetric algorithm or build a private encryption algorithm to encrypt source codes of websites before uploading on the network. This way guarantees the privacy of websites because nodes on the IPFS network and normal web browsers cannot understand the contents of the websites. In order to access websites, the SP has to provide an application as common web browsers, which contains the secret key of the SP to decrypting websites that hosted by the SP, and limits the view source code feature. When a user uses this app to access a certain website, it will download objects of the website, and then decrypt and show results on the application window. In order to improve security, customers can also use cryptography functions in their websites to guarantee the privacy for storage data.

3 Proposed Protocol

The proposed protocol for web hosting satisfies the privacy requirements mentioned in Sect. 2.2. Transactions and related information are transparent on a blockchain. Figure 3 describes the general diagram of the protocol for web hosting. We explain the symbols in the diagram as follows: P/O represents a person or an organization who owns a website and wants to register the web hosting service, each P/O has a public key and a private key denoted by P_PO and p_PO respectively. The web hosting service provider is denoted by SP, has a public key P_SP and a private key p_SP.

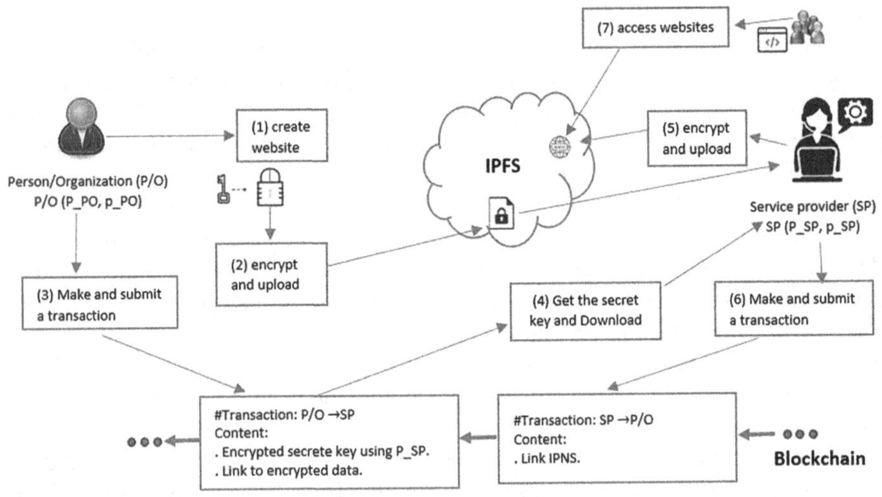

Fig. 3. The protocol diagram for web hosting.

The process of registering the service and accessing websites of the protocol as the following steps:

1. The P/O builds a website.
2. The P/O encrypts the website with a secret key (denoted by *k1*) and the selected symmetric cryptography. The P/O can use the symmetric cryptosystem provided

by the SP or in case of using a private cryptosystem, the P/O must provide this cryptosystem or the source code of the cryptosystem to the SP for decryption. Then the P/O uploads the ciphertext of the website to the public IPFS network. After the upload process has completed, the P/O receives the hash value of the ciphertext from the IPFS node.

3. The P/O performs a transaction to the blockchain network: In this transaction, besides the information about the sender and receiver addresses, it also has the following necessary information:

    - *k1*, for decrypting the original data, is encrypted by the SP's public key.
    - The link of the ciphertext on the public IPFS network.

4. The SP get *k1* and download the data: When receiving a transaction from the P/O, the SP uses p_SP to decrypt ciphertext of *k1*. Then, the SP downloads the ciphertext of the website on the IPFS network and uses *k1* to decrypt to get the source codes of the website.
5. The SP uses a symmetric cryptography and a secret key (denoted by *k2*) to encrypt the website. The output is uploaded to a cluster node of the SP. The returned hash value is mapped to a certain public key by the IPNS service of IPFS.
6. The SP creates and submits a transaction to the blockchain network to return this IPNS value to the P/O. The P/O can use this value to configure a domain name for the website by using the TXT record of the domain name system service [16].
7. In order to access the website, users have to download an application provided by the SP. This application can show websites hosted by the SP because it contains *k2* for decrypting websites.

In case the P/O wants to update the source codes of the website, the P/O has to perform steps similar to the service registration process.

4 Experiment Results

To deploy this protocol, we use the *ipfs.io* network which is a public IPFS network. For the blockchain network, we use Ganache to create a virtual ethereum blockchain. By default, Ganache generates ten different accounts, each account wallet contains 100 ETH.

We use the solidity language to build smart contracts for storing information of the web hosting service as shown in Fig. 4.

We write a simple website using HTML, CSS, and JavaScript programming languages, all files related to the website are placed in a folder named *Website*. For cryptography algorithms, we use the AES-256 algorithm in CBC mode to encrypt the *Website* as described at step 2 of our protocol. We use the RSA cryptosystem with 1024-bit keys to

```
pragma solidity >=0.4.21 <0.6.0;
import "github.com/Arachnid/solidity-
stringutils/strings.sol";
contract DataSharing {
    mapping (address => string) message;
    using strings for *;
     event AddData(string,string);
     event ReturnLink(string);
function SubmitData(address recipient, string  Key, string
link) public {
        string memory data;
        data=Key.toSlice().concat("|".toSlice());
        data=data.toSlice().concat(link.toSlice());
        data=data.toSlice().concat("|".toSlice());
        message[recipient] = data;
        emit AddData(Key, link);
    }
function SubmitIPNS(address recipient, string  ipns) public
{
        message[recipient] = ipns;
        emit ReturnLink(ipns);
    }
function GetData() public view returns (string ) {
        return (message[msg.sender]);
    }
}
```

Fig. 4. Code script of the smart contracts for web hosting.

generate key pairs for the P/O and the SP. We install the *IPFS-Desktop-Setup-0.10.4.exe* application on a computer to join the *ipfs.io* network [17], and then upload the encrypted form of the *Website* to the network. The secret key is encrypted by the public key of the SP and then sent together with the link of the file on the IPFS network to the blockchain network. The result of the transaction is shown in Fig. 5.

Fig. 5. The decoded result of the transaction.

We use two virtual machines (OS Windows 7, RAM 1 GB, CPU 2.6 GHZ) joined to the public IPFS network as the nodes of the SP, Fig. 6 shows the node's information on the IPFS network. These nodes are enabled the clustering service and the pinning feature. The *Website* is encrypted by the AES algorithm and *k2*, and then the output is uploaded to the *ipfs.io* network as shown in Fig. 7, all files of the *Website* are also encrypted as shown in Fig. 8. The *Website* is stored long-term in these two nodes.

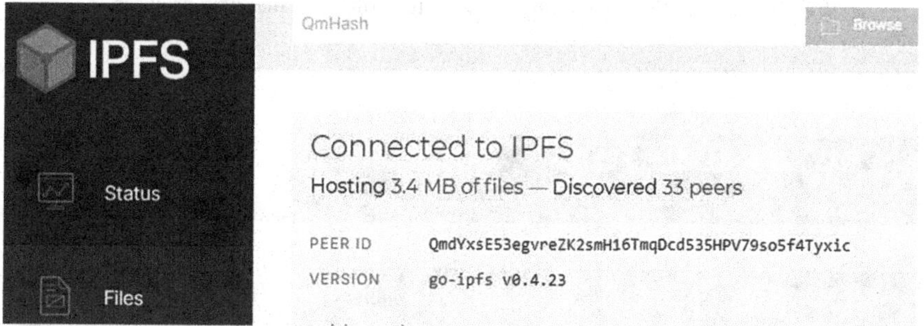

Fig. 6. The information of the SP' node on the IPFS network.

Fig. 7. The hash value of the Website folder on the IPFS network.

Fig. 8. Some files of the Website folder on the IPFS network.

Users can access the *Website* through the gateway address (https://www.ipfs.io/ipfs/<hash_value>) or can also access to the localhost of the devices joined to the network (http://127.0.0.1:8080/ipfs/<hash_value>). As shown in Fig. 9, the normal web browsers can access the *Website* but cannot understand its contents. Similarly, other nodes on the public IPFS network will also receive the ciphertext when accessing the hash value of the website, hence the privacy of websites is guaranteed. In order to show the *Website*, we use the C# programming language to create an application named *Web_Client* which contains *k2* for decrypting the source codes of the *Website*, as shown in Fig. 10.

Fig. 9. The result of accessing the website from a normal web browser

Fig. 10. The website is accessed from the Web_Client application.

5 Conclusions

Websites hosted on IPFS can access faster and more secure than the traditional model because objects of websites are identified by the cryptographic hash of their contents and objects of a certain website can retrieve from peer-to-peer nodes. In this paper, we have built the protocol for web hosting presented on a previous paper [8]. Our protocol can protect the privacy of websites on the public IPFS network. The protocol is built from the advantages of the IPFS platform for storing distribution, the blockchain technology for providing anonymous, and cryptography for protecting confidentiality and authentication. The experimental results show that our protocol operates efficiently, can easily implement in practice.

Acknowledgements. This research is funded by Vietnam National University Ho Chi Minh City (VNU-HCM) under grant number NCM2019-18-01.

References

1. Xiao, Z., et al.: A new architecture of web applications-the widget/server architecture. In: 2010 2nd IEEE International Conference on Network Infrastructure and Digital Content, pp. 866–869. IEEE (2010)

2. Shafiq, M.Z., Liu, A.X., Khakpour, A.R.: Revisiting caching in content delivery networks. In: The 2014 ACM International Conference on Measurement and Modeling of Computer Systems, pp. 567–568. ACM (2014)
3. Mokhtarian, K., Jacobsen, H. A.: Caching in video CDNs: building strong lines of defense. In: Proceedings of the 9th European Conference on Computer Systems, pp. 1–13 (2014)
4. Hosanagar, K., Krishnan, R., Smith, M., Chuang, J.: Optimal pricing of content delivery network (CDN) services. In: 37th Annual Hawaii International Conference on System Sciences, 2004. Proceedings of the IEEE. IEEE (2004)
5. Sengupta, A., Tandon, R., Clancy, T.C.: Fundamental limits of caching with secure delivery. IEEE Trans. Inf. Forensics Secur., **10**(2), 355–370. IEEE (2014)
6. Fan, Q., et al.: Video delivery networks: challenges, solutions and future directions. Comput. Electr. Eng. **66**, 332–341 (2018)
7. Dias, D., Benet, J.: Distributed web applications with IPFS, tutorial. In: Bozzon, A., Cudre-Maroux, P., Pautasso, C. (eds.) ICWE 2016. LNCS, vol. 9671, pp. 616–619. Springer, Cham (2016). https://doi.org/10.1007/978-3-319-38791-8_60
8. Huynh, T.T., Nguyen, T.D., Tan, H.: A decentralized solution for web hosting. In: 2019 6th NAFOSTED Conference on Information and Computer Science (NICS), pp. 82–87. IEEE (2019)
9. Benet, J.: IPFS-content addressed, versioned, P2P file system. arXiv preprint arXiv:1407.3561 (2014)
10. Steichen, M., Fiz, B., Norvill, R., Shbair, W., State, R.: Blockchain-based, decentralized access control for IPFS. In: 2018 IEEE International Conference on Internet of Things (iThings) and IEEE Green Computing and Communications (GreenCom) and IEEE Cyber, Physical and Social Computing (CPSCom) and IEEE Smart Data (SmartData), pp. 1499–1506. IEEE (2018)
11. Huynh, T.T., Nguyen, T. D., Tan, H.: A survey on security and privacy issues of blockchain technology. In: 2019 International Conference on System Science and Engineering (ICSSE), pp. 362–367. IEEE (2019)
12. Zheng, Z., Xie, S., Dai, H. N., Wang, H.: Blockchain challenges and opportunities: a survey. In: International Journal of Web and Grid Services, pp. 352–375 (2018)
13. Conti, M., Kumar, S., Lal, C., Ruj, S.: A survey on security and privacy issues of bitcoin. IEEE Commun. Surv. Tutorials, **20**(4), 3416–3452. IEEE (2018)
14. Buterin, V.: A next-generation smart contract and decentralized application platform. White paper 3(37) (2014)
15. The IPFS network. https://ipfs.io. Accessed Apr 2020
16. DNS txt record. https://tools.ietf.org/html/rfc6763. Accessed Apr 2020
17. The client application of the ipfs.io network. https://github.com/ipfs-shipyard/ipfs-desktop. Accessed Apr 2020

A Novel Secure Protocol for Mobile Edge Computing Network Applied Downlink NOMA

Dac-Binh Ha[1,2(✉)], Van-Truong Truong[1,2], and Duy-Hung Ha[3]

[1] Faculty of Electrical-Electronic Engineering, Duy Tan University, Da Nang 550000, Vietnam
truongvantruong@dtu.edu.vn

[2] Institute of Research and Development, Duy Tan University, Da Nang 550000, Vietnam
hadacbinh@duytan.edu.vn

[3] Faculty of Electrical Engineering and Computer Science, VSB-Technical University of Ostrava, Ostrava, Czech Republic
haduyhung@tdtu.edu.vn

Abstract. In this paper, we study a mobile edge computing (MEC) network based on non-orthogonal multiple access (NOMA) scheme, in which a user can offload its tasks to two MEC servers through downlink NOMA. Due to security constraint, the confidential tasks must be computed on the trusted server, the remain tasks can be offloaded to other server if needed. In this scenario, we propose a novel secure protocol, namely APS-NOMA MEC, based on access point selection (APS) scheme to guarantee the security constraint. The exact closed-form expression of successful computation probability for this proposed system protocol is derived. We further study the impact of the network parameters on the system performance to confirm the effectiveness of deployment of NOMA in MEC network. The numerical results show that our proposed protocol outperforms the conventional NOMA MEC scheme in terms of computation efficiency. Finally, the simulation results are also provided to verify the accuracy of our analysis.

Keywords: Mobile edge computing · Non-orthogonal multiple access · Downlink NOMA · Successful computation probability · Trusted server

1 Introduction

Cloud computing has been seen emerging as a new paradigm of computing in the last decade. In the last few years, mobile edge computing (MEC) is a new trend in computing with the function of clouds moving towards the network edges to support the intensive computation needs of the next-generation wireless communication networks (NGWCN). In this MEC network, the edge servers serve as the computational access points to help in accomplishing the computation tasks of users through wireless links [1–6]. Meanwhile, NOMA technique has

N.-S. Vo and V.-P. Hoang (Eds.): INISCOM 2020, LNICST 334, pp. 324–336, 2020.
https://doi.org/10.1007/978-3-030-63083-6_25

been recognized as a strong candidate for NGWCN due to its ability of serving multiple users using the same time and frequency resources, so that it can improve network capacity [7–11]. Naturally, the application of NOMA technique in MEC is considered in a few of works to improve the performance of MEC networks [12–15]. In [12], a NOMA MEC network, in which two users may partially offload their respective tasks to a single MEC server through uplink NOMA, was studied. The offloading scheme in three different modes, namely the partial computation offloading, the complete local computation, and the complete offloading was proposed for this considered model. The optimal solutions for an optimization problem to maximize the successful computation probabilty were obtained by jointly optimizing the parameters of the proposed scheme. In [13], a computation efficiency maximization framework was proposed for wireless-powered MEC networks based on uplink NOMA according to both partial and binary computation offloading modes. Two algorithms, namely iterative algorithm and alternative optimization algorithm, were proposed to solve the computation efficiency non-convex problem. The authors in [14] proposed the efficient algorithms to solve the weighted sum-energy minimization problems under both cases with partial and binary offloading for multi-antenna NOMA multiuser MEC system. The work of [15] studied NOMA MEC networks for both uplink and downlink transmissions. The studied results have shown that the use of NOMA can efficiently reduce the latency and energy consumption of MEC offloading compared to their conventional orthogonal multiple access (OMA) counterparts.

Motivate by the work of [15], in this work we consider the scenario that the security requirement of MEC network is deployed based on trusted server. The confidential tasks must be computed on the trusted server, the remain tasks can be offloaded to other server if needed by using downlink NOMA scheme. The main contributions of our paper are as follows.

- A novel secure APS-NOMA MEC scheme for mobile edge computing system applied downlink NOMA network based on trusted server is proposed.
- We derive the exact closed-form expression of successful computation probability for this scheme.
- Numerical results are provided to investigate the impact of the network parameters, i.e., transmit power, power allocation ratio, on the system performance to verify the effectiveness of deployment of NOMA in MEC network.
- Simulation results show that our proposed protocol can achieve better successful computation probabilty compared with other NOMA scheme.

The rest of this paper is organized as follows. Section 2 presents the system model. The performance of this considered system is analyzed in Sect. 3. The numerical results and discussion are shown in Sect. 4. Finally, we conclude our work in Sect. 5.

Notation: g_i denotes the channel power gain of link $S - AP_i$ ($i = \{1, 2\}$), $g_i \sim \mathcal{CN}(0, N)$. B denotes the channel bandwidth. P_s stands for the transmit power of S. $\gamma_s = \frac{P_s}{N}$ is the average transmit signal-to-noise ratio (SNR).

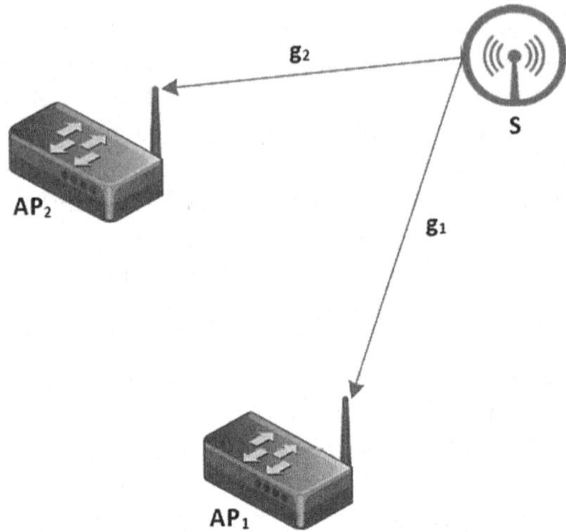

Fig. 1. System model for downlink NOMA MEC network

2 System Model

The Fig. 1 depicts the system model for a secure downlink NOMA mobile edge computing network, in which a single user partially offloads its tasks to two (trusted and untrusted) MEC access points (APs) through downlink NOMA. Without loss of generality, we denote user, trusted AP and untrusted AP as S, AP_1 and AP_2, respectively. They are assumed to have a single antenna and operate in the half duplex mode. Assuming that S has a L-bit task with L_1 confidential bits ($L_1 < L$) to be executed and it may not be able to execute its tasks locally within the latency budget due to the limited computational ability. Therefore, S needs the help from APs through wireless links subject to quasi-static Rayleigh fading. However, due to security constraint the trusted server (AP_1) must be prioritized for selecting to compute the tasks of user, meanwhile, the untrusted server (AP_2) is only used to guarantee the performance of this considered system if needed. We assume that the task-input bits are bit-wise independent and can be arbitrarily divided into different groups [12]. Therefore, L-bit task can be divided into a confidential L_1-bit task (Task 1) and a non-confidential L_2-bit task (Task 2).

We propose a new secure protocol, called APS-NOMA MEC scheme, for MEC network as follows.

- In the first phase, the user S estimates the channel parameters at the beginning of each transmission in duration τ_0.

- In the second phase, according to the channel state information (CSI), S assigns the servers and offloads its tasks to the corresponding APs during duration τ_1 according to APS Algorithm 1. Notice that, in APS Algorithm 1, Task 1 is confidential task with L_1 bits, Task 2 is non-confidential task with L_2 bits and $L_1 + L_2 = L$.

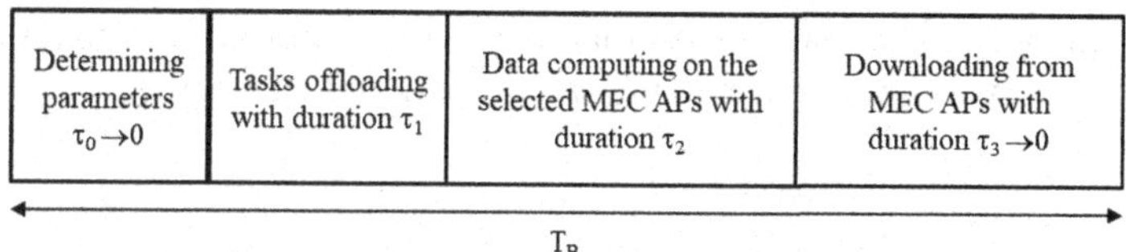

Fig. 2. Time flow chart for APS protocol

- In the third phase, data is executed on the selected MEC APs in duration τ_2.

- Finally, in the fourth phase S downloads the results from APs during duration τ_3.

The time flow chart for APS protocol is as Fig. 2, in which τ_0 and τ_3 are assumed very small and thus are neglected [12], and T_B denotes the transmission block time.

Algorithm 1. Access Point Selection (APS) Algorithm

```
1: procedure SELECTION(AP1, AP2)
2:     if g1 > g2 then AP1 selected
3:     offload L-bit task to AP1
4:     goto 9
5:     else divide L-bit task into L1-bit task (Task 1) and L2-bit task (Task 2)
6:     applying NOMA, S offloads Task 1 to AP1 and Task 2 to AP2
7:     goto 9
8:     end if
9:     download results from corresponding APs
```

The i.i.d. quasi-static Rayleigh channel gains g_i, $i \in \{1,2\}$, follows exponential distributions with parameters λ_i. Therefore, the cummulative density function (CDF) and probability density function (PDF) of g_i $(i = 1,2)$ are respectively given by

$$F_{g_i}(x) = 1 - e^{-\frac{x}{\lambda_i}}, \tag{1}$$

$$f_{g_i}(x) = \frac{1}{\lambda_i} e^{-\frac{x}{\lambda_i}}. \tag{2}$$

3 Performance Analysis

In order to characterize the performance of a MEC system, the successful computation probability, called Pr_s, is used as an important performance metric [12]. It is defined as the probability that all tasks are successfully executed within a given time $T > 0$, which is expressed as

$$Pr_s = \Pr(\tau_1 + \tau_2 \leq T) = \Pr(\tau \leq T), \tag{3}$$

where τ_1 and τ_2 are the transmission latency and computation latency, respectively; $\tau = \tau_1 + \tau_2$. The execution time τ is calculated as follows.

$$\tau = \begin{cases} \frac{L}{B\log(1+\gamma_s g_1)} + \frac{\rho L}{f_1}, & g_1 > g_2 \\ \max\left\{t_1 + \frac{\rho L_1}{f_1}, t_2 + \frac{\rho L_2}{f_2}\right\}, & g_1 < g_2 \end{cases} \quad (4)$$

where $t_1 = \frac{L_1}{B\log\left[1+\frac{a\gamma_s g_1}{(1-a)\gamma_s g_1+1}\right]}$ is the transmission latency from S to AP_1, $t_2 = \frac{L_2}{B\log[1+(1-a)\gamma_s g_2]}$ is the transmission latency from S to AP_2, a $(0 < a < 1)$ is the power allocation ratio when applying NOMA scheme, ρ denotes the number of required CPU cycles for each bit, and f_i stands for the CPU-cycle frequency at the AP_i, $i = \{1, 2\}$.

Therefore, according to APS Algorithm 1, the successful computation probability can be rewritten as

$$\begin{aligned} Pr_s = &\Pr\left(g_1 > g_2, \frac{L}{B\log(1+\gamma_s g_1)} + \frac{\rho L}{f_1} \le T\right) \\ &+ \Pr\left(g_1 < g_2, \max\left\{t_1 + \frac{\rho L_1}{f_1}, t_2 + \frac{\rho L_2}{f_2}\right\} \le T\right). \end{aligned} \quad (5)$$

In order to evaluate the performance of this considered NOMA MEC system, we obtain the following theorems.

Theorem 1.

Under quasi-static Rayleigh fading, the exact closed-form expression of the successful computation probability Pr_s for this considered downlink NOMA MEC system based on proposed APS-NOMA MEC scheme is given by

$$Pr_s = \begin{cases} e^{-\frac{\beta}{\lambda_1}}\left[1 - e^{-\frac{\beta}{\lambda_2}}\right] + e^{-\frac{\beta_1}{\lambda_1} - \frac{\beta^*}{\lambda_2}} + \\ \qquad \frac{\lambda_1}{\lambda_1+\lambda_2}\left[e^{-\left(\frac{1}{\lambda_1}+\frac{1}{\lambda_2}\right)\beta} - e^{-\left(\frac{1}{\lambda_1}+\frac{1}{\lambda_2}\right)\beta^*}\right], & a > \frac{\Phi}{1-\Phi} \\ e^{-\frac{\beta}{\lambda_1}}\left[1 - e^{-\frac{\beta}{\lambda_2}}\right] + \frac{\lambda_1}{\lambda_1+\lambda_2} e^{-\left(\frac{1}{\lambda_1}+\frac{1}{\lambda_2}\right)\beta}, & a < \frac{\Phi}{1-\Phi} \end{cases} \quad (6)$$

where $\beta = \frac{2^{\frac{L}{\Omega B}}-1}{\gamma_s}$, $\beta_1 = \frac{\Phi}{\gamma_s[a-(1-a)\Phi]}$, $\beta^* = \max\{\beta_1, \beta_2\}$, $\beta_2 = \frac{2^{\frac{L_2}{\Omega_2 B}}-1}{(1-a)\gamma_s}$, $\Phi = 2^{\frac{L_1}{\Omega_1 B}} - 1$, $\Omega = T - \frac{\rho L}{f_1}$, $\Omega_1 = T - \frac{\rho L_1}{f_1}$, $\Omega_2 = T - \frac{\rho L_2}{f_2}$.

Proof. See Appendix A.

In the work [15], the application of downlink NOMA transmission to MEC was investigated. In this scenario, the user S always offloads Task 1 to AP_1 and Task 2 to AP_2 by using NOMA scheme. In order to verify the effectiveness of this proposed protocol, we derive the exact closed-form expression for successful computation probability of conventional downlink NOMA MEC system similar to [15] as following theorem.

Theorem 2.

Under quasi-static Rayleigh fading, the exact closed-form expression of the successful computation probability $Pr_s^{(ref)}$ for the conventional downlink NOMA MEC system is expressed as

$$Pr_s^{(ref)} = \begin{cases} e^{-\left(\frac{\beta_1}{\lambda_1}+\frac{\beta_2}{\lambda_2}\right)}, & a > \frac{\Phi}{1-\Phi} \\ 0, & a < \frac{\Phi}{1-\Phi} \end{cases} \tag{7}$$

where $\beta_1 = \frac{\Phi}{\gamma_s[a-(1-a)\Phi]}$, $\beta_2 = \frac{2^{\frac{L_2}{\Omega_2 B}}-1}{(1-a)\gamma_s}$, $\Phi = 2^{\frac{L_1}{\Omega_1 B}} - 1$, $\Omega_1 = T - \frac{\rho L_1}{f_1}$, $\Omega_2 = T - \frac{\rho L_2}{f_2}$.

Proof. See Appendix B.

4 Numerical Results and Discussion

In this section, we provide the numerical results in terms of successful computation probability Pr_s to reveal the impact of key system parameters to system performance. The simulation parameters used in this work are provided in Table 1.

Table 1. Simulation parameters

Parameters	Notation	Typical values
Environment		Rayleigh
Number of antennas of each device	N	1
Transmit power	P_s	0–30 dB
CPU-cycle frequency of AP_1	f_1	5 GHz
CPU-cycle frequency of AP_2	f_2	10 GHz
The number of CPU cycles for each bit	ρ	10
Channel bandwidth	B	100 MHz
The threshold of latency	T	0.5 s
The number of data bits	L	80 Mbits

4.1 Impact of Average Transmit SNR and the Length of Task 1

In order to study the impact of the length of confidential task to the performance, we let $L_1 = \epsilon L$, $L_2 = (1-\epsilon)L$, where ϵ is denoted as data allocation coefficient, $0 \leq \epsilon \leq 1$. We set the power allocation ratio as $a = 0.75$ for our simulation. The impact of average transmit SNR (γ_s) and the length of Task

1 (L_1) on system performance in terms of successful computation probability Pr_s is shown in Fig. 3 and Fig. 4. We can observe from these figures that Pr_s increases when γ_s increases. In other words, the performance can improve by inceasing the transmit power of user. Meanwhile, the impact of the length L_1 on the performance is quite different, i.e., the variation of L_1 does not affect to the performance much. However, there exists an optimal value of L_1^* corresponding to ϵ^* that Pr_s achieves a maximum value.

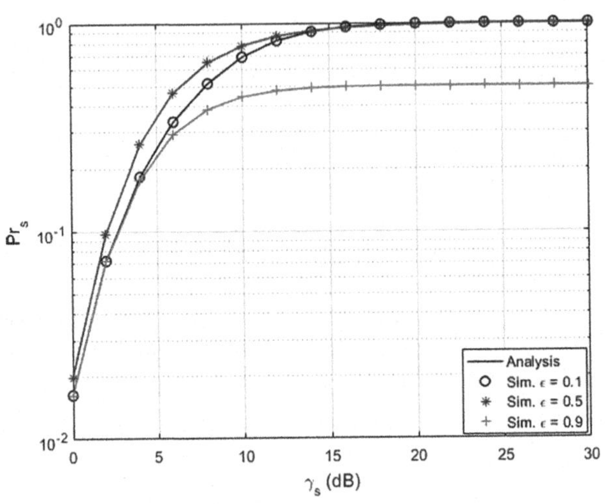

Fig. 3. Pr_s vs. average transmit SNR γ_s with different ϵ.

4.2 Impact of Transmit Power Allocation Ratio and the Length of Task 1

The impact of transmit power allocation ratio (a) and data allocation coefficient (ϵ) on system performance in terms of successful computation probability Pr_s is shown in Fig. 5 and Fig. 6. For a given value of ϵ, Pr_s increases when $a > \frac{\Phi}{1-\Phi}$ and it does not change when $a \leq \frac{\Phi}{1-\Phi}$. It means that in order to improve the performance of this considered system we can allocate the more transmit power the better. However, when a moves closer to 1, the performance degrades. Therefore, there exists an optimal value of $a*$ that Pr_s achieves a maximum value. Figure 7 and Fig. 8 also depict this conclusion.

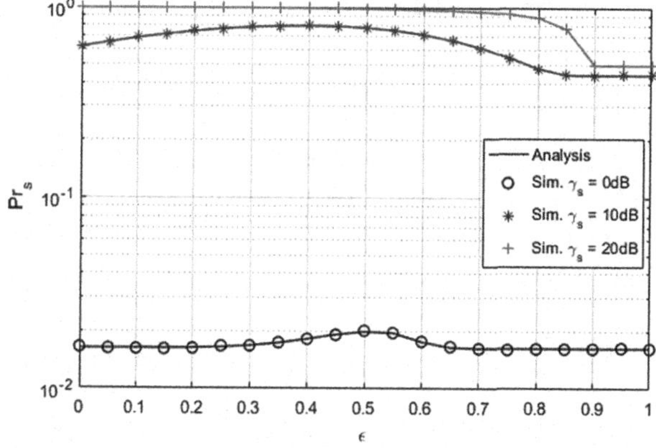

Fig. 4. Pr_s vs. data allocation coefficient ϵ with different γ_s.

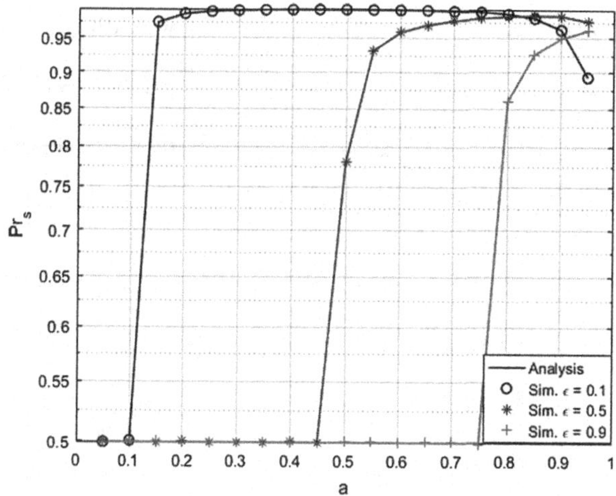

Fig. 5. Pr_s vs. power allocation ratio a with different ϵ.

4.3 Comparison to Conventional NOMA Scheme

Figure 9 and Fig. 10 depict the comparison results between APS-NOMA scheme and conventional NOMA scheme in terms of Pr_s vs. γ_s with different ϵ and with different a. We can see from these figures that our proposed protocol outperforms the conventional NOMA scheme in terms of successful computation probability.

Finally, from above Figs. 3, 4, 5, 6, 7 and 8 we can see that the superior match between analytical and simulation results. This verifies the correctness of our analysis.

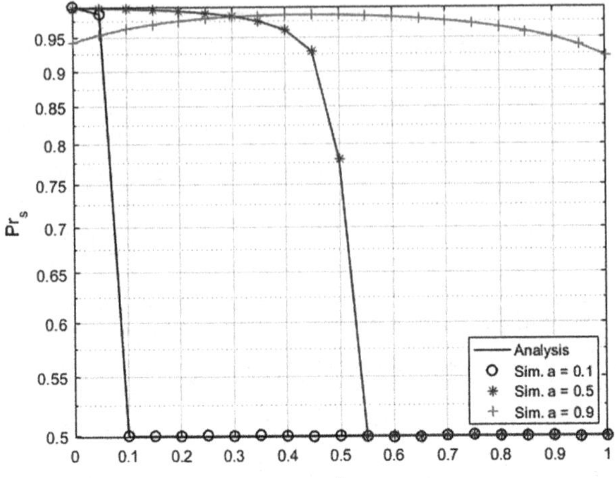

Fig. 6. Pr_s vs. data allocation coefficient ϵ with different a.

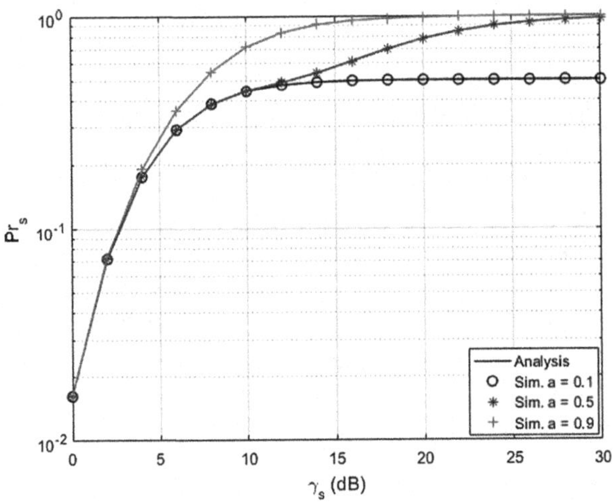

Fig. 7. Pr_s vs. γ_s with different a.

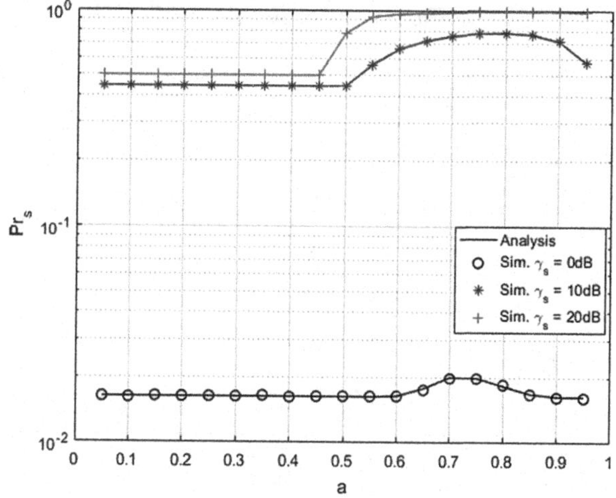

Fig. 8. Pr_s vs. power allocation ratio a with different γ_s.

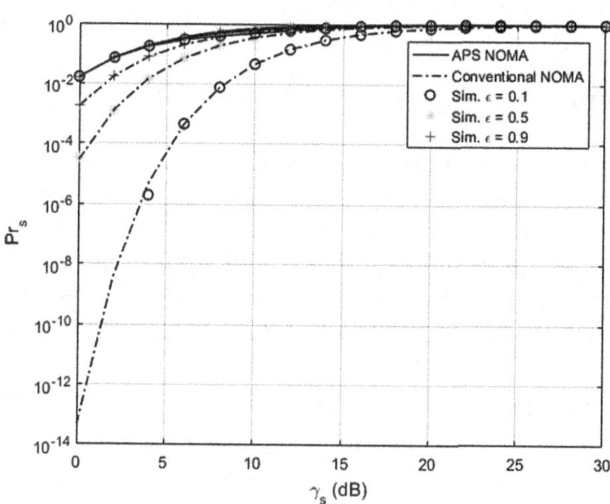

Fig. 9. The comparison between APS-NOMA scheme and conventional NOMA scheme in terms of Pr_s vs. γ_s with different ϵ.

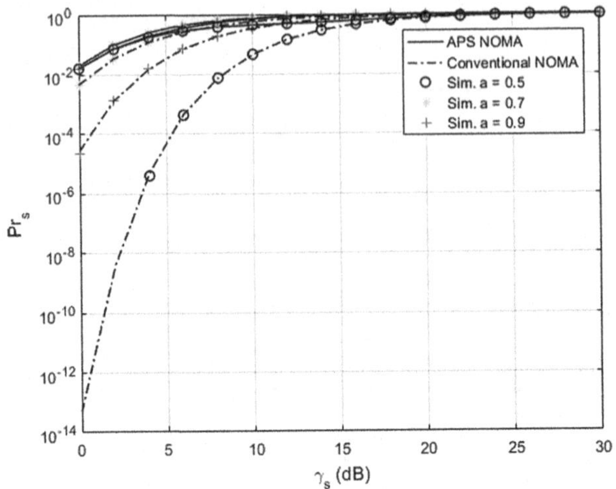

Fig. 10. The comparison between APS-NOMA scheme and conventional NOMA scheme in terms of Pr_s vs. γ_s with different a.

5 Conclusion

In this paper, we have proposed a new offloading APS-NOMA MEC scheme for a secure NOMA MEC network. We have also derived the exact closed-form expressions of successful computation probabilty for this proposed system. The numerical results have been provided to reveal that our proposed scheme achieves better performance than conventional NOMA scheme. Moreover, the performance of this system can be improved by increasing the transmit power and/or by select the optimal value of power allocation ratio.

APPENDIX A: PROOF OF THEOREM 1

Here, from (5) we derive the closed-form expression of Pr_s as follows.

$$\begin{aligned} P_s &= \Pr\left(g_1 > g_2, g_1 > \frac{2^{\frac{L}{\Omega B}} - 1}{\gamma_s}\right) \\ &+ \Pr\left(g_1 < g_2, \left[1 - \frac{(1-a)(2^{\frac{L_1}{\Omega_1 B}} - 1)}{a}\right] g_1 > \frac{2^{\frac{L_1}{\Omega_1 B}} - 1}{a\gamma_s}, g_2 > \frac{2^{\frac{L_2}{\Omega_2 B}} - 1}{(1-a)\gamma_s}\right) \\ &= \begin{cases} \underbrace{\Pr\left(g_1 > g_2, g_1 > \beta\right)}_{I_1} + \underbrace{\Pr\left(g_1 < g_2, g_1 > \beta_1, g_2 > \beta_2\right)}_{I_2}, & a > \frac{\Phi}{1-\Phi} \\ \underbrace{\Pr\left(g_1 > g_2, g_1 > \beta\right)}_{I_1}, & a < \frac{\Phi}{1-\Phi} \end{cases} \end{aligned} \quad \text{(A-1)}$$

where $\beta = \frac{2^{\frac{L}{\Omega B}}-1}{\gamma_s}$, $\Omega = T - \frac{\rho L}{f_1}$, $\beta_1 = \frac{\Phi}{\gamma_s[a-(1-a)\Phi]}$, $\Phi = 2^{\frac{L_1}{\Omega_1 B}} - 1$, $\beta_2 = \frac{2^{\frac{L_2}{\Omega_2 B}}-1}{(1-a)\gamma_s}$, $\Omega_1 = T - \frac{\rho L_1}{f_1}$, $\Omega_2 = T - \frac{\rho L_2}{f_2}$,

$$\begin{aligned} I_1 &= [1 - F_{g_1}(\beta)]\, F_{g_2}(\beta) + \int_{\beta}^{\infty} [1 - F_{g_1}(y)]\, f_{g_2}(y) dy \\ &= [1 - F_{g_1}(\beta)]\, F_{g_2}(\beta) + \frac{1}{\lambda_2} \int_{\beta}^{\infty} e^{-\frac{y}{\lambda_1} - \frac{y}{\lambda_2}} dy \\ &= [1 - F_{g_1}(\beta)]\, F_{g_2}(\beta) + \frac{\lambda_1}{\lambda_1 + \lambda_2} e^{-\left(\frac{1}{\lambda_1} + \frac{1}{\lambda_2}\right)\beta}. \end{aligned} \tag{A-2}$$

$$\begin{aligned} I_2 &= [1 - F_{g_1}(\beta_1)]\, [1 - F_{g_2}(\beta^*)] - \int_{\beta^*}^{\infty} [1 - F_{g_1}(y)]\, f_{g_2}(y) dy \\ &= [1 - F_{g_1}(\beta_1)]\, [1 - F_{g_2}(\beta^*)] - \frac{\lambda_1}{\lambda_1 + \lambda_2} e^{-\left(\frac{1}{\lambda_1} + \frac{1}{\lambda_2}\right)\beta^*}, \end{aligned} \tag{A-3}$$

where $\beta^* = \max\{\beta_1, \beta_2\}$.

From (1), (A-2), (A-3) and (A-1), we obtain the result as (6). This concludes our proof.

APPENDIX B: PROOF OF THEOREM 2

Here, we derive the closed-form expression of $Pr_s^{(ref)}$ as follows.

$$\begin{aligned} Pr_s^{(ref)} &= \Pr\left(t_1 + \frac{\rho L_1}{f_1} < T, t_2 + \frac{\rho L_2}{f_2} < T\right) \\ &= \Pr\left(\left[1 - \frac{(1-a)(2^{\frac{L_1}{\Omega_1 B}} - 1)}{a}\right] g_1 > \frac{2^{\frac{L_1}{\Omega_1 B}} - 1}{a\gamma_s}, g_2 > \frac{2^{\frac{L_2}{\Omega_2 B}} - 1}{(1-a)\gamma_s}\right) \\ &= \begin{cases} [1 - F_{g_1}(\beta_1)]\, [1 - F_{g_2}(\beta_2)], & a > \frac{\Phi}{1-\Phi} \\ 0, & a < \frac{\Phi}{1-\Phi} \end{cases} \\ &= \begin{cases} e^{-\left(\frac{\beta_1}{\lambda_1} + \frac{\beta_2}{\lambda_2}\right)}, & a > \frac{\Phi}{1-\Phi} \\ 0, & a < \frac{\Phi}{1-\Phi} \end{cases} \end{aligned} \tag{B-1}$$

This concludes our proof.

References

1. Mao, Y., You, C., Zhang, J., Huang, K., Letaief, K.B.: A survey on mobile edge computing: the communication perspective. IEEE Commun. Surv. Tutorials **19**(4), 2322–2358 (2017)
2. Zhou, F., Wu, Y., Hu, R.Q., Qian, Y.: Computation rate maximization in uav-enabled wireless powered mobile-edge computing systems. IEEE J. Sel. Areas Commun. **36**(9), 1927–1941 (2018)
3. Sun, H., Zhou, F., Hu, R.Q.: Joint offloading and computation energy efficiency maximization in a mobile edge computing system. IEEE Trans. Veh. Technol. **68**(3), 3052–3056 (2019)

4. Zhang, Y., Lan, X., Li, Y., Cai, L., Pan, J.: Efficient computation resource management in mobile edge-cloud computing. IEEE Internet of Things J. **6**(2), 3455–3466 (2019)
5. Li, Q., Zhao, J., Gong, Y.: Computation offloading and resource allocation for mobile edge computing with multiple access points. IET Commun. **13**(17), 2668–2677 (2019)
6. Lu, W., Yin, B., Huang, G., Li, B.: Edge caching strategy design and reward contract optimization for uav-enabled mobile edge networks (2020). EURASIP J. Wireless Commun. Networking **38**, 1–10 (2020). https://doi.org/10.1186/s13638-020-1655-2
7. Men, J., Ge, J.: Performance analysic of non-orthogonal multiple access in downlink cooperative network. IET Commun. **9**(18), 2267–2273 (2015)
8. Islam, S.M.R., Avazov, N., Dobre, O.A., Kwak, K.S.: Power-domain non-orthogonal multiple access (NOMA) in 5G systems: potentials and challenges. IEEE Commun. Surv. Tutorials **19**(2), 721–742 (2017)
9. Ha, D.-B., Nguyen, S.Q.: Outage performance of energy harvesting DF relaying NOMA networks. Mob. Netw. Appl. **23**(6), 1572–1585 (2017). https://doi.org/10.1007/s11036-017-0922-x
10. Tran, D.D., Tran, H.V., Ha, D.B., Kaddoum, G.: Cooperation in NOMA networks under limited user-to-user communications: Solution and analysis. In: IEEE Wireless Communications and Networking Conference (WCNC), 15–18 April 2018, Barcelona, Spain (2018)
11. Tran, D.D., Ha, D.B.: Secrecy performance analysis of QoS-based non-orthogonal multiple access networks over nakagami-m fading. In: The International Conference on Recent Advances in Signal Processing, Telecommunications and Computing (SigTelCom), HCMC, Vietnam (2018)
12. Ye, Y., Lu, G., Hu, R.Q., Shi, L.: On the performance and optimization for MEC networks using uplink noma. In IEEE International Conference on Communications Workshops (ICC Workshops), Shanghai, China. IEEE (2019)
13. Zhou, F., Wu, Y., Hu, R.Q., Qian, Y.: Computation efficiency in a wireless-powered mobile edge computing network with NOMA. In: IEEE International Conference on Communications (ICC), Shanghai, China, 20–24 May 2019 (2019)
14. Wang, F., Xu, J., Ding, Z.: Multi-antenna noma for computation offloading in multiuser mobile edge computing systems. IEEE Trans. Commun. **67**(3), 2450–2463 (2019)
15. Ding, Z., Fan, P., Poor, H.V.: Impact of non-orthogonal multiple access on the offloading of mobile edge computing. IEEE Trans. Commun. **67**(1), 375–390 (2019)
16. Ha, D.H., Ha, D.B., Zdralek, J., Voznak, M.: A new protocol based on optimal capacity for energy harvesting amplify-and-forward relaying networks. In: 5th NAFOSTED Conference on Information and Computer Science (NICS) (HCMC, Vietnam) (2018)
17. Gradshteyn, I., Ryzhik, I.: Table of Integrals Series and Products. Elsevier Academic Press, Brulington (2007)

Author Index

Printed in the United States
By Bookmasters